PROPERTY OF U. S. GOVERNMENT
U. S. DEPARTMENT OF ENERGY

PLASTICITY AND HIGH TEMPERATURE STRENGTH OF MATERIALS

Combined Micro- and Macro-mechanical Approaches

PROPERTY OF U. S. GOVERNMENT
U. S. DEPARTMENT OF ENERGY

PLASTICITY AND HIGH TEMPERATURE STRENGTH OF MATERIALS

Combined Micro- and Macro-mechanical Approaches

MASATERU OHNAMI
*Professor of Mechanical Engineering,
Ritsumeikan University, Kyoto, Japan*

ELSEVIER APPLIED SCIENCE
LONDON and NEW YORK

ELSEVIER APPLIED SCIENCE PUBLISHERS LTD
Crown House, Linton Road, Barking, Essex IG11 8JU, England

Sole Distributor in the USA and Canada
ELSEVIER SCIENCE PUBLISHING CO., INC.
52 Vanderbilt Avenue, New York, NY 10017, USA

WITH 37 TABLES AND 344 ILLUSTRATIONS

© ELSEVIER APPLIED SCIENCE PUBLISHERS LTD 1988

British Library Cataloguing in Publication Data

Ohnami, Masateru
  Plasticity and high temperature strength
  of materials—combined micro- and macro-
  mechanical approaches.
  1. Plasticity 2. Materials 3. Fracture
  mechanics 4. Materials at high temperature
  I. Title
  620.1′1233   TA418.14

Library of Congress Cataloging in Publication Data

Ohnami, Masateru, 1931–
  Plasticity and high temperature strength of materials.

  Bibliography: p.
  Includes index.
  1. Plasticity. 2. Fracture mechanics.
  3. Materials at high temperatures. I. Title.
  TA418.14.O36   1987   620.1′1233   87-15533

  ISBN 1-85166-119-0

No responsibility is assumed by the Publisher for any injury and/or damage to persons or property as a matter of products liability, negligence or otherwise, or from any use or operation of any methods, products, instructions or ideas contained in the material herein.

Special regulations for readers in the USA

This publication has been registered with the Copyright Clearance Center Inc. (CCC), Salem, Massachusetts. Information can be obtained from the CCC about conditions under which photocopies of parts of this publication may be made in the USA. All other copyright questions, including photocopying outside of the USA, should be referred to the publisher.

All rights reserved. No part of this publication may be reproduced, stored in a retrieval system, or transmitted in any form or by any means, electronic, mechanical, photocopying, recording or otherwise, without the prior written permission of the publisher.

Photoset in Malta by Interprint Ltd
Printed in Great Britain by Page Bros. (Norwich) Limited

I should like to dedicate this book, with respect, to Ritsumeikan University, Kyoto, and my colleagues

Contents

Chapter 1 **Introduction**... 1

Chapter 2 **Concepts and Methodologies for the Study of Plasticity and Fracture of Solids**..................... 5
2.1 Concepts and Methodologies Used for the Study of Strength and Fracture of Materials...................... 5
2.2 Variety and Interaction................................... 7
2.3 Discontinuity in 'Space and Time' on Deformation and Fracture of Materials..................................... 14
2.4 Irreversible Process....................................... 29
2.5 Fluctuation in Strength and Fracture of Materials (SFM) and Factor of Safety....................................... 34
2.6 Modeling and Scale-up.................................. 43
References... 54

Chapter 3 **History of the Combined Micro- and Macro-mechanical Approach to the Study of Plasticity and Fracture of Materials** 59
3.1 Interdisciplinary Approach............................... 59
3.2 History of the Study of Plasticity of Heterogeneous Materials 61
3.3 History of the Theory of Distributed Dislocations and the Continuum Approximation Approach..................... 74
3.4 Advances in Fractology................................... 90
References... 111

Chapter 4 **Plasticity Laws of Polycrystalline Bodies**............ 120
4.1 Crystal Plasticity under Simple Tension 120
4.2 Crystal Plasticity under Cyclic Stressing.................. 130
4.3 Influence of Strain History on Plastic Deformation of Polycrystalline Metals Subjected to Cyclic Stressing...... 137

4.4	Continuum Mechanical and Micromechanical Studies of Cyclic Strain-hardening Behavior in a Biaxial Stress State at Elevated Temperatures.................................	145
4.5	Plasticity Laws in Creep of Polycrystalline Metals at Elevated Temperatures...................................	162
	References...	180

Chapter 5 **Plasticity Laws of Imperfect Crystals and Misorientation of Metals**........................... 184

5.1	Study of Misorientation in Metallic Plasticity Relevant to the Geometrical Aspects of Continuously Dislocated Continuum...	184
5.2	Microstructural Study of Static Flow Stress in Metals by Means of CDC..	199
5.3	Microstructural Study of the Cyclic Constitutive Relation in Metals by Means of CDC..............................	210
5.4	Microstructural Study of the Influence of Pressure Soaking on Plastically Deformed Metals by Means of CDC	220
5.5	Microstructural Study of Effects of Hydrostatic Pressure on Flow Stress and High-temperature Creep in Metals by Means of CDC	228
	References...	237

Chapter 6 **Fracture Laws in a Continuum Mechanical Approach**. 240

6.1	Effect of Hydrostatic Stress on Crack Behavior and Fracture Life under Creep, Low-cycle Fatigue and the Interaction of These Conditions at Elevated Temperatures	241
6.2	Effect of Stress Biaxiality on Crack Behavior and Failure Life in High-temperature Low-cycle Fatigue of Heat-resisting Metals and Alloys in the Creep Range	250
6.3	Effect of Load Waveform and Sequence on High-temperature Low-cycle Fatigue of Heat-resisting Metals and Alloys from the Viewpoint of Crack Behavior	281
6.4	Effect of Strain Wave Shapes and Varying Principal Stress Axes on High-temperature Biaxial Low-cycle Fatigue	293
6.5	Notch Effect on Low-cycle Fatigue under Creep–Fatigue Interaction Conditions.....................................	304
6.6	Life Prediction of Notched Specimens under Creep–Fatigue Interaction Conditions	327
6.7	Thermal Fatigue of a Circular Disk Specimen in	

Connection with the Strength of Gas Turbine Disks...... 340
References 353

Chapter 7 **Fracture Mechanics of Solids with Microstructure and the *J*-Integral** 358
7.1 Influence of Hydrostatic Pressure on the Fracture Toughness of Polycrystalline Metals and *J*-Integral 359
7.2 Interaction Between Crack, Plasticity and Environment in the Crack Behavior of Polycrystalline Metals............. 368
7.3 Effect of Load Wave Shape on the High-temperature Low-cycle Fatigue of Pure Metals 400
7.4 X-Ray Study of Damage Evaluation for Structural Metals under Creep–Fatigue Interaction Conditions.............. 409
References 424

Chapter 8 **Technology Development of Metal Plasticity under High Hydrostatic Pressure and of Life Assessment/Remaining Life Prediction Methods for Structural Materials with High-temperature Applications.** 426
8.1 Plastic Drawing of Pure Metals under High Hydrostatic Pressures.. 426
8.2 Plastic Compression of Pure Metals under High Hydrostatic Pressures.................................... 436
8.3 Research and Development of Nuclear Steelmaking in Japan and the Life Prediction Method of Structural Materials for Very High Temperature Gas Reactor (VHTGR) Applications 443
8.4 Collaborative R and D Efforts on the Reliability of Structural Materials for High-temperature Applications in Japan.. 470
8.5 Collaborative R and D Efforts Examining the Corrosive and Mechanical Behavior of Heat-resisting Alloys for 50 MW (Thermal) VHTGR Applications in Japan 482
References 485

Appendix I... 488

Appendix II 502

Index 505

Chapter 1

Introduction

A combined micro- and macro-mechanical research of plasticity and fracture of solids started with the formation of the study of the strength and fracture of materials (SFM) in the 1940s and 1950s. Formation of the study of SFM is characterized as follows: (i) development of the boundary field in research; (ii) formation of a new integrated field of research, where an interdisciplinary approach between various research fields is necessary. A variety of phenomena has been found in relation to SFM, and the study of SFM concerns a wide field in which both science and technology today are firmly interconnected and are related closely to social assessment. Thus it is desirable for scientists and engineers to cooperate actively with each other in studying it.

This book was written with the following main objectives:

1. To present fundamental information concerning the plasticity and high-temperature strength of structural materials from an integrated viewpoint of a combined micro- and macro-mechanical approach.
2. To present this information as an organized, logical and systematic body of knowledge, providing the fundamental concepts and methodologies, the history of thought, the unification of theories and experiments, and the applications to technology today, in the research field of plasticity and high-temperature strength of materials.
3. To provide the following people with the means of assessing and improving upon the reliability of the material evaluations, design and maintenance of machines and structures: researchers of the strength of materials, mechanical engineers concerned with the design and assessment of machines and structures, and researchers

and engineers engaged in materials science and metallurgy.
4. To provide the readers, the researchers and engineers mentioned above, graduates and undergraduates of mechanical engineering, materials science and metallurgy, with a comprehensive understanding of the important ideas and theories in the study of plasticity and high-temperature strength of materials, and illustrate their relevance with applications to technological situations.
5. To provide the reader with a concentrated means by which to understand unique theories and experimental data with reference to the authors' papers and the collaborative efforts in Japan. The majority of the data used for reference in this book has been studied in the laboratory of the author and is the result of collaborative Japanese efforts in which the author has participated.

'Many parameters have been identified, but there is not enough service data or understanding to rank them for applications.' 'Many individual experiments or testing programs have been reported, but they are seldom integrated to provide a rigorous understanding.' 'There is too much emphasis on testing and reliance on large safety factors (short-term economics); there is insufficient recognition of the power and cost-effectiveness of suitably blending testing with mechanism research.' (see Section 2.1 in Chapter 2). These points outline some of the critical problems in the study of the strength and fracture of materials. This book aims at answering these problems and providing an integrated understanding extending from conceptual and theoretical studies to technological practice concerning the strength and fracture of solids.

The format of this book was designed to meet the needs of the researcher and student as well as practising engineers. A general outline of the basic concepts, methodologies and a brief history of the thought which comprises the evolution of the study of plasticity and fracture of solids are described in Chapters 2 and 3, from the viewpoint of a combined micro- and macro-mechanical approach. Both theoretical and experimental studies on plasticity and the high-temperature strength of materials are systematically described in the following chapters. Plasticity laws of polycrystalline bodies constructed from inhomogeneous crystals are described in Chapter 4 from the viewpoint of crystal plasticity. In addition, the plasticity laws of imperfect crystals are described in Chapter 5 by means of the geometrical aspect of the continuously dislocated continuum and the important concept of misorientation in crystals. In Chapter 6, to illustrate the fracture laws in a continuum mechanical approach, we describe some important mechanical aspects which demonstrate the drastic difference between the crack behavior and failure life of metallic creep and that of metallic low-cycle fatigue at high temperatures. In Chapter 7 we describe the relationship between the fracture mechanics of heterogeneous materials having microstructure and the J-integral in continuum mechanics. At the end of the book, in Chapter 8, we describe the following two problems in technology today: the plastic drawing of brittle metal under high hydrostatic pressures and the life assessment/remaining life prediction of structural metals with high-temperature applications.

It is my great pleasure to write my thanks to colleagues in my laboratory in Ritsumeikan University, Kyoto, Japan. First mention goes to Emeritus Professor Sotoo Endo, of Ritsumeikan University, Professor Yoshiteru Awaya, of Maizuru National Technical College, Professor Katsuhiko Motoie, Professor Hiroshi Umeda and Professor Naomi Hamada, of Hiroshima-Denki Institute of Technology, Professor Masaru Ohmura, of Setsunan University, Professor Masahiro Shikida, of Osaka Industrial University, Professor Kazuaki Shiozawa, of Toyama University, Assistant Professor Masao Sakane, of Ritsumeikan University, Dr Ryuzo Imamura, of Ishikawajima–Harima Heavy Industries Co., Ltd, and Dr Seiichi Nishino, Babcock–Hitachi Co., Ltd. It is also with pleasure that I acknowledge the help of many graduate and undergraduate students in my laboratory. It is also my great pleasure to write my thanks to the faculty members in my department for useful discussions concerning research in this field.

It is my great honor to write my thanks to Emeritus Professor Takeo Yokobori, of Tohoku University, to whom I owe a great personal debt for his encouragement to publish this book. As the author, it is also my great honor to write my thanks to the many distinguished researchers and engineers who provided truly enlightening discussions which found their origin in the collaborative studies performed by the following academic committees in which I participated, in the capacity noted: the Research Committee on the High Temperature Strength of Materials, the Society of Materials Science, Japan (Committee Chairman), the Research Committee on the High Temperature Strength of Materials, the Iron and Steel Institute of Japan (Subcommittee Chairman), the Special Committee on the Structural Strength of High Temperature Heat Exchangers in the National Project of Nuclear Steelmaking, sponsored by the Ministry of International Trade and Industry, Japan (Subcommittee Chairman), the Expert Committee on the National

Project based on the Fund for Promotion of Science and Technology, the Science and Technology Agency, Japan (Subcommittee Chairman), the Expert Committee on a Very High Temperature Gas Reactor for Multi-Purpose Use, the Japan Atomic Energy Research Institute (Member), the Editorial Committee on Comprehensive Lectures on the Strength and Fracture of Materials, 8 volumes (Chief Editor), and the Scientific Committee of the International Conference on Creep, 1986, Tokyo (Committee Chairman). Finally, I wish to thank Mr Charles Rotbart for his excellent revision of the manuscripts.

Chapter 2

Concepts and Methodologies for the Study of Plasticity and Fracture of Solids

This chapter describes basic concepts and methodologies used for the study of the strength and fracture of materials. Seven concepts are adopted: interaction, continuum approximation, interface structure, irreversible process, stochastic process, scattering and safety margin, and modeling and scale-up. These will be described with stratificational and historical cognitions from the viewpoint of unification of 'variety' and 'monotony' and that of 'continuity' and 'discontinuity'.

2.1 CONCEPTS AND METHODOLOGIES USED FOR THE STUDY OF STRENGTH AND FRACTURE OF MATERIALS

It is said that there is no *general methodology* or *algorithm* in studying the *strength and fracture of materials* (abbreviated as *SFM*). This is due to the variety of phenomena associated with SFM. But this does not mean that there are not any specific methodologies, as we will point out in the following [1].

(1) *There are a variety of phenomena involved in the strength and fracture of materials.* It may be true that there are no uniformly applicable methodologies, but various research approaches are useful. In other words, this variety of phenomena indicates that Nature is intricate and unceasingly dynamic. Moreover, the concept of 'motion' has led to the use of various scientific methods such as theory, instrument, and norm.

(2) *In this field it is necessary to have an understanding of modern technology and basic science; an interdisciplinary approach is important.* The study of SFM concerns a wide field in which both science and technology are firmly interconnected. Thus it is desirable for scientists and engineers to cooperate actively with each other.

(3) *In accumulating data for the study it will be useful to decide on the research approach that will eliminate redundancies.* For example, the number of papers on fatigure of materials during the period from 1830 to 1980 was over 20 000 [2]. It was estimated that the number of papers of fatigue of materials published by the American Society for Testing and Materials (ASTM) would increase from 4000 in 1962 to 10 000 in 1970 [3]. In fact, this figure expands to over 13 000 when publications besides those of ASTM are included. The publication Trends of Study on Fatigue of Materials, edited by the Fatigue Committee of the Society of Materials Science, Japan (JSMS), reported that the annual number of papers was 564 in 1965, 896 in 1970, 1252 in 1975, 1336 in 1980 and 1653 in 1984.

It was reported by Fong, a member of the ASTM E-9 Committee on Fatigue [4], that the annual cost for research into the fatigue of materials in the United States runs to billions of dollars. Faulty study methodology may result in inadequate solutions for scientific and technical problems. Fong reported the following reasons [5]: (i) Many parameters have been identified, but there is not enough service data or understanding to rank them for applications. (ii) Many individual experiments or testing programs have been reported, but they are seldom integrated to produce a rigorous understanding. (iii) There is too much emphasis on testing and reliance on large safety factors (short-term economics); there is insufficient recognition of the power and cost-effectiveness of suitably blending testing with mechanism research.

Considering the above, the following features have been recognized:

(1) *Variety in deformation and fracture of materials.* The unavoidable variety of phenomena is related to the *interaction* between the many phenomenal processes.

(2) *The effect of discontinuity in 'space' and 'time' on deformation and fracture of materials.* Both of the concepts 'discontinuity in space', in *stratificational cognition*, and 'discontinuity in time', in *recognition of stochastic processes*, are useful in bridging the gap between 'macro' and 'micro'.

(3) *Structure-sensitive properties of deformation and fracture of materials.* Both materials with higher and lower grades of structure have discontinuity in quality, but there is continuity where the former is based on the latter.

(4) *Scattering of fracture of materials.* Fracture of materials is essentially a statistical phenomenon.

(5) *Modeling.* Stratificational cognition may connect with various levels of theoretical development in the study of modeling.

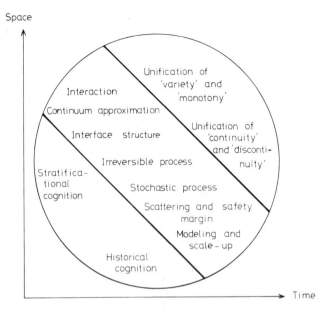

Fig. 2.1. Basic concepts used for strength and fracture of materials. (Reprinted from ref. 1 by permission of OHM-Sha)

(6) *Spanning 'model' with 'original'*. Both the structure and the function of the model must be related to those at full-size by taking 'scale-up' into consideration.

Concepts (1)–(6) are interrelated, and each item represents a single definite concept. These concepts are illustrated in Fig. 2.1 [1]. The concepts used in the SFM study will be overviewed with reference to several historic papers.

2.2 VARIETY AND INTERACTION

2.2.1 Variety of Fracture of Materials

The fracture of solids is not a deterministic physical phenomenon of single or homogeneous processes such as the evaporation of liquids, liquefaction of vapor and melting of solids, but a common phenomenon of 'fracture' occurs throughout a very considerable variety of these processes [6]. The liquefaction from vapor to liquid is a physical process of nucleation, and the process has been applied to the fracture of solids [7]. An analogy between the plastic flow or fracture of solids and the

Table 2.1
Classification of Characteristic Modes in Fracture Surface of Metals (Reprinted from ref. 9 by permission of BAIFUKAN)

(A) Transgranular Fracture
   (1) Ductile fracture
      (a) Microcavity coalescence fracture
          Dimple $\begin{cases} \text{Equiaxial dimple} \\ \text{Elongated dimple} \end{cases}$
      (b) Glide plane fracture
          Serpentine glide, Ripple, Stretching
   (2) Brittle fracture
      (a) Cleavage fracture
          River pattern, Cleavage step, Cleavage facet, Tongue
      (b) Quasi-cleavage fracture
          River pattern, Cleavage step, Quasi-cleavage facet, Tongue
   (3) Fatigue fracture
         Striation $\begin{cases} \text{Ductile striation, Plateau} \\ \text{Brittle striation, River pattern} \end{cases}$
         Tire truck
(B) Intergranular Fracture
   (1) Ductile fracture (due to coalescence with microcavities)
         Granular fracture surface with dimples
   (2) Brittle fracture
         Granular fracture surface without shell marks (or with few marks)
   (3) Fatigue fracture
         Striation, granular fracture surface with striation pattern

---

melting of solids has also been studied by many researchers in the USSR [8]. However, these are not studies of homogeneous processes (see Section 2.4). Therefore modes of fracture of solids vary according to the respective fracture processes. Scanning electron microscopy (SEM) has clarified the characteristic modes of fracture on the surface of metals, as shown in Table 2.1 [9].

The modes are broadly divided into *transgranular* and *intergranular fractures*. These are further classified into (i) *ductile fracture* after plastic deformation, (ii) *brittle fracture* without marked plastic deformation and (iii) *fatigue fracture* with gradual crack propagation under repeated loading.

In the static tension test of technical ductile metal specimens, *cap and cone fracture* is often observed. *Fibrous fracture surfaces* occur in the central portion of the specimens and *shear fracture surfaces* on the perimeter. SEM observation shows a large number of small cavities due to both microvoid nucleation and coalescence. In the tension test of pure metal specimens, *slip-off* or *chiselpoint fracture* may occur. SEM observation shows *serpentine glide* due to glide plane fracture.

An example of the elongated *dimple* of commercial copper in a tear test in air at room temperature is shown in Fig. 2.2(a) [10]. An example of the serpentine glide of the same specimen is shown in Fig. 2.2(b) [10]. The specimen is under hydrostatic pressure which inhibits the nucleation of minute voids. Under hydrostatic pressure, the fracture mode of the metals changes according to the fracture processes such as microvoid nucleation, coalescence, necking and shear fracture of the ligaments between cavities.

In the impact test for low-carbon steel, as shown in Fig. 2.2(c) [11], the specimen breaks on a specified crystalline plane of *cleavage* without marked plastic deformation. Since the fracture does not occur on one cleavage plane but extends over parallel planes, a *cleavage step* or *river pattern* is observed by SEM. This results from the confluence of cleavage steps as well as rivers joining together. The direction of crack propagation can be determined from the river pattern. In the cleavage fracture of metal a *tongue-like* projection can be observed.

On the other hand, a stripe pattern of *striations* is observed in fatigue fracture of metal, which is formed in each repeated load cycle. An example of the high-temperature low-cycle fatigue of austenitic stainless steel is shown in Fig. 2.2(d) [12]. Striations will be formed perpendicularly to the direction of crack propagation on the striped portions of *plateau* or *patch*. Although these portions appear flat they consist of a considerable number of uneven slip lines. Under other load conditions, the fractured surface of the striated area in Fig. 2.2(d) changes to the intergranular fractured surface in Fig. 2.2(e) [12], even if the material and temperature are the same as those in Fig. 2.2(d). This is caused by the interaction of creep and fatigue with a slower rate of tension than of compression. In general, the high-temperature fatigue life of the materials for technical use is reduced more in Fig. 2.2(e) than in 2.2(d) (see Chapter 6) due to the *creep/fatigue interaction* in (e).

Thus there is a wide variety of fracture modes, and SEM observation will extend from a microscopic scale of $0.1\ \mu m$ to a macroscopic scale of mm. Variable data may be obtained by taking a large number of measurements. This is understandable provided that the phenomenon is

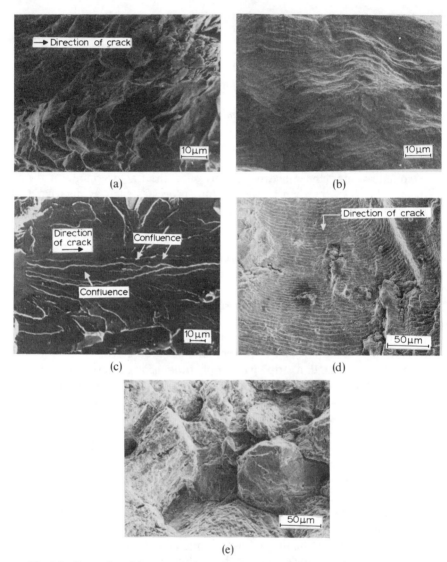

Fig. 2.2. Examples of fracture surface of metals by scanning electron microscopy. (a) Static tear fracture of a commercial pure copper in air [10]. (b) Static tear fracture of a commercial pure copper under hydrostatic pressure [10]. (c) Impact fracture of low-carbon steel in air (Charpy test) [11]. (d) Low-cycle fatigue fracture of 304 type stainless steel at 823 K in air (total strain range controlled push-pull; frequency 0·1 Hz) [12]. (e) Low-cycle fatigue fracture of type 304 stainless steel at 823 K in air, under 0·002%/s of tension and 0·2%/s of compression [12].

transitional and successive in 'time' and 'space', and that each process of the fracture is momentary and is changeable with respect to the others.

2.2.2 Interaction

The interaction between various fracture processes accounts for the variety of material fracture modes. The crack behavior of metallic fatigue and creep will be described with this in consideration.

Historically, the phenomenon of metallic fatigue fracture had been known early in the development of technical fields after the Industrial Revolution of the 18th and 19th centuries in Europe. Nevertheless, the first series of scientific studies on iron and steel was performed by August Wöhler (1819–1914), a Prussian mechanical engineer [13]. For a number of years his 'Burchgrenze' (*fatigue limit*) was regarded as describing the general behavior of metallic fatigue. It was later determined that there is no fatigue limit for the majority of nonferrous metals and some kinds of alloys, and it is rare to observe a fatigue limit of metals in a corrosive environment or at high temperatures. The existence of a fatigue limit suggests that, after a considerably large number of load cycles, fatigue crack growth will be checked by the healing of fatigue damage and the two processes will balance one another. The metallic fatigue limit has been studied as shown in the following example.

It is now considered that the fatigue limit of steels is definitely affected by various factors and is related to the crack behavior and structural change. The Japanese physicist T. Yokobori, founder of the International Conference on Fracture (ICF) [14], classified microcracks of metallic fatigue into two types according to the propagation process. One is a microcrack that does not become a fatal fracture, here referred to as type A. Another is a microcrack or short crack that contributes to final rupture, here referred to as type B. It may be related to *threshold stress* which is the lower limit at which fatigue crack propagation begins. Non-propagating cracks belong to type A and are related to the existence of a fatigue limit.

The majority of the nucleation sites of steel fatigue cracks are in the nonmetallic inclusions in hard steel with high tensile strength. However, the predominant crack nucleation sites of ductile steel are in grain boundaries, slip bands, boundaries between pearlite and ferrite and other structural boundaries; these fatigue cracks belong to type B. The B-type crack contributes to the fatal rupture of the material through short crack growth or coalescence with microcracks. The A-type crack is regarded as having the same propagation process as the B-type. Therefore the B-type contains the A-type 'as a moment'. In fatigue crack growth a

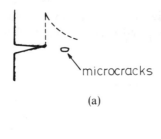

(a)

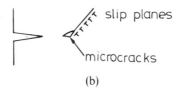

(b)

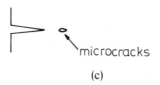

(c)

Fig. 2.3. Some models of the micromechanism of metallic fatigue crack. (Reprinted from ref. 14 by permission of IWANAMI-Shoten) (a) Microcracks may nucleate by stress concentration at the main crack tip. (b) Microcracks may nucleate when atomic bonding is broken by stress concentration at slip planes. (c) Microcracks may nucleate by cohesion of atomic vacancies.

characteristic local plasticity zone forms in front of the crack tip. Crack propagation is not only affected by local plasticity, but also by coalescence with microcracks ahead of the main crack. As shown in the model in Fig. 2.3 [14], fatigue crack behavior is determined by stress concentration at the crack tip, concentration due to slip bands in front of the crack tip and coalescence with macrocracks due to the cohesion of atomic vacancies. Therefore fatigue crack propagation shows a discontinuous curve with an *incubation period*, not only on a microscopic scale but also on a macroscopic scale.

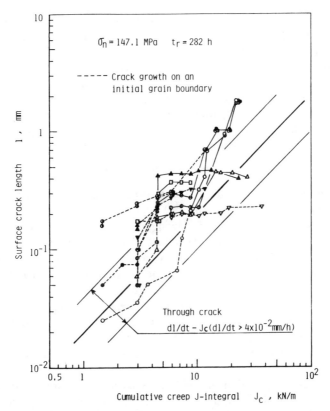

Fig. 2.4. Surface crack length, l, versus cumulative creep J-integral, J_c, curves. Test conditions: 316 SS, 923 K, in vacuum (10^{-2} Pa). (Reprinted from ref. 15 by permission of Soc. Mater. Sci., Japan)

Similar behavior is observed in metallic creep at high temperatures. The surface intergranular crack growth in a smooth specimen of austenitic stainless steel is shown in Fig. 2.4 [15]. The ordinate shows the half-crack length l and the abscissa shows the cumulative creep J-integral, J_c, which was calculated by integrating the *creep J-integral*, $J_c = \sigma_n V$, where σ_n is the nominal applied stress and V the surface crack opening displacement rate. The respective grain boundary cracks grow discontinuously but l increases nearly proportionally to J_c.

On the other hand, empirical laws of continuous propagation rate hold for the through-crack propagation rate. Propagation rate is a concept that eliminates the contradiction that a crack exists at a certain place in the material at a given time but at the same time does not exist at that place. Both concepts of 'negation' and 'affirmation' apply here. The unification of these concepts is attained with the concept of 'motion'. This concept guarantees continuity through the interaction between the microscopic factors which control crack propagation. The discontinuity in microscopic surface crack propagation indicates that each process occurs while interacting with other processes.

The solid line in Fig. 2.4 is the through-crack propagation rate curve for specimens of the same material with a center-crack. It is written as

$$\mathrm{d}l/\mathrm{d}N = CJ_c \qquad (C = \text{constant}) \tag{2.1}$$

Since discontinuous surface crack growth curves are formed along a continuous curve based on eqn (2.1), the unity between 'discontinuity' and 'continuity' is due to the following 'motions': (i) interaction between the crack and viscoplasticity ahead of the crack; (ii) interaction between the main crack and microcracks or voids in front of the crack; (iii) interaction between the crack and the oxidation environment at the crack tip.

Because J_c in eqn (2.1) is regarded as being proportional to the local creep rate ahead of the crack, the concept of 'crack propagation rate' generally suggests that the *fracture occurs continuously when the local deformation rate in front of the crack comes up to the intrinsic deformation rate.*

The crack propagation of solids consists of (i) continuous processes which are controlled by law at various levels, and (ii) discontinuous processes in 'space and time'. Unification of (i) and (ii) occurs due to the *interaction* of these processes. Such interaction is broken down into the following three categories [16]: succession; occurrence; inevitability. Once the category is determined, we can effectively recognize the variety of phenomenon. The crack propagation of solids is not a linear summation of the processes but a *nonlinear* summation.

2.3 DISCONTINUITY IN 'SPACE AND TIME' ON DEFORMATION AND FRACTURE OF MATERIALS

2.3.1 Stratificational Cognition

The stratificational concepts of elasticity, plasticity and fracture of crystalline solids are shown in Tables 2.2–2.4 [17, 18].

Material space on level I in Tables 2.2 and 2.3 is the so-called

Table 2.2
Stratificational Concepts of Elasticity of Crystalline Solids (Reprinted from ref. 17 by permission of OHM-Sha)

Class	Level	Structure of material space	Mechanical subject and method of continuum approximation	Discipline
I	Macro-scopic	Averaging discontinuous structures; body is continuous and homogeneous	Formation of Hooke's law and equation of motion in level of continuum mechanics	Theory of macro elasticity
II	Sub-macro-scopic	Body is continuous but not homogeneous, and is composed of elements of the order of grain size in polycrystals	Derivation of macroscopic Hooke's law and equation of motion, from both the law of single crystals and information on texture	Theory of crystal elasticity Statistical continuum mechanics
III	Sub-macro-scopic	Body is continuous but not homogeneous, and is composed of elements of crystal lattice	Derivation of macro equation of motion from both knowledge of atomic interactions and formal theory of statistical mechanics	Lattice theory
IV	Micro-scopic	Body is discontinuous and composed of particles (atoms)		

continuum. The physical properties of small pieces subdivided from the continuum are independent of the shape and size of the pieces. Pieces can be subdivied until they reach a *material point* or a *particle.* Thus the continuum consists of an aggregate of particles. Conversely, on level IV the body loses continuity and consists of discrete particles. There is a great dimensional gap between levels I and IV. Studies on various levels have been done to bridge this gap. Typical examples are levels II and III in Tables 2.2–2.4. The majority of materials for technical use, such as polycrystalline metals, are changeable in their physical properties and microstructure distribution. These are *heterogeneous materials.*

Materials for technical use which have microstructure have boundaries on which physical properties change when the subdivision exceeds

Table 2.3
Stratificational Concepts of Plasticity of Crystalline Solids (Reprinted from ref. 17 by permission of OHM-Sha)

Class	Level	Structure of material space	Mechanical subject and method of continuum approximation	Discipline
I	Macro-scopic	Averaging discontinuous structures; body is continuous and homogeneous	Formation of plasticity law in continuum mechanics	Theory of macro-plasticity
II	Sub-macro-scopic	Body is continuous but not homogeneous, and is composed of elements of the order of grain size in polycrystals	Derivation of macro-scopic plasticity law of polycrystals from both law of single crystals and information on texture	Theory of crystal plasticity Statistical continuum mechanics
III	Sub-macro-scopic	Body is continuous but not homogeneous, and is composed of elements of the order of subgrain size in crystals	Continuum approximation in which imperfect crystal is regarded as Euclidean connection space	Theory of continuous distribution of disloca-tions Statistical continuum mechanics
IV	Micro-scopic	Body is discontinuous and composed of particles (atoms)	Fields of crystal dislocations and interaction between dislocations	Crystal dislocations

that boundary. The boundary is not always distinct and there are many cases in which physical properties change gradually. But in crystalline solids the structural boundaries between the various levels are distinct. Boundaries are a problem of so-called *stratificational cognition*, and research at new levels will develop new advances. The stratificational concept will bridge the gap between 'macro' and 'micro'. For this it is important to clarify the submacroscopic concept which bridges the gap between macro and micro and to find the *measure* or *parameter* which concretely represents the stratificational concept. Those mentioned previously are examples of such parameters. First, there are various fracture modes of material as shown in Table 2.1. There are also various measures such as dislocation density, cell size, misorientation and void

Table 2.4
Stratificational Concepts of Fracture of Crystalline Solids (Reprinted from ref. 17 by permission of OHM-Sha)

Class	Level	Fracture region in material space	Fracture criterion	Discipline
I	Macro-scopic	$10^{-2}-10^{-1}$ cm, i.e. the order of macro-scopic cracks	For macroscopic cracks in continuum where discontinuous microscopic structures are neglected, fracture may occur under the control of singularity of stress/strain at the crack tip and global energy condition	LEFM EPFM DM PFM
II	Sub-macro-scopic	About 10^{-3} cm, i.e. the order of grain size	Fracture of this level may occur when submacroscopic cracks grow up in material with discontinuous structures of the order of grain size	Crystallo-graphy of fracture (fracto-graphy)
III	Sub-macro-scopic	10^{-4} cm, i.e. the order of subgrain size	In material with discontinuous structures of the order of grain size, fracture of this level may occur when microscopic cracks nucleate and grow up or stop or coalesce under the control of interaction with crystal lattice defects; both global energy and localized stress conditions are necessary in the crack growth	Micro-mechanics fracture Physical fracture mechanics Electron fractography X-ray studies on mechanical behavior of materials
IV	Micro-scopic	10^{-8} cm, i.e. the order of atomic distance	Fracture of this level may occur when atomic bonding is broken and the internal interfaces are created; crystal lattice defects are influential to fracture strength of solid	Crystal dislocation theory of fracture

LEFM, Linear elasticity fracture mechanics
EPFM, Elastic-plasticity fracture mechanics
DM, Damage mechanics
PFM, Phenomenological fracture mechanics

density detected by the X-ray diffraction method and transmission electron microscopy (TEM). The model of interaction between the micro and macro cracks shown in Fig. 2.3 is also an important measure.

At each level of Table 2.4 there are one or more *criteria* of material fracture. At level IV, the fracture strength of real single crystals is only one-tenth or one-hundredth of the *ideal fracture strength*, $\sigma_{th} \simeq E/10$ (see Section 2.4), where E is Young's modulus of the crystal. This discrepancy results mainly from the stress concentration due to crystal lattice defects, and the decrease in atomic bonding due to environmental effects. The criterion of solid fracture at level IV is theoretically formulated considering the above. On the other hand, the fracture of polycrystalline metal is dependent on the structure and mechanical behavior of each grain and those of the grain or subgrain boundaries. If plastic deformation affects the fracture process, metal behavior will become even more complicated. This behavior is referred to as *structure-sensitive properties*. Therefore the theoretical formulation of fracture criterion on levels II and III is more intricate than that at the atomic level of IV. This is why study on a new level is necessary.

Thus there are various characteristic laws or disciplines for each level in Tables 2.2–2.4. These indicate one phase of 'discontinuity'. However, the material structure of the higher level is based on that of the lower level, and the former contains the latter 'as a moment'. This indicates another phase 'continuity' in which the two levels cannot be separated. Therefore, when higher and lower levels exhibit different characteristics they can be correlated with each other. A study which considers both higher and lower levels is necessary.

2.3.2 Continuum Approximation

Considering the *continuum approximation* as another example, it is the methodology which describes the mechanical behavior of materials with microstructure by extending the concept of 'continuum'.

As shown in Fig. 2.5 [17], the distributed surface force on a definite area, $\Delta S^{(v)}$, of the minute elements, which make up the macroelement, can be transferred into both surface stress vector **t** and couple stress vector **m** as well as the force vector in 'particle mechanics'. When $\Delta S^{(v)}$ approaches zero, only the surface stress vector arises because there is no distribution of surface force. Rotation of the minute elements is related to the couple stress. The *scale parameter* which has the dimension of length appears in the constitutive equation. The physical meaning of the scale parameter has not yet been clarified, but it is one of the stratifi-

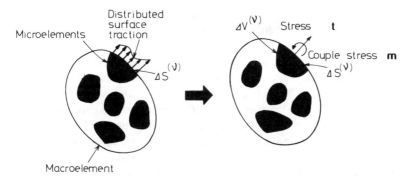

Fig. 2.5. Distributed surface traction on microelement $\Delta S^{(v)}$ can be replaced by both stress vector, **t**, and couple stress vector, **m**, on $\Delta S^{(v)}$. (Reprinted from ref. 17 by permission of OHM-Sha)

cational concepts concerning the mechanical behavior of materials with microstructure.

As shown in Table 2.5 [17], the dimension of the scale parameter is small when there is little spatial fluctuation of the physical properties of the minute elements in an elastic body with microstructure. Such a body is a statistically homogeneous elastic body (here abbreviated to A) which is called a *simple body* by W. Noll, an American researcher of rational mechanics (1958) [19]. On the other hand, when there is much fluctuation, hence a statistically inhomogeneous body (here abbreviated to B), the scale parameter is comparatively large. In this case, the body is

Table 2.5
Spatial Fluctuation of Physical Properties of Microscopic Constituent Elements in Elastic Bodies with Microstructure (Reprinted from ref. 17 by permission of OHM-Sha)

Case of small fluctuation [A]	*Case of large fluctuation* [B]
(1) Small scale parameter l	(1) Large scale parameter l
(2) 'Simple body' (material of lower grade) 　(i) Locality 　(ii) Uniform deformation	(2) 'Complex body' (material of higher grade) 　(i) Non-locality 　(ii) Polarity (including microscopic deformation with particle rotation)

considered a *complex body* which has qualities different from those of the simple body. Table 2.5 has been taken from a paper by I.A. Kunin, a physicist in the USSR [20].

When two particles in a body have the same frame, mass and physical properties, the body is called *materially isomorphic* [21]. If all particles are isomorphic, the body is called *materially uniform*. If a *configuration* with uniform physical properties can be chosen, it is called *materially homogeneous*. Therefore a material in which the physical properties of particle X are dependent on the nearest local configuration X and do not correlate with the configuration of the whole is considered a *simple body*.

However, when the physical properties of non-isomorphic material are influenced by distant particles, *nonlocality* of the material appears. In such a material, additional strain occurs due to particle rotation, and *polarity* of the material appears. A and B differ widely in Table 2.5. However, there is a relation between large- and small-scale parameters which indicates a connection between A and B. The quantitative change brings a transition to the qualitative change. Therefore nonlocality and polarity play an important role in the continuum approximation on levels II and III in Tables 2.2 and 2.3.

Another important concept concerned with the continuum approximation is the *complexity of configurations*. Because it is essentially impossible to unify an imperfect crystal with a part of the perfect crystal without lattice defects, a 'smeared' imperfect crystal is considered *non-Euclidean space of a higher grade*, provided that the smeared perfect crystal is Euclidean space. Thus the imperfect crystal loses the following two important properties of Euclidean space: (i) Euclidean topology is constant; (ii) Euclidean metric is constant. *Crystal dislocation* exhibits the qualities mentioned above.

It is possible to analyze the plasticity of imperfect crystals and the lattice defects on the atomic level, but the relation between perfect and imperfect crystals cannot be geometrically represented even though the locus of change in each atom is physically analyzed. Thus we can recognize the necessity for a geometrical formulation of imperfect crystals based on a high order of material space. Geometric study on level III in Table 2.3 is useful not only in determining the parameter which represents the behavior of the lattice defect, but also in describing the structural change and plasticity of heterogeneous material. The importance of this geometrical study was first pointed out in 1955 by K. Kondo, a Japanese researcher of solid mechanics and a founder of the

Research Association of Applied Geometry (RAAG) in Japan [22], and B.A. Bilby, a British physicist [23].

The continuum approximation of imperfect crystals with lattice defects corresponds to the study of the *continuous distribution of dislocations* [22–28]. From the viewpoint of the differential geometry of Euclidean connected manifolds, the dislocations are considered 2nd order *torsion tensors* [22, 23]. The global lattice defects are considered *Riemann–Christoffel curvature tensors* [22]. Using Laue X-ray diffraction, we [28] clarified that both tensors are related to *misorientation* which represents a continuous change of the orientation of plastically deformed metal grains (see Chapter 5). The 4th order curvature tensor is a particularly important parameter for the continuum approximation which represents *inhomogeneity of strain* [29]. On the other hand, S. Minagawa, a Japanese researcher of solid mechanics (1968) [30], suggested that there is a *dual relation* in which torsion tensors represent dislocation density in 'strain space', but represent couple stress in 'stress space'. It has also been clarified that couple stress is related to crystal dislocation [29, 30] (see Chapter 3).

The simple point correlation function of the *dislocation density tensor* $\{\alpha_{kl}\}$ of the 2nd order shows the average dislocation density as first proposed by J. Nye (1955) [24]. However, simple tensile deformation occurs with an equal number of positive and negative dislocations. In order to clear up this problem, a *multi-point correlation function tensor*, which contains the two-point tensor of the 4th order, $\{\alpha_{ij}(x)\,\alpha_{kl}(x_1)\}$, has been studied by E. Kröner, a German physicist [31]. It was discovered that this type of tensor provides for a more accurate description of the continuous distribution of dislocations [32] (see Chapter 3). We [33] pointed out that the multi-point correlation function tensor plays the important role of the 'circle' in the continuum approximation, where the three concepts of polarity, nonlocality and complexity of configuration are interconnected, as illustrated in Fig. 2.6 [17].

2.3.3 Structural Boundary and Surface Energy

The structural discontinuity of a 'boundary' is composed of an internal interface having a thickness of one molecule or atom. This interface is caused by the fracture of liquids and solids. The structural boundary and the internal interface play an important role in the study of SFM, but this has yet to be clarified completely.

The structural boundary between differently oriented grains is con-

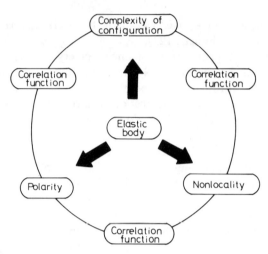

Fig. 2.6. Three approaches in 'continuum approximation' and the role of 'correlation function'. Arrows show the direction for the precision and historical development in 'Theory of continuum approximation' or 'Generalized continuum mechanics'. (Reprinted from ref. 17 by permission of OHM-Sha)

sidered to be the portion of misfit where planes of atoms come to an end. This type of boundary is called the *crystallographic boundary*. The existence of lattice imperfections is necessary for grains to make contact, and they give *interface energy* to the grain boundaries. When impurities are contained in the crystal, the elements will gather around the boundary to decrease the energy, complicating the boundary. This is a *real grain boundary*.

Misorientation occurs because of differently oriented grains or subgrains. The interface of misorientation is not only composed of edge and screw dislocations, but also a network of lattice defects which contain atomic vacancies as well as others. This type of crystallographic boundary is only applicable to the *small-angle boundary* where the interface angle is 15° at most.

However, the *large-angle boundary* has not yet been completely studied due to difficulties in investigating the interface structure, especially the atomic configuration. The development of electron microscopy (EM) and of high resolution and field-ion microscopy (FIM) made for easier direct observation at the atomic level. With such observations the grain boundary was discovered to be a region of atomic distance and is composed of both 'good' and 'bad' portions of the crystal lattice. This direct observation had an impact on the theoretical study of boundaries and made computer simulation possible. An analysis of the impurities on the grain boundaries is also performed by the use of the techniques of Auger electron spectrography, electron spectroscopy for chemical analysis (ESCA) and others.

Research on the large-angle boundary has been supported not only by direct observations but also by new theories. Before the 1930s there were two contradistinctive ideas. One is a model in which the grain boundary is composed of an amorphous structure with a thickness of a hundred atomic distances; the other is a model in which the misfit is one atomic distance and is located where the crystal lattice planes intersect. Thus the discussion continues.

The coincidence boundary model proposed by D. G. Brandon, a British metallurgist, and his colleagues (1964) is shown in Fig. 2.7 [34]. The A–B and C–D portions are regions in which there is a good contact interface. They are called *coincidence boundaries* and atom points on the boundaries agree with the lattice points of both grains. Therefore the coincidence boundary is in the covalent form of both grains. Since atoms are arranged regularly on the boundary and the boundary energy is lower than that of irregular regions, a coincidence boundary will appear when it is arranged in a specific direction. In the figure the coincidence points are arranged in a ratio of one per eleven points. This ratio is called the 'coincidence boundary of $\Sigma 11$'. According to this definition, $\Sigma 3$ is a dual crystal and $\Sigma 1$ a single crystal. The value of Σ is calculated when both the rotation axis and the angle are given, and the structure of the coincidence lattice is determined [34]. Not all real grain boundaries are explained by this regular interface structure, but the coincidence boundary model and the associated Σ value are important stratificational concepts in the study of SFM.

The following concerns *surface energy*, which requires the creation of an internal interface in solids.

True surface energy γ_s is calculated as the negative of work per unit area done by separating neighboring crystal lattice planes. These planes extend from the equilibrium distance a to the infinite distance where the atomic interaction force σ vanishes (see Fig. 2.8). The equation to calculate the true surface energy γ_s is as follows:

$$2\gamma_s = \int_0^\infty \sigma(x)\,dx \qquad (2.2a)$$

The atomic interaction force per unit area, $\sigma(x)$, is derived from the negative derivative of potential energy $U(x)$ with respect to atom

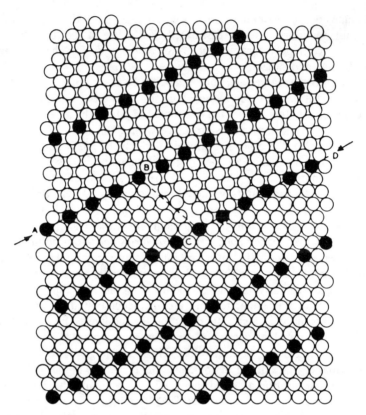

Fig. 2.7. Two-dimensional model of a bcc bicrystal with coincidence boundary of $\Sigma 11$. (Reprinted from ref. 34 by permission of Pergamon Press)

position x if the *state concerned is assumed to be the 'eigenstate' in quantum mechanics*. In other words, $\sigma(x)$ is given by the expectation value of microforce and is derived from the 'eigenvalue' of potential energy.

Provided that $\sigma(x)$ is approximated as $\sigma = \sigma_{th} \sin(2\pi x/\lambda)$, the following equation holds:

$$2\gamma_s = \lambda \sigma_{th}/\pi \tag{2.2b}$$

Furthermore, assuming that $\sigma(x)$ is approximated as $\sigma_{th}(2\pi x/\lambda)$ and Hooke's law $\sigma = Ex/a$ holds in the vicinity of a, the ideal fracture strength, σ_{th}, of perfect solids in eqn (2.2b) is written as

$$\sigma_{th} = \sqrt{\gamma_s E/a} \tag{2.3a}$$

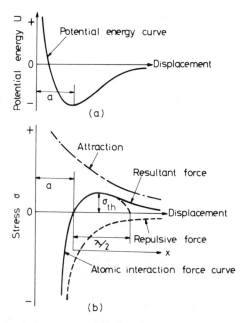

Fig. 2.8. Derivation of (b) atomic interaction force versus displacement curve from (a) potential energy curve.

where E is Young's modulus. Since γ_s is estimated to have nearly the same value as $Ea/20$, eqn (2.3a) is also written as

$$\sigma_{th} \simeq E/10 \tag{2.3b}$$

Many thermodynamics textbooks state that equilibrium surfaces are treated in isolation, but this is decidedly untrue. It is questionable whether equilibrium interface thermodynamics has a valid application to fracture, because fracture does not usually occur in an equilibrium system. But thermodynamics of surfaces has played an important role in the development of fracture criteria. Many of these classic papers seem to have been based on simplistic assumptions whose validity needs to be reexamined [35]. From this point of view we will describe the following example.

With a knowledge of the *energy momentum tensor*, as formulated by J. D. Eshelby (1916–1981), a British physicist [36], the so-called *J-integral*, J, of cracked specimens is written below from the negative of the potential energy U dissipated by the growing crack increment l. This integral was independently proposed in 1967–8 by J. R. Rice [37] and G. P. Cherepanov [38], researchers of solid mechanics in America and the USSR, respectively:

$$J = (1/B)(-\delta U/\delta l) \tag{2.4a}$$

where l is the crack length and B is the thickness of the specimen. Therefore the total input energy E is written as

$$BE = 2\gamma_s l - \int_0^l J\,dl \tag{2.4b}$$

From the equilibrium condition the maximum value of E is obtained when

$$J_c = 2\gamma_s \tag{2.5}$$

This is in agreement with the *brittle fracture theory* (1920) proposed by Alan Arnold Griffith (1893–1963), a British aeronautical engineer [39]. However, local plastic deformation occurs at the interface even when iron and steel fracture is in the cleavage mode, and the plastic work γ_p per unit area on the interface is remarkably larger than γ_s in eqn (2.5). Griffith studied solid fracture at the Royal Aircraft Factory, later known as the Royal Aircraft Establishment (RAE), and is called the 'father of the science of fracture'. He is also well known as a pioneer in the development of a jet engine with an axial compressor. Once the crack begins to propagate, the increment dl is no longer an equilibrium displacement, and J_c loses its energy interpretation. Thus the conditions under which crack growth begins can be determined using the so-called J_c *criterion*. For example, ASTM E813 and JSME S001 are the standard test methods for a J_{Ic} measure of fracture toughness. In this case the evaluation of γ_s and γ_p for solid materials must be made by taking into consideration the stratification factors [1]. However, the effect of the spreading local plasticity zone ahead of the crack tip is not included in γ_p. The equation $J_c = 2\gamma_p$ yields the necessary and sufficient condition for crack growth with small or large scale yielding.

2.3.4 Stochastic Process

Because the fracture of solids involves probabilistic processes, it is a dynamic or time-statistical phenomenon and is discontinuous in time. Therefore the fracture of solids is regarded as a *stochastic process*. The study of stochastic processes arose from the physics of the fracture of solids previous to other fields of physics. The first fracture test of glass under constant loading was performed by M. Hirata (1948) [40] and others [41]. Hirata, a Japanese physicist, is the pioneer of the theory of stochastic processes. After this first test, many brilliant papers were presented by T. Yokobori (from 1949) on the various types of fracture (Table 2.1) and the yielding of steel [42, 43]. Studies by other researchers were also done on high polymers, concrete and wood. We will describe the essential concepts of the stochastic process along the lines of Yokobori's research [43].

The *transition probability* can be written for that instant of time in which fracture originates. No fracture occurs before time t and fracture occurs during an infinitesimal increment of time dt, as $\mu(\sigma_1, t)dt$, where σ_1 is the local stress which controls the nucleation of cracks. In linear fracture mechanics (abbreviated as *LFM*), when the material with crack length l is subjected to stress σ, the *stress intensity factor*, $K = \sigma\sqrt{\pi l}$, is also used. 'Taking account of t' means the transition of the material properties in a high-temperature or corrosive environment; it also means the change of material structure in front of the crack tip.

As described above, the fracture phenomena of fatigue, creep, delayed fracture, etc. are represented by the process of *successive transition probability* of the lifetime. This lifetime is composed of many steps such as crack initiation and propagation as well as multiple successive steps of crack growth and coalescence with microcracks. Considering the processes of the $n+1$ state and the n step, the average fatigue life, $\bar{N}$, is represented by $\bar{N} = 1/\mu_{i,i+1}$. Moreover, the average creep rupture life or the delayed fracture life, t, is represented by $\bar{t} = 1/m_{i,i+1}$, where $\mu_{i,i+1}$ is the transition probability per cycle from the i state to the $i+1$ state and m is the transition probability per time interval. The *Shanon diagram* of the transition probability of the fracture of solids is shown in Fig. 2.9 [43]. In SFM there are many cases where the transition probability m has the physical and mechanical properties of 'rate' in the *thermal activation process of stress-dependent type*. This was first pointed out by K. Ohmori, a Japanese physicist (1941) [44], and was developed by T. Yokobori [7].

Considering that fatigue crack growth is the stochastic process of the $n+1$ state and the n step, the relation for the propagation rate, $dl/dN = \varepsilon_A/\bar{N}_i$, can be obtained by substituting both $dl/dn = \varepsilon_A$ and $dN/dn = N_i$ into $dl/dN = (dl/dn)/(dN/dn)$. Here ε_A is the increment of crack length at each step, and $\bar{N}_i$ is the average life at the i step, written $\bar{N}_i = 1/\mu_A$. When the crack does not grow at each loading cycle but is regarded as growing by the stochastic process, the propagation rate will be represented

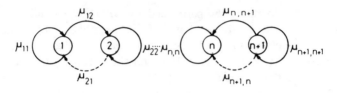

⎡ ① : No fracture, ② ············ ⓝ : Discontinuous growth of crack, ⎤
⎢ ⓝ⁺¹ : Fatal fracture, μ_{ij} : Transition probability from 'i' state ⎥
⎣ to 'j' state, where $\mu_{21} \approx \mu_{32} \approx \cdots \approx \mu_{n+1,n} \approx 0$ ⎦

Fig. 2.9. Transition probability diagram of $n+1$ state and n-step process. (Reprinted from ref. 43 by permission of IWANAMI Shoten)

as [43]

$$dl/dN = \varepsilon_A \mu_A \qquad (2.6a)$$

Here the propagation rate is of the A type.

However, when the crack is forced to increase at each cycle and the increment ε_B is regarded as an average of the stochastic process, $dl/dN = \varepsilon_B$ and $\varepsilon_B = \lambda \mu_B$ can be obtained, where λ is a constant and μ_B is the transition probability. In this case ε_B corresponds to the *striation space* as shown in Fig. 2.2(d), and the propagation rate dl/dN is written as [43]

$$dl/dN = \lambda \mu_B \qquad (2.6b)$$

Here the propagation rate is of the B type.

A similar representation for metallic creep fracture is given as [43]

$$dl/dt = \varepsilon_c \mu_c \qquad (2.6c)$$

where ε_c is the increment of growth at each step and μ_c is the transition probability per unit time.

The time- or cycle-dependent fracture of solids caused by the stochastic process in 'velocity space' can be understood by drawing an analogy between the flow of solids and the turbulence of fluids. This means that 'time' and 'space' statistics on the fracture process of solids must be taken into consideration. This also corresponds to the idea of there being a mixture of both A and B, where μ_B of type B is multiplied by a transition probability of the occurrence of plastic flow ahead of the crack tip [43].

2.4 IRREVERSIBLE PROCESS

2.4.1 Irreversible Process and Crack Nucleation

The fracture of solids is a so-called irreversible phenomenon.

Much unproductive discussion has taken place due to confusion concerning the concepts 'reversible change' and 'irreversible change'. In classic thermodynamics or statistical mechanics there is no such thing as an irreversible process in the physical picture (physikalische Bild), but irreversible change is observed in the physical phenomenon. In other words, irreversible change is only observed in phenomena which are experienced daily. This illustrates the *distinction between 'physical picture' and 'physical phenomenon'*. Physical picture is an abstraction of physical phenomenon and thus does not always include all causes of irreversible change in the phenomenon. However, in quantum mechanics an irreversible process already exists in the microscopic process which includes the cause of irreversible change, i.e. the unstable state causes a change in a quantum mechanically stable state.

The phenomenon of irreversible change in natural science has been interpreted using the 'entropy production' of the second law of thermodynamics, the 'unidirectional change in the H-function' of statistical mechanics and the 'uncertainty principle' of quantum mechanics. The thermodynamic approach is to apply the *heterogeneous nucleation theory* to the fracture of solids. It was developed by T. Yokobori in the 1950s [7,45]. It was his important idea that, since cracks mainly occur at the portion of stress concentration, $q\sigma$, this portion must be the nucleation site in solids. He called it a *stress-dependent heterogeneous process* [7], where q is the stress concentration factor and σ the external stress. This was the first new development since the application of the *theory of absolute reaction rates* to the fracture of solids and liquids in the 1940s [46–50]. The theory of absolute reaction rates was first made by Henry Eyring, an American chemist of Mexican birth, and his colleagues (1941) [51].

The nucleation theory, rather than the theory of absolute reaction rates, seem to have the characteristic that the fundamental law determining the direction of phase transformation is taken into consideration through the 'decrease in free energy'. As shown in Fig. 2.10, even if an embryo is initiated by thermal fluctuation, an increase in free energy F is necessary to form the nucleus and the embryo probably vanishes before it reaches the *critical nucleus size*, r^*. However, if the embryo does happen to reach r^*, it will continue growing in the irreversible process

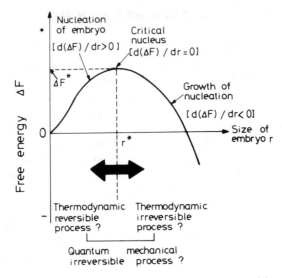

Fig. 2.10. Relation between size of embryo and change of free energy.

$d(\Delta F)/dr < 0$. Strictly speaking, nucleus growth is not an irreversible process prior to nucleation. To avoid contradiction, the *phenomenological rate process theory* has been applied to the whole process of crack nucleation in which the difference between both reaction rates in positive and negative directions is effective under applied stress.

In the stress-dependent heterogeneous nucleation theory, the maximum free energy required to reversibly create a crack with the shape of a circular disk is written as $\Delta F^* = \pi^3 \gamma_s^3 E^2 /[6(1-v^2)^2 \sigma^4]$, where γ_s is surface energy, σ is external stress, E is Young's modulus and v is Poisson's ratio. Therefore, for instance, the rate of nucleation of brittle fracture is formally represented as [7]

$$I_F = ZV(k/h)(q\sigma/\sigma_0)^{1/n_b kT} \exp(-\Delta f^*/kT) \qquad (2.7)$$

where Z is the number of atoms in the portion of stress concentration, V is the crack tip neighborhood volume, Δf^* is the activation energy required to separate atoms at the crack tip, k is Boltzmann's constant, h is Planck's constant, T is absolute temperature, and n_b and σ_0 are material constants. T. Yokobori [7] showed that eqn (2.7) was applicable to the stress velocity dependence in brittle fracture of solids. In the equation, $q\sigma$ is also replaced by the stress intensity factor K in LFM.

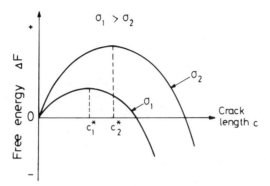

Fig. 2.11. Relation between latent Griffith crack length and change of free energy.

Thus it seems that the stress-dependent heterogeneous nucleation theory can overcome the deficiencies in Griffith's brittle fracture theory (1920), in which no irreversible process of fracture was taken into consideration [39]. In fact, when the external stress σ is kept constant, the relation between Griffith's latent crack length c and the change in free energy ΔF is represented in Fig. 2.11 [52] as well as in Fig. 2.10. But there is a distinctive difference. In Griffith's theory, a latent crack at the critical size, c^*, exists as a stable crack under no load, but not in the nucleation theory [52].

The stress-dependent nucleation theory was first proposed in the study of fatigue crack propagation in the 1950s. It was developed in the 1960s and 1970s. According to the models in Figs 2.3 (a) and (b), using eqn (2.7) as well as eqn (2.6a), the fatigue crack propagation rate was derived as [7]

$$dl/dN = (\varepsilon_A V Z k T / f h) M_1 \exp(-\Delta f^*/kT) \Delta K^{2\beta/[1+\beta)(n_f k T)]} \quad (2.8)$$

Here the nucleation rate of cracks per cycle is written as $\mu = \int_0^{1/f} I_F \, dt$, where f is the frequency, β is the cyclic strain hardening coefficient, $\Delta K = (\pi l)^{1/2} \sigma_a$, l is the crack length, σ_a is the stress amplitude, and M_1 and n_f are material constants. A similar equation was formulated for the model in Fig. 2.3(c). These equations are known as *Yokobori's theory*.

2.4.2 Analogy of Crack and Electron

In spite of these successes in the nucleation theory of the fracture of solids, some questions remain. S. Yamamoto, a Japanese metallurgist, pointed out the following fundamental questions [53]:

(1) Because embryo nucleation requires an increase in free energy, it cannot be detected by only a decrease in free energy. Moreover, since the concept 'time' is not included in the change of free energy, the reaction rate cannot be interpreted by using only the so-called nucleation theory.

(2) Thermodynamic equilibrium is assumed in the nucleation theory to determine the distribution of embryos. Can embryos be formed in an equilibrium state? Can equilibrium be maintained even when the nucleus at r^* has vanished?

(3) Since the nucleation theory is described using classic thermodynamics, the theory is based on the macroscopic concept. Can nucleation be interpreted from the macroscopic viewpoint? The nucleation theory will have to be uniformly interpreted from microscopic and dynamic irreversible processes. The important concept 'critical nucleus' has not yet been experimentally examined.

The so-called rate process theory was formally adopted by T. Yokobori [17] in order to solve the above (1). He emphasized the importance of the *kinetic theory of fracture of solids* from the viewpoint of the combined nucleation theory and the dynamic theory of distributed dislocations or the stochastic theory. Here (2), the *principle of local equilibrium* [54] is assumed, by which the state function at a particular point in the system can be considered. This consideration can be made when there is a small degree of nonequilibrium state even if the system is in the irreversible process. However, because the propriety of the principle has not been proved, it is necessary to ascertain the agreement between theoretical and experimental results based on the principle of local equilibrium. Item (3) is a problem of stratificational cognition, and it is desirable to uniformly interpret the mechanism of crack nucleation at the microscopic level.

In Fig. 2.4 it might be interesting to note whether or not the continuous crack propagation rule, $dl/dt \propto J_c$, may be extrapolated to the region of crack initiation. However, generally it will be difficult to find a direct connection between macroscopic and microscopic crack behavior because of stratificational recognition. In this example, the critical curvature radius, r^*, of the spherical nucleus at the grain boundary calculated by $r^* = 2\gamma_s/\sigma$ [7] is nearly 100 Å at most. Here σ is the external stress, and the true surface energy, γ_s, is estimated to be nearly 1 N/m. On the other hand, the J_c value estimated from eqn (2.5) is nearly 10^{-3} N/m. This is with the provision that the cavity nucleus is formed due to interface exfoliation when the local concentrated stress at the interface reaches the ideal fracture strength. In this sense, the cavity type of grain boundary crack with a size of nearly 200 Å may be considered the *microcrack threshold*. But it is necessary to examine this experimentally.

The first proposal of an *analogy of cracks and electrons* was made in 1931 by T. Terada, a Japanese physicist and one of the discoverers of the Laue X-ray diffraction spot [55]. He wrote: 'According to the uncertainty principle as it is called, there are n-electrons in the field, but their exact positions or velocities cannot be pointed out. The number of cracks can be foretold from the condition of strain, but not the exact positions or velocities of their end-points' [55]. Terada concluded: 'This analogy is only a qualitative one, and not a complete one at that. However, it may at least serve as a first step towards an understanding of the ultimate structure of matter reviewed in the light of modern physics'.

One of the structure-sensitive properties, in terms of lattice imperfection, is the phenomenon of fracture or plastic deformation of crystalline solids. Quantum mechanical studies on imperfections in crystal lattices have not been completely successful because of the mathematical complexity of the problem. For instance, the atomic interaction force versus the displacement curve in Fig. 2.8(b) for covalent crystal material such as diamond is determined by the quantum mechanical approach, but there is still too much difficulty in drawing such a curve for metallic crystals for technical use.

Sir Alan H. Cottrell, a British physicist [56], wrote: 'The strength and plastic properties of solids, for example, are not determined by the average behavior of the atoms but by the exceptional behavior of those relatively few atoms situated at lattice irregularities. The theories of such properties could not be developed without providing some intermediate concepts, to by-pass the mathematically formidable and to some extent physically irrelevant problem of solving *Schrödinger's equation* for lattice irregularities, and to enable the theory to work directly in terms of the atoms and their movements. One of the most useful of these has been the representation of the *cohesion of a solid* in terms of 'atomic bonds' i.e. the force–displacement relation between pairs of atoms such as that in Fig. 2.8(b)'.

This is important considering that the higher level of the strata of atomic cohesion will be more effective in studying metal fracture than the lower one of electrons.

2.5 FLUCTUATION IN STRENGTH AND FRACTURE OF MATERIALS (SFM) AND FACTOR OF SAFETY

Scattering is generally observed in the strength and fracture of solids. This results from the stochastic process of plastic flow and fracture of solids as well as the inhomogeneity of materials for technical use. Therefore the fluctuation of SFM is related to many fields of research, from *reliability physics to statistical continuum mechanics* and *reliability-based design*. The essential concepts are described as follows.

2.5.1 Scattering and Distribution Function

Tables A and B [57] in Appendix I show examples of the fluctuation of fracture strength and lifetime of solid materials for technical use through discussions of the *probability distribution function*.

The *coefficient of variation* (c.o.v.) is generally employed to measure scattering. If random variables of an event with a sample size of N are $x_1, x_2, \ldots, x_N$, then the expectation $E(x)$ and the variance $V(x)$ are written as

$$E(x) = \sum_{i=1}^{N} x_i P(x_i)$$

$$V(x) = \sum_{i=1}^{N} [x_i - E(x)]^2 P(x_i) = \sum_{i=1}^{N} x_i^2 P(x_i) - E(x)$$

Therefore the c.o.v. is obtained from $[V(x)]^{1/2}/E(x)$.

When the distribution in the tensile strength σ of structural steel is represented by a normal distribution, as shown in Fig. A of Table A, a c.o.v. of 4·33% indicates that 68·3% of the sample size is in the region $1·043\bar{\sigma} > \sigma > 0·957\bar{\sigma}$, where $\bar{\sigma}$ is the mean. When there is rupture time of solids under a constant load, the distribution function is frequently shown by an exponential distribution, and the c.o.v. will be nearly 100%. As seen from Tables A and B, the c.o.v. of the tensile strength of ductile metal is small. However, this is large compared with the fracture strength of brittle metal and low-temperature brittle fracture strength of ductile metal. Moreover, the c.o.v. decreases as sample size increases. These facts are understood from the distribution of random variables.

In inductive statistics, let the random variable Z of the function of variables $x_1, x_2, \ldots, x_n$ be

$$Z = g(x_1, x_2, \ldots, x_n) \tag{2.9}$$

and expand it by the means $E(x_1), \ldots, E(x_n)$, respectively. Thus the following equation is obtained:

$$Z = [g]_0 + \sum_{i=1}^{n} [x_i - E(x_i)][\partial g/\partial x_i]_0$$

$$+ \left(\frac{1}{2}\right) \sum_{i=1}^{n} \sum_{j=1}^{n} [x_i - E(x_j)][x_j - E(X_j)][\partial^2 g/(\partial x_i \partial x_j)]_0 \quad (2.10)$$

In eqn (2.10), $[g]_0$, $[\partial g/\partial x_j]_0$, etc. are the values when $X_1, X_2, \ldots, X_n$ are equal to $E(X_1), E(X_2), \ldots, E(X_n)$, respectively. Therefore the mean $E(Z)$ and the variance $V(Z)$ are given by

$$E(Z) = [g]_0 + \left(\frac{1}{2}\right) \sum_{i=1}^{n} \sum_{j=1}^{n} [\partial^2 g/\partial X_i \partial X_j]_0 \operatorname{Cov}[X_i, X_j] + \cdots$$

$$V(Z) = \sum_{i=1}^{n} \sum_{j=1}^{n} C_i C_j \operatorname{Cov}[X_i, X_j] + \cdots \quad (2.11a)$$

where $C_i = [\partial g/\partial x_i]_0$. Since $\operatorname{Cov}[X_i, X_j]$ is the *covariance* of both X_i and X_j, $\operatorname{Cov}[X_i, X_j] = V[X_i]$ when $i = j$. Because of this, and if the random variables are independent of each other, the following relations are obtained and the second and remaining terms of the right-hand side of eqn (2.10) are omitted:

$$E(Z) = [g]_0 = g[E(X_1), E(X_2), \ldots, E(X_n)]$$

$$V(Z) = \sum_{i=1}^{N} C_i^2 V(X_i) \quad (2.11b)$$

If Z is in its most simple form $\sum_{i=1}^{n} X_i/n$, where X_i is an independent random variable, it is known from the central limit theorem that the distribution becomes nearly normal as n increases. In this case, both equations $E(Z) = E(X)$ and $V(Z) = V(X)/n$ are obtained.

If the tensile strength Z of a ductile material is determined from the mean of the deformation characteristic X of the material microstructure, the distribution of Z will be represented by the normal distribution, shown in an example in Fig. A in Table A, derived from the central limit theorem. In general, X varies with inhomogeneities in critical shear stress and crystal grain, grain size, grain boundary strength, etc. When the material is produced in a well controlled manufacturing process, there is

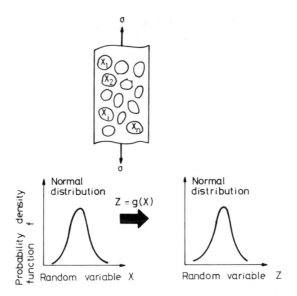

Fig. 2.12. Normal distribution function, Z, of random variables, X, for the ductile fracture of materials.

comparatively little variation in X and the distribution of X is almost normal. Therefore Z is approximated by the linear equation with respect to X, when all but the first two terms of eqn (2.10) are omitted. Here the distribution of Z is almost normal, as shown in the schema of Fig. 2.12. In this case the c.o.v. is comparatively small.

However, the brittle fracture of a solid material is determined by the weakest constitution in the material. If the cumulative distribution function of strength x of each defect is $G(x)$, then the cumulative distribution function of the *resistance* $F(x)$ of a metal with n cracks is equal to the *extreme distribution* of n realizable random variables controlled by the distribution $G(x)$. A material with tensile strength x is composed of n links, if the cumulative distribution function is $F(x)$ and the probability density distribution function is $f(x) = dF(x)/dx$. Since the tensile strength x will be determined by the strength of the weakest of the n links, the resistance of the links will be represented by the probability distribution $f(x)$. However, in the schematic representation of Fig. 2.13, $f(x)$ appears as the extreme distribution $f_1(x)$ as n increases. When n becomes large, the extreme distribution function is written as

$$1 - F_1(x) = [1 - F(x)]^n = \exp[\log[(1 - F(x))^n] = \exp[1 - nF(x)] \quad (2.12)$$

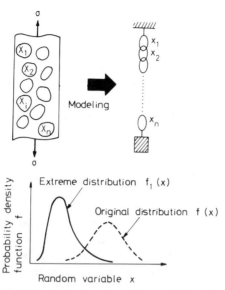

Fig. 2.13. Extreme distribution in 'weakest link model' for the brittle fracture of materials.

This distribution was studied by E. J. Gumbel, an American statistician of German birth [58]. Three types of distribution function have been clarified: exponential, Cauchy, Weibull. The first and the third types are well known in the study of SFM. The latter was first studied by Waloddi Weibull (1887–1979), a prominent Swedish scientist, famous for his distribution function (1939) [59] shown in Table A.

As described in Section 2.3, the fracture strengths of solids have structure-sensitive properties. Since the brittle fracture of a metal is determined by the weakest property of the metal, fracture strength is represented by the extreme distribution and the fluctuation certainly becomes large, as shown in Fig. B of Table A. For the rupture time of glass, metallic creep rupture life and metallic fatigue fracture life, there is a similar situation in which microcracks begin at the surface or the interior of the material and fracture occurs when the local critical condition is satisfied through the interaction in Section 2.2 or the stochastic process in Section 2.3. This situation accounts for the fact that delayed fracture, stress corrosion cracking, metallic creep rupture and metallic fatigue are phenomenally represented by extreme distribution functions such as the exponential, logarithmic normal and Weibull distributions.

2.5.2 Safety Margin

The *factor of safety* does not mean a measure of safety but should be understood as a *safety margin* between the external load and resistance of a material. This safety margin should maintain reliability [60]. The *allowable stress* x_R^* is smaller than the strength basis x_R^0 such as the mean or the mode. When it is employed to cover unreliability in structural design, the safety margin between x_R^* and x_R^0 is defined as the *resistance factor*, $S_R = x_R^0/x_R^* > 1$. The *load factor* S_L, which represents the margin between applied stress basis x_L^0 and design stress x_L^*, is defined as $S_L = x_L^*/x_L^0 > 1$.

To evaluate the strength of materials, $x_L^* < x_R^*$ is often employed as a basis. For instance, considering the history of research since Carl von Bach, a German mechanical engineer (1847–1931) [61], S_L has been considered a minor contribution in mechanical design compared with S_R. The safety margin S is generally defined as $S = x_R^0/x_L^0 > 1$.

Consider the margin which shows the difference between the minimum value x_R^* in the distribution of x_R, and the maximum value x_L^* in that of x_L at a start time t_0 in Fig. 2.14. When deterioration of the material occurs in service, x_R^* will be equal to x_L^* at time t_1, and x_L^* becomes larger than x_R^* at time t_2. Thus the probability of fracture, P, increases when both distributions overlap with each other at t_2.

Regarding the fatigue limit of metals, the safety margin S is written as $S = \bar{x}_R/\bar{x}_L$, and the distribution with random variable x_{R-L} is represented as shown in Fig. 2.15, where $x_{R-L} \equiv x_R - x_L$. Because fracture occurs in

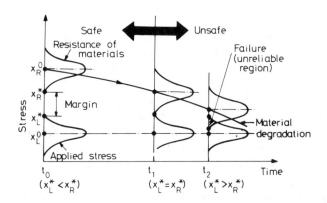

Fig. 2.14. Material degradation of structural material in service and occurrence of unreliable region.

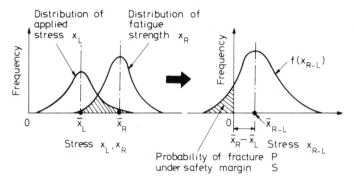

Fig. 2.15. Safety margin for metallic fatigue strength.

the region of $x_{R-L} < 0$, the probability of fracture, P, under the safety margin S is written as $f(x_{R-L}) \, dx_{R-L}$.

Assuming that the distributions of both x_L and x_R are represented by the normal distribution, the following equation can be obtained:

$$P = \int_0^\infty (2\pi)^{-1/2} \exp(-t^2/2) \, dt \tag{2.13}$$

where

$$t = (x_{R-L} - \bar{x}_{R-L})/(s_R^2 + s_L^2)^{1/2}$$

and

$$t_V = \bar{x}_{R-L}/s_{R-L}$$

In eqn (2.13) s_R, s_L and s_{R-L} are the standard deviation of x_R, x_L and x_{R-L}, respectively. Substituting $S = x_R/x_L$ into the equation for t_V, and solving the equation for S, the following equation is obtained:

$$S = \{1 + [1 - (1 - t_V^2 \delta_R^2)(1 - t_V^2 \delta_L^2)]^{1/2}\}/(1 - t_V^2 \delta_R^2) \tag{2.14}$$

where δ_R and δ_L show the c.o.v. of x_R and x_L, respectively. Table 2.6 [62] shows the values of S calculated by eqn (2.14) with $P = 1/100$ or $R = 99.9\%$. It is found from the table that S becomes larger as δ_L or δ_R increases. The safety margin $S = x_R/x_L$ corresponds to *partial safety factor design* in structural design. It was first applied to the limit design in the USSR (1955) and therefore was incorporated into the ISO (1973), BS (1978) and other standards.

2.5.3 Safety Life, Fail-safe and Damage of Materials

After the 1950s, an important change started in the structural design of

Table 2.6
Factor of Safety for Reliability, R, of 99·9% (From ref. 62 by courtesy of Soc. Mater. Sci., Japan)

L	R				
	0·01	0·03	0·05	0·10	0·15
0	1·03	1·10	1·18	1·45	1·86
0·1	1·30	1·32	1·37	1·58	1·98
0·3	1·94	1·95	1·99	2·15	2·50
0·5	2·58	2·60	2·62	2·78	3·15

airplanes, nuclear power plants, etc. This was the implementation of the new concept 'fail-safe'.

Safety life is usually considered to be the duration before cracks begin forming in a structural material in service. However, fail-safe is based on the following postulation. Crack initiation might be limited to a small portion of the structure, but it should be inspected in service and failed members repaired before integrity declines.

In normal safe-life design, the structure will not fail until the integrity of the majority of structural members has declined considerably, but the progressive deterioration of the material in service can be unavoidable because of the statistical character of crack initiation and propagation. If deterioration of the material is identified in service before such failure, interruption of operation will be disadvantageous, thus the fail-safe design was started. ASIP (1975) [63] and ASME Code Section III (1972) [64] are the so-called damage tolerance designs.

According to a report by the International Civil Aviation Organization (ICAO) on aircraft accidents in commercial airlines of the world, in the 1950s accidents occurred at a rate of about one to 100 000 flying hours, but in 1978 it was one to 800 000 flying hours. Data for test flights, war, and air piracy are not contained in the report. Commercial flights have become eight times safer in about 20 years. Due to the fact that failure of a structure in this case may bring about loss of human life and/or damage to property, the concepts 'safety' and 'reliability' are considered synonymous. However, the former is a broad concept compared with the latter. Therefore the study of safety is an interdisciplinary field involving social as well as natural sciences.

In general, damage tolerance design is performed with the following considerations:

1. There is inevitably a defect which is not detected.
2. There are various degrees of inspection.
3. The degree of 'damage' detectable varies according to the inspection method employed.

It is not easy to define *materials damage* accurately. The type of damage that is difficult to detect is that which degrades the material due to distributed microcracks. Such cracks will bring about the fatal fracture of the material due to crack propagation; they decrease the rigidity, toughness and lifetime of the materials.

Requirements for the evaluation methods concerning the material damage of structural materials with high-temperature applications are becoming more sophisticated in response to the *remaining life assessment* in the long-term service of high-temperature power plants. Many methods have been proposed [65] but an effective method has been difficult to find. Typical measures of metal damage at elevated temperatures are shown in Table 2.7 [66]. These measures are determined from both non-destructive and destructive examinations, and various types of non-destructive inspection have been proposed. Seven methods are employed in the *in-service inspection* of ASME Code Section V (1971) [67]. On the other hand, the destructive examination has the advantage of being a direct evaluation of the strength and remaining life of a structural material. A flow chart of the evaluation of defects in ASME Section IX is shown in Fig. 2.16 [68] (1978). In the figure, a_f is the maximum size of defect at the final period of life of nuclear power plant components. It is estimated (ASME Section III) to be a semi-elliptically shaped crack with depth $t/4$ and length $1.5t$, where t is the thickness of membrane components. In Code Case Section IX, the size is nearly one-tenth that in Code Section III. The critical size a_c and the maximum size a_f in nuclear power components are determined by means of LFM.

It is also necessary to clarify the relationship between local evaluation of damage and total evaluation of integrity, through investigation into the actual conditions of damage in machine structures, analysis of damage factors and study of evaluation methods of integrity in machine structures as shown in Fig. 2.17 [69].

2.5.4 Statistical Law

The law employed in reliability engineering is based on quantitative study; it does not clarify the transition from quantitative to qualitative

Table 2.7

Measure of Damage of Metals at Elevated Temperatures (Reprinted from ref. 66 by permission of Iron and Steel Institute, Japan)

Measure of damage, D	Characteristics	Applications
Cavitation detriment $D=f(V)$ (V, cavity density)	(1) D is composed of nucleation of boundary cavities and cracks; V is represented by function of applied stress σ, temperature T, and strain ε (2) D is not necessarily constant for the life fraction given	Creep Creep/fatigue
Aging detriment $D=1-t_R/t'_R$ (t_R, life of non-aged metal; t'_R, life of aged metal)	(1) Life of metals is changeable by factors such as precipitation density, dislocation density and others (2) D is not necessarily constant for the life fraction given	Creep Fatigue
Macroscopic mechanical parameter $D=1-(\sigma/\bar{\sigma})$ $D=1-(\varepsilon_p/\bar{\varepsilon}_p)^n$ ($\bar{\sigma}$, net stress of damaged metal; $\bar{\varepsilon}_p$, maximum plastic strain of damaged metal; n, cyclic work-hardening coefficient)	(1) This is the damage mechanical approach and is represented by the development equation which controls the development of damage variables (2) Net stress, $\bar{\sigma}$, means the magnified stress which resulted from the decrease in effective cross-sectional area due to material degradation	Creep Fatigue Creep/fatigue
Residual life $D=1-t_R/t'_R$ (t_R, life of virgin metal; t'_R, life of damaged metal)	(1) D will be determined from tests of the specimen sampled from structure in service (2) Procedure is easy and applicable to nonlinear damage rule	Creep Fatigue Creep/fatigue

change in intrinsic inevitability. However, this is not to underestimate the importance of statistical law. The inevitability observed in the processes of plastic flow and fracture of solids is objectively carried through by the medium of many contingencies.

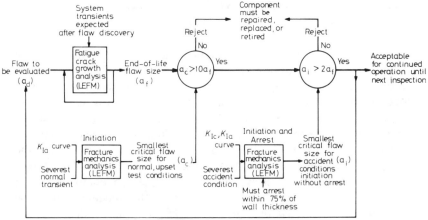

Fig. 2.16. Flow chart of evaluation of defects in ASME Code Section XI. (Reprinted from ref. 68 by permission of Electric Power Research Institute)

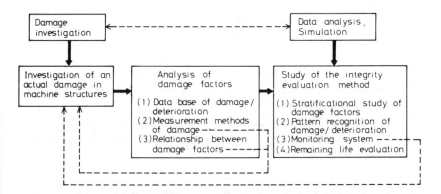

Fig. 2.17. Study of integrity evaluation method for a machine structure. (Reprinted from ref. 69 by permission of Science and Technology Agency, Japan)

2.6 MODELING AND SCALE-UP

2.6.1 Modeling [70]

A *model* is conceptually different from an *original* or *prototype*. It is employed as a means of understanding the original.

The following are *analogues* which were devised to understand the objective: (i) models of the crack-tip mechanism taking into consideration interaction and stochastic processes; (ii) models of the continuum approximation in the mechanics of material with microstructure; (iii) models of the structural boundary etc., as previously mentioned in Sections 2.2 and 2.3. Therefore the model represents the 'structure' and 'function' of the objective. As a matter of fact, a model is different from the original, but if it resembles the original too closely it will lose the capability of representing the structure and function of the original in a simplified form. Thus *modeling* is a kind of abstraction by which the objective is clarified. Modeling is a useful research tool, but it is a mistake to believe that it will replace the need to comprehend the objective. To avoid this mistake it is important that modeling should be recognized as a *stratificational structure* or *hierarchy* and understood as a means of theoretical study. It is also important that the limits of application are clarified. Thus modeling plays a useful role when one correctly recognizes the objective.

2.6.2 Scale-up

Iwasaki and Miyahara, a Japanese philosopher and a physicist respectively [71] (1976), classified the concept of the model as follows:

1. It is a scale representation of the original.
2. Both structure and function of the original are represented in an essential way to study their respective problems independent of the size of the original.
3. Attention is mainly focused on the function of the original.
4. Both structure and function of the objective are ideally represented in philosophical thinking.

Items 2 and 3 are as mentioned previously in Sections 2.2 and 2.3, and item 4 is employed in theoretical physics. Item 1 will now be described regarding the *scale-up* which is very important in the study of SFM.

The etymological origin of the word 'model' is in French where it means a scaling down of the original, according to C. Iwasaki and S. Miyahara [71]. Models often involve a scaling down of the shape of the original. There are cases where only the shape is a scaled down model of the original, but in many cases a model is an analogue of the structure and function as well as of the original.

In the study of SFM, a laboratory test for miniature [65] or small specimens, a simulation test for the model and a to-scale mock-up test

STUDY OF PLASTICITY AND FRACTURE OF SOLIDS   45

are performed for predicting remaining life of structural components in service and for designing machine structures. However, the *law of similitude* between the data for miniature or small specimens and those for to-scale does not necessarily hold due to various influencing factors. This is the reason scale-up is not a simple matter in the study of SFM.

About five centuries ago Leonardo da Vinci (1425–1519) recorded his observations on 'testing the strength of iron wires of various length'; folio 82, recto-b, *Codex Atlanticus* (property of the Ambrosiano library, Milan). An illustration from this work is shown in Fig. 2.18 [72].

He wrote (author's additions in parentheses): 'the object of this test was

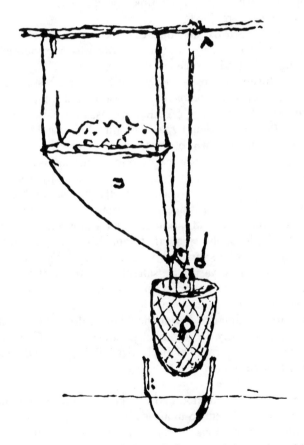

Fig. 2.18. Leonardo da Vinci (1425–1519) first tested iron wire with different lengths. (From ref. 72)

to determine the load an iron wire can carry. Attach an iron wire 2 braccia (1 braccia = 60 cm) long to something that will firmly support it, then attach a basket (Leonardo called it a 'cavagno', using a term from the Milanese dialect) or a bag ('sporta', from a Tuscan word) to the wire and feed into the basket some fine sand through a small hole placed at the end of a hopper. A spring is fixed so that it will close the hole as soon as the wire breaks. The basket is not upset while falling, since it falls only a very short distance. The test is repeated several times to check the results. Then a wire of 1/2 the previous length is tested and the additional weight it carries is recorded; then a wire of 1/4 the original length is tested and so forth, noting each time the ultimate strength and the location of the fracture' [73].

This may have been the first time that the tensile fracture strength of metal, wood, stone or rope was examined. But why he proposed wires of different lengths instead of different diameters is not clear, as the varied weights of such short pieces of wire would have no measurable effect on the results. 'The only reasonable explanation is that Leonardo made an error in the use of words', wrote W. B. Parsons, an American historian of science [73]. Leonardo began by taking a wire 2 braccia long for the first test. Having the measurement of length in mind, he wrote 'length' instead of 'thickness (diameter)' when he described the succeeding step, and 'additional' instead of 'lesser' for the weight carried by a smaller wire. Thus he must have learned that the tensile fracture strength of iron wire was changed by the length instead of the diameter. George R. Irwin, an American researcher of solid mechanics and a pioneer of linear fracture mechanics (1961) [74], also wrote that Leonardo discovered the increase in fracture strength of a short iron wire. In Leonardo's day the strength of iron wire must have varied considerably. In those days people were sure to have learned by experience that the strength of materials varied, but to devise a qualitative test was the work of genius.

'We can now summarize the method Leonardo seems to have used, doubtless unconsciously, in scientific questions', wrote C. Truesdell, an American researcher of rational mechanics (1968) [75], and he indicated the following three points:

1. Observe the phenomenon and list the quantities having numerical magnitude that seem to influence it.
2. Set up linear relations among pairs of these quantities as are not obviously contradicted by experience.
3. Propose 'rules of three for trials in experiments'.

On the other hand, M. J. Barba, a French civil engineer (1880) [76], proposed that a decrease in tensile fracture strain ε is inversely proportional to the gage length L. This was the first step toward *Oliver's law* (1928), $\varepsilon = K(A/L)^n$, where A is the cross-sectional area of the specimen, and K and n are material constants. The Iron and Steel Institute of Japan (ISIJ) (1973) [77] reported that Oliver's law ($n = 0.4$) was applicable to specimens of hot rolled steel in the region from -0.53 to -1.15 of $\log(A/L)$. This judgement was made based on data for 532 specimens.

2.6.3 Law of Similitude

The following concerns the *law of similitude* in the study of SFM. Sedov, a researcher of continuum mechanics in the USSR (1957) [78], pointed out the following:

1. Phenomena with constant 'π-factors' in *dimensional analysis*, and the same *criterion of similitude*, are called '*similar*' to each other.
2. The solution of the control equation of similar phenomena is written as a homogeneous equation with parameters in the criterion of similitude.

Therefore the following similitudes are necessary for performing simulation tests:

1. Geometrical similarity.
2. Similarity of both composition and structure of material.
3. Similarity of mass distribution.
4. Similarity of both input and output (response), i.e. mechanical and physical similarities such as load, temperature, environment, and their distribution.
5. Similarity of timing.

To apply the law of similitude to the analysis of SFM, the following are employed [78]:

1. Local criterion of similitude.
2. Similarity condition of deformation or fracture process.
3. Extraction of an important and stable growing process in mechanical behavior.
4. Linear fracture mechanics (LFM) approach.

For instance, apply the stress intensity factor (SIF) K, which represents the mechanical intensity at the crack tip, to the local criterion of similitude. The similarity condition of K at the crack tip for both model and original is written as

$$K(\text{model})/K(\text{original}) = \text{constant} \tag{2.15}$$

In the local criterion of similitude of eqn (2.15), because K is determined to be the ratio of crack length l to the representative length L of the cracked material, it is necessary to maintain a constant ratio. Size effects will probably have some influence when the parameter in the local criterion of similitude contains L.

In the plate with double side notches shown in Fig. 2.19, (b) is geometrically n times as large as (a). Because the *form factor* α (non-dimensional) of (a) is equal to that of (b), the maximum stresses at the notch root are given as $\sigma_{\max} = \alpha\sigma$ and $\sigma'_{\max} = \alpha\sigma'$, respectively. On the other hand, in the plate with geometrically similar double sided cracks, shown in Fig. 2.20, $K = \sigma\sqrt{C\pi l}$ and $K' = \sigma'\sqrt{C\pi n l}$, where C is the only constant determined by the size ratio. If $\sigma = \sigma'$, then $K' = \sqrt{n}\,K$. In order to obtain the same K, the condition $\sigma' = \sigma/\sqrt{n}$ should be satisfied. If a cracked material fractures when K reaches the critical magnitude, such as *fracture toughness* K_c under monotonic loading and *fatigue fracture toughness* K_f under repeated loading, the local criterion of similitude of the critical K is more appropriate than the maximum stress, $\sigma_{\max}$. Since K_c is connected with the *energy release rate*, G_c, eqn (2.15) also shows the local criterion of similitude at certain energy levels. Furthermore, because K_c or G_c determines the *critical crack length*, l_c, at which

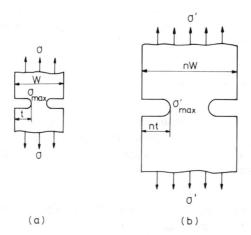

Fig. 2.19. Two plates with geometric similar double-sided notches.

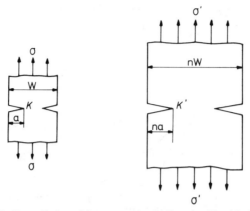

Fig. 2.20. Two plates with geometric similar double-sided cracks.

unstable fracture occurs, eqn (2.15) shows the geometrical criterion of similitude.

2.6.4 Size Effect

Tensile elongation, brittle fracture strength, fatigue strength and other mechanical properties of metallic materials are influenced by the specimen size, and the so-called *size effect* is observed.

In spite of many experimental and theoretical studies since the 1950s, there are still many unknown factors concerning the size effect in metallic fatigue. However, the following factors have been confirmed:

(1) The more the specimen size increases, the more the fatigue limit, σ_w, of the specimen decreases. With specimen diameters from 10 mm to 25–50 mm, σ_w decreases greatly but becomes constant with larger diameters.

(2) For the same kind of steel, the more the tensile strength or fatigue strength is increased by heat treatment, the more conspicuous is the effect. This conspicuous effect appears as the fluctuation of strength of material increases.

(3) The size effect is greater for notched specimens than for smooth specimens. The *fatigue strength reduction factor*, β, (nondimensional) is equal to the form factor, α, when α is comparatively small, but there is a general tendency for β to be smaller than α when α is comparatively large. When α is large there are unusual cases in which β decreases as α increases, or β becomes larger than α.

(4) The size effect is greater in the rotating bending fatigue test but less

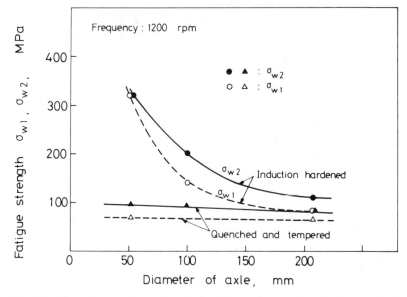

Fig. 2.21. Size effect on the fatigue limit of induction-hardened railway axle for the Japanese bullet train. σ_{w1}: Failure was indicated at the press-fitting portion; σ_{w2}: fatigue cracks were indicated by magnetic particle inspection. (Reprinted from ref. 79 by permission of ASTM)

in push–pull and reversed torsion tests. Fatigue data are shown in Fig. 2.21 [79] for induction-hardened railway axles composed of structural steel S38C press-fitted into the hub, with diameters of 52, 100 and 209 mm. A series of full-sized rotating cantilever bending fatigue tests were performed in the development of the Japanese *bullet train*. A rotation speed of 1200 rpm was used in these tests, which corresponds to 200 km/h, the service speed of the bullet train. For the induction-hardened axles, the size effect was conspicuously great at a diameter of 50 mm. The reason was clarified to be the decrease in the superficial compressive residual stress due to the large relative slip at the fretting portions of the larger axles [79].

The following factors have been studied to explain these facts:

1. *Statistical factors* such as fluctuation in the strength of materials, and the existence of a surface layer (volume) subjected to large stress.
2. *Mechanical factors* such as the stress gradient.

3. *Energy factors* which have more elastic energy as the specimen size increases.
4. *Deviation from the similitude condition.*

Factor 1 is explained by using the *weakest link model* mentioned in Fig. C in Table A of Appendix I. For instance, if both model and original have the same length to diameter ratio of 1:4, and are made from the same material with homogeneously distributed defects, the ratio of both tensile fracture strengths σ_c is given by eqn (4) in Table A as [74]

$$\sigma_c(\text{original})/\sigma_c(\text{model}) = (1/4)^{1/m} = (1/4)^{\delta/1 \cdot 2} = 1 - \delta < 1 \qquad (2.16)$$

Here δ is the relative variance, $\delta = \sqrt{\Sigma_{i=1}^{N}[(\sigma_i - \sigma)/\bar{\sigma}]^2/N}$, and the relation $\delta = (1 \cdot 2)/m$ was used. As seen from eqn (2.16), the size effect is connected with fluctuations in the strength of materials. When there is such a large fluctuation, σ_c will decrease greatly. However, it is not so simple for the fatigue fracture of metals under stress alternation.

The following concerns factor 2. This mechanical factor is explained as follows. If both surface stresses are equal, the stress gradient in specimens having large diameters is smaller than that of small diameter specimens.

The influential region of stress gradient within a finite length normal to the specimen is denoted as R^* in Fig. 2.22. The *stress mean value*, σ_1^*, in volume V_1 over the region R^* is larger for a specimen having a larger diameter than that seen in volume V_2 for a specimen having a small diameter, even though both surface stresses are equal. Therefore the probability of fracture for a specimen having a large diameter is greater than that for a specimen having a small diameter. The stress mean value, σ^*, is defined as $\sigma^* = (1/R^*) \int_{R-R^*}^{R} \sigma_{eq} \, dr$, where σ_{eq} is the comparative stress in a state where multiaxial stresses are present.

This type of size effect for smooth specimens is observed in reversed bending or reversed torsion tests. On the other hand, two geometrically similar round notched bar specimens have equal maximum stresses at the notch root, but the difference between the stress mean values σ_1^* and σ_2^* is larger for notched bar specimens than for smooth bar specimens because of the influence of a steep gradient of stress in the notched bar.

The theory of stress mean value is based on the *micro-support effect* proposed by Heinz Neuber, a German researcher of solid mechanics, in 1936 [80]. He was born in Stettin in 1906, and studied at technical universities in Berlin and Munich. (He started his research as an assistant of Professor Ludwig Foppl in Munich and was appointed professor of the Dresden Technical University and held many other

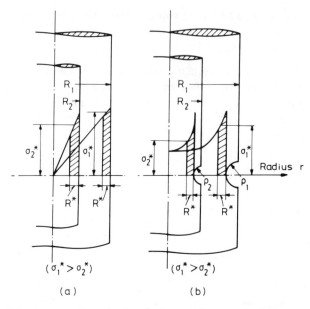

Fig. 2.22. The stress gradient in the specimen influences the fatigue strength. (a) Smooth round bar subjected to cyclic bending or cyclic torsion. (b) Round notched bar subjected to push-pull and cyclic bending or cyclic torsion.

academic posts in Germany.) Certain domains of a material have strong bonds between particles; therefore they are supported as quasi-rigid blocks by their environment. Thus, in the vicinity of the point of high stress, it is assumed that the material has a stress mean value over a finite length normal to the surface within the range of high stress concentration, the so-called *fictive length of structure*, R^*. The fictive radius of curvature R_F can be determined for a sharply curved notch. This curvature differs from the real radius of curvature R by the small length R^*:

$$R_F = R + R^* \qquad (2.17)$$

Considering a new radius R_F, the new stress concentration factor, α^*, is written as $\alpha^* = \alpha(R/R_F)^{1/2}$. This replaces the ideal (Hooke's) stress concentration factor α. If the theory of stress mean value is applied to the stress concentration problem at holes and notches in the Cosserat continuum (mentioned in Section 2.3), then the fictive radius of curvature can be represented in the form [81]

$$R_F = R\left(1 + \sum_{i=1}^{\infty} \sum_{j=0}^{\infty} S_{ij} R^{*i} l^j R^{-i-j}\right) \qquad (2.18)$$

where l is the Cosserat structural length and S_{ij} the nondimensional quantity, and $R_F = R + R^*$ when $i=0$ and $j=0$, where $R^* = S_{10}$. Generally speaking, the stress concentration at holes and notches in the Cosserat elastic body is relatively small compared with that in a linear elastic body [82]. Furthermore, the fracture toughness of the Cosserat elastic body is smaller than that of a linear elastic body [82].

A similar problem with *surface layer* δ was independently proposed in 1953 [83] by Fujio Nakanishi (1897–1964), a professor at the University of Tokyo. He considered the layer weaker than the interior and that the maximum concentrated stress in notched material occurred a short distance δ normal from the surface. Thus he estimated the yield strength of the sharp notched material with large scale from that of a sharp notched small specimen if δ was given. Nakanishi also discovered the so-called *two types of phenomenological yielding of mild steel* in 1928–9 [84, 85]. One is a kind of stability problem for the whole material, and the other is the criterion for local yielding to occur when the yield stress reaches the critical value before yielding due to the above-mentioned stability. The two yield conditions are written $\bar{\sigma} = \sigma_y$ and $\sigma_s = 1 \cdot 5\sigma_y$, respectively. Here σ_y is the so-called yield stress of the material, $\bar{\sigma}$ is the mean yield stress with weight on the whole section yielded in the material, and σ_s is the stress at which local yielding occurs. If σ_s was replaced by σ_s' at the interior distance δ, the criterion of $\sigma_s = 1 \cdot 5\sigma_y$ will be applicable to cases where the surface is influential in material yielding such as in the simple tension of a small wire, the bending of a thin plate specimen, and the simple tension of a sharp notched specimen. (Nakanishi also studied rocket thrust, rocket mechanics, and fuel injection and combustion, at the Institute of Aeronautics, University of Tokyo, first opened in 1909 as the Japanese Military Balloon Research Association and changed to an academic institute in 1921).

Historically, many studies have been made in which the fatigue strength reduction factor (nondimensional), β, is estimated from the elastic stress concentration factor (nondimensional), α. For instance, if α^* is assumed to be equivalent to β, the effect of the fictive length of structure R^* or stress gradient on β can be considered.

Lastly, the energy factor 3 mentioned above is also considered from the point of view that the fracture toughness, K_c (energy release rate G_c) itself contains the size ratio.

Thus it is necessary to consider the effect of size on the fracture strength of solids in machine structure design. In fact, this consideration is of the utmost technical importance. In this theory the relation between parameters in the local criterion of similitude is generally written by a homogeneous equation of the nondimensional quantities. In the theory of scale-up, it is desirable to contrive a new kind of parameter from a stratificational point of view. A computer has successfully performed a scale-up by using the concept 'hierarchy' in the computer memory and the network. However, this is not always successful in the study of SFM.

REFERENCES

1. M. Ohnami, in *Science and Technology on Strength and Fracture of Materials*, Vol. 1 in A Series of Comprehensive Lectures on Strength and Fracture of Materials (ed. M. Ohnami), OHM-Sha, Tokyo, 1984, pp. 33–6 (in Japanese).
2. T. Yokobori, in *Science and Technology on Strength and Fracture of Materials* (ed. M. Ohnami; see ref. 1), p. 12 (in Japanese).
3. S. S. Manson, Fatigue: a complex subject—some simple approximations (Lecture), *Experimental Mechanics*, **5** (1956), 193, Soc. Exp. Stress Anal.
4. J. T. Fong, Statistical aspect of fatigue at microscopic, specimen and component levels, *ASTM Special Technical Publication*, **675** (1979), 730, Amer. Soc. Test. Mater.
5. J. T. Fong, Purpose of a research briefing, ASTM Committee E-9 (1983).
6. Ref. 1, pp. 36–42.
7. T. Yokobori, *An Interdisciplinary Approach to Fracture and Strength of Solids*, 2nd Edn, IWANAMI-Shoten, Tokyo, 1974, pp. 256–71 (in Japanese) (Russian Edn, Dumka, Kiev, 1978).
8. V. S. Ivanova, V. F. Terent'ev, *Privoda Ustalost Metallov*, Metallurgiya, Moskva, 1975, pp. 6–11 (Japanese Edn, GENDAI RIKOHGAKU-Sha, Tokyo, 1979).
9. H. Kitagawa and R. Koterazawa, Fractography, in *Fracture Mechanics and Strength of Materials* (Series of Monographs, ed. H. Miyamoto *et al.*), Vol. 15, BAIFUKAN, Tokyo, 1977, pp. 3–9 (in Japanese).
10. M. Ohnami, Influence of hydrostatic pressure on fracture toughness of polycrystalline metallic materials, *Journal of Japan Society for Technology of Plasticity*, **15** (1974), 769 (in Japanese), CORONA-Sha, Tokyo.
11. M. Ohnami, unpublished.
12. M. Ohnami, M. Sakane and N. Hamada, Effect of changing of principal stress axes on low-cycle fatigue life in various strain shapes at elevated temperatures, *ASTM Special Technical Publication*, **853** (1985), 622.

13. A. Wöhler, Uber die Festigkeitsversuche mit Eisen und Stahl, *Zeitschrift fur Bauwesen*, **20** (1870), 73, Mitwirkung der Konigl. Technischen Bau-Deputation und des Architeckten-Vereins zu Berlin.
14. Ref. 7, pp. 185–6.
15. R. Ohtani, M. Okuno and R. Shimizu, Microcrack growth in grain boundaries at the surface of a smooth specimen of 316 stainless steel at high temperature creep, *Journal of the Society of Materials Science, Japan*, **31** (1982), 505 (in Japanese).
16. C. Iwazaki and S. Miyahara, *Modern Natural Science and Materialism Dialectic*, OHTSUKI-Shoten, Tokyo, 1972, p. 266 (in Japanese).
17. Ref. 1, pp. 42–51.
18. M. Ohnami and K. Shiozawa, Strength and fracture of polycrystalline body, in *Fracture Mechanics and Strength of Materials* (ed. H. Miyamoto et al.) Vol. 14, 1976, pp. 2–8 (in Japanese).
19. W. Noll, A mathematical theory of the mechanical behavior of continuous media, *Archives for Rational Mechanics and Analysis*, **2** (1958), 197, Springer Verlag, Berlin.
20. I. A. Kunin, The theory of elastic media with microstructure and the theory of dislocations, in *Mechanics of Generalized Continua* (ed. E. Kröner), Springer Verlag, Berlin, 1968, p. 197.
21. C. Truesdell and R. Toupin, The classical field theories, in *Handbuch der Physik* (ed. S. Flugge), III-I (1960), 226–793, Springer Verlag, Berlin.
22. K. Kondo, *RAAG Memoirs of the Unifying Study on Basic Problems in Engineering and Physical Science by Means of Geometry*, Vols 1–4, Gakujitsubunken-Fukyukai, Tokyo, 1955–68.
23. B. A. Bilby, R. Bullough and E. Smith, Continuous distributions of dislocations: a new application of the methods of non-Riemannian geometry, *Proceedings of the Royal Society*, **A231** (1955), 263, The Royal Society, London.
24. J. F. Nye, Some geometrical relations in dislocations in dislocated crystals, *Acta Metallurgica*, **1** (1953), 153.
25. E. Kröner, *Kontinuumstheorie der Versetzungen und Eigenspannungen* Springer Verlag, Berlin, 1958, pp. 1–177.
26. T. Mura, Continuum theory of dislocations and plasticity, in *Mechanics of Generalized Continua* (ed. E. Kröner), Springer Verlag, Berlin, 1968, p. 269.
27. S. Minagawa, Dislocation theory as a field theory, *Journal of the Materials Science Society of Japan*, **9** (1972), 271 (in Japanese), Soc. Mater. Sci., Japan.
28. M. Ohnami and K. Shiozawa, Definition of misorientation and the application to evaluation of strength of metallic materials, *Transactions of the Japan Society of Mechanical Engineers*, **39** (1973), 769 (in Japanese).
29. M. Ohnami (ed.) *Introduction to Micromechanics*, OHM-Sha, 1980, pp. 205–54 (in Japanese) (Russian Edn, *Vedenie v Mikromekhaniki*, Metallurgiya, Moskva, 1986).
30. S. Minagawa, On some generalized features of the theory of stress function space with references to the problems of stress in a curved membrane, *RAAG Memoirs* (see ref. 22), **4C** (1968), 99.
31. E. Kröner, The rheological behavior of metals, *Rheology Acta*, **12** (1973), 37.
32. E. Kröner and C. Teodosiu, Lattice defects approach to plasticity and

viscoplasticity, in *Problems of Plasticity* (ed. A. Sawczuk), Noordhoff Publishers, Leyden, 1974, pp. 45–88.
33. Ref. 29, pp. 91–6.
34. D. G. Brandon, B. Ralph, S. Ranganathan and M. S. Wald, A field ion microscopic study of atomic configurations at grain boundaries, *Acta Metallurgica*, **12** (1964), 813.
35. J. W. Cahn, Nonequilibrium surface and interface thermodynamics, in *Atomistics of Fracture* (ed. R. M. Latanision and J. R. Pickens), 1981, p. 427, published in co-operation with NATO Sci. Affairs Div., New York.
36. J. D. Eshelby, The force on an elastic singularity, *Philosophical Transactions of the Royal Society*, **A244** (1951), 87, The Royal Society, London.
37. J. R. Rice, A path-independent integral and the approximate analysis of strain concentration by notches and cracks, *Transactions of the American Society of Mechanical Engineers*, **E35** (1968), 379.
38. G. P. Cherepanov, Cracks in solids, *International Journal of Solids and Structures*, **4** (1968), 811, Pergamon Press.
39. A. A. Griffith, The phenomena of rupture and flow in solids, *Philosophical Transactions of the Royal Society*, **A221** (1920), 163.
40. M. Hirata, Statistical phenomena in science and engineering (Lecture), *Science of Machine*, **1** (1949), 231 (in Japanese), YOKENDO, Tokyo.
41. H. Kubota, Time of rupture of glass, *Journal of Physics, Japan*, **17** (1948), 286 (in Japanese), Physical Soc. of Japan.
42. T. Yokobori, *The Strength, Fracture and Fatigue of Materials*, GIHODO, Tokyo, 1955, pp. 6–10, 178–84 (in Japanese).
43. Ref. 7, Chapter 8.
44. K. Ohmori, Application of the theory of the rate of chemical reaction to the process of distribution, *Memoirs of Faculty of Engineering, Kyushu University*, **16** (1941), 239 (in Japanese).
45. T. Yokobori, Failure and fracture of metals as nucleation processes, *Journal of Physical Society of Japan*, **7** (1952), 44.
46. S. Glasstone, K. J. Laidler and H. Eyring, *The Theory of Rate Processes* McGraw-Hill, 1941. New York.
47. A. Tobolsky and H. Eyring, Mechanical properties of polymeric materials, *Journal of Chemical Physics*, **11** (1943), 125, Amer. Institute of Physics.
48. E. S. Machlin and A. S. Nowick, Stress rupture of heat-resisting alloys as a rate process, *Trans. AIME*, **172** (1943), 386.
49. T. Yokobori, Delayed fracture in creep of copper, *Journal of Physical Society of Japan*, **6** (1951), p. 78.
50. J. C. Fisher, J. H. Hollomon and D. Turnbull, Nucleation, *Journal of Applied Physics*, **19** (1948), 775, Amer. Institute of Physics.
51. For example, J. C. Fisher, The fracture of liquids, *Journal of Applied Physics*, **19** (1948), 1062.
52. Ref. 7, p. 120.
53. S. Yamamoto, The present situation and fundamental problems in reaction rate theory (Review), *Bulletin of the Japan Institute of Metals*, **19** (1980), 866 (in Japanese).
54. I. Prigogine, *Introduction to Thermodynamics of Irreversible Process*, 3rd Edn, Interscience Publishers, New York, 1961.

55. T. Terada, Analogy of crack and electron, *Proceedings of the Imperial Academy*, **7** (1931), 215, The Imperial Academy of Japan.
56. A. H. Cottrell, Fracture, in *The Physics of Metals* (ed. P. B. Hirth), Vol. 2, Defects, Cambridge University Press, 1975, p. 247.
57. Ref. 1, pp. 51–65.
58. E. J. Gumbel, *Statistics of Extreme*, Columbia University Press, New York, 1958.
59. W. Weibull, The phenomenon of rupture in solids, *Ingeniors Vetenskaps Akademien Hundlinger*, **153** (1939), Stockholm.
60. H. Okamura and H. Itagaki, Statistics of strength and fracture of material, in *Fracture Mechanics and Strength of Materials*, (ed. H. Miyamoto et al.; see ref. 9), Vol. 6, 1979, pp. 34–6 (in Japanese).
61. Carl von Bach, *Elastizität und Festigkeit*, 2nd Edn, Berlin, 1894.
62. Society of Materials Science, Japan (Ed.), *Design Manual on Fatigue of Metallic Materials*, YOKENDO, Tokyo, 1978, p. 13 (in Japanese).
63. MIL-STD-1530(A), Aircraft Structural Integrity. Airplane Requirements (1975).
64 American Society of Mechanical Engineers, ASME Boiler and Pressure Vessel Code, Section III, Rules for Construction of Nuclear Power Plant Components (1972).
65. JSME/IMechE/ASME/ASTM, Proceedings of the panel discussion on remaining life assessment of fossil fuel power plants: current situation and activities (Chairman, M. Ohnami; Co-chairman, M. Kitagawa), in *International Conference on Creep* (April 14–17, 1986, Tokyo), pp. 1–33.
66. M. Ohnami, Development of remaining life prediction system for high temperature plant components: example of steam turbine rotors (Review), *Journal of the Iron and Steel Institute of Japan*, **69** (1983), 1549 (in Japanese).
67. Amer. Soc. Mech. Engrs, ASME Boiler and Pressure Vessel Code, Section V, Non-destructive Examination (1971).
68. T. U. Marson, Flow evaluation procedures: Background and applications of ASME Section XI, Appendix A, NP-719-SR, p.B.–14 (1978), Electric Power Research Institute.
69. Science and Technology Agency, Japan, Report on Development of Techniques of Reliability Evaluation of Structural Materials (1987) (in Japanese).
70. Ref. 1, pp. 65–71.
71. C. Iwasaki and S. Miyahara, *Theory of Scientific Recognition*, OHTSUKI-Shoten, Tokyo, 1976 (in Japanese).
72. Leonardo da Vinci, *An Artabras Book*, Reynal and Co., New York, p. 274.
73. W. B. Parson, *Engineers and Engineering in the Renaissance*, MIT Press, Boston, 1968, pp. 72–3.
74. G. R. Irwin, Effects of size and shape on fracture of solids, *ASTM Special Technical Publication*, **283** (1961), 120.
75. C. Truesdell, *Essays in the History of Mechanics*, Springer Verlag, Berlin, 1968, p. 38.
76. M. J. Barba, *Memoires de la Société des Ingénieurs Civils*, Part I, Paris, 1880, p. 682.

77. Iron and Steel Institute of Japan (Ed.), Specimen size effect on tensile elongation, *Data Series*, **2** (1973), pp. 1–39 (in Japanese).
78. Ref. 8, pp. 67–9.
79. K. Nishioka, K. Ishii and H. Komatsu, Fatigue strength of induction hardened railway axles, *Journal of Materials*, **4** (1969), 413, Amer. Soc. Test. Mater.
80. H. Neuber, Zur Theorie der technischen Formzahl, *Forschung a. d. Geb. d. Ingenieurwessen*, **7** (1936), 271.
81. H. Neuber, On the effect of stress concentration in Cosserat continua, in *Mechanics of Generalized Continua* (ed. E. Kröner), Springer Verlag, Berlin, 1968, p. 109.
82. Ref. 29, pp. 169–92.
83. F. Nakanishi, M. Hanada *et al.*, Yield point of mild steel under sharply concentrated stress, *Transactions of the Japan Society of Mechanical Engineers*, **19** (1953), 14 (in Japanese).
84. F. Nakanishi, On the yield point of mild steel, *Journal of the Japan Society of Mechanical Engineers*, **31** (1928), 39.
85. F. Nakanishi, Strength of mild steel beams under uniform bending, *Journal of the Japan Society of Mechanical Engineers*, **32** (1929), 171 (in Japanese).

Chapter 3

History of the Combined Micro- and Macro-mechanical Approach to the Study of Plasticity and Fracture of Materials

This chapter describes perspectives of work done concerning a combined micro- and macro-mechanical approach to the strength and fracture of materials (SFM). We will chiefly pay attention to giving a view of the study of the plasticity of heterogeneous materials, the theory of distributed dislocations and the continuum approximation approach, and advances in fractology, from the standpoint of an interdisciplinary approach to SFM. In addition, the men who were responsible for these studies and work will be reviewed.

3.1 INTERDISCIPLINARY APPROACH

Combined micro- and macro-mechanical research into the strength and fracture of materials started with the formation of the study of SFM in the 1940s and 1950s. Formation of the study of SFM is characterized as follows:

1. Development of the boundary field in research.
2. Formation of a new integrated field of research, where an interdisciplinary approach between various research fields is necessary.

'Interdisciplinary' is used to denote that two or more research fields are related. Table 3.1 [1] shows the interdisciplinary phase of the study of SFM from the viewpoint of *specialization* and *integration*. The study of SFM has been created and advanced so as to combine the fields into a different hierarchy of sciences. Advances in interdisciplinary research itself appear to be progressing towards specialization, but this can also

Table 3.1
Specialization and Integration in the Study of Strength and Fracture of Materials (Reprinted from ref. 1 by permission of OHM–Sha)

Research fields \ Level	Atom, molecule, lattice defect (10^{-8})	Subgrain cell (10^{-4})	Polycrystal (10^{-2}) micro cracks / inclusions	Macro defect and notch (1)	Machine structures (10^{2})
Solid physics	├────┤				
Metallurgy	├──────────┤				
Generalized continuum theory[a]		├──────────┤			
Solid mechanics[b]			├──────────────┤		
Mechanical study of structures					├────┤
Study of SFM	├──┤				

Scale [cm]

[a] Contains the theory of the continuum approximation and of the statistical continuum regarding the continuous distribution of dislocations and polycrystalline bodies and others.
[b] Contains the so-called fracture mechanics.

be regarded as a sort of connection between respective studies through specialization toward the boundaries of knowledge in each field. For instance, the *theory of the continuous distribution of dislocations* constitutes the boundary between the crystal dislocation theory and crystal plasticity theory. In turn, a characteristic field of study has formed, known as the physical theory of plasticity, which led to physical fracture mechanics where both lattice defects and macro defects such as cracks and notches are connected. Takeo Yokobori (1974), an emeritus professor at Tohoku University [2], first proposed the 'micro–macro connected mechanical, kinetic and statistical approach to the fracture of solids', *fractology*, and he pointed out that characteristically it was the field most often studied in SFM. He received the Academy Award from the Japanese Academy of Sciences, an annual award for outstanding achievement and contribution, for his research into SFM.

Interdisciplinary studies are not really new. We can trace the first sources back to ancient Greece and Rome. Nevertheless Yokobori first published his book on the strength, fracture and fatigue of materials in 1955 [3], and founded the forum of the Japan National Symposium on Strength, Fracture and Fatigue of Materials (JNSSFM) in 1955. Afterwards an American symposium similar to JNSSFM was held in 1963 [4]. In 1965 he founded the International Conference on Fracture (ICF). These were important for establishing what was to be the driving force for the interdisciplinary study of SFM.

I had the privilege of presenting a short paper, dealing with what we can learn from the history of study on the fracture of materials, at the 30th JNSSFM in 1985 [5]. We know that there are seemingly infinite kinds of fatigue failure within structural materials, and it has been important to predict failure accurately. Nevertheless there are many questions remaining.

A. J. Kennedy, a former professor at the College of Aeronautics, Britain, in his short 1958 report on the century of research into fatigue since Wöhler [just one hundred years after the first of a series of scientific papers on metallic fatigue had been written by August Wöhler (1819–1914), a Prussian mechanical engineer], wrote: 'The study of the so-called laws of fatigue, as it now stands, is akin to attempting a study no less wide than that of the laws of economics or the laws of human illness. These cannot be fruitful scientific studies until we define what economic activity, or what illness, is under analysis—or to use the proper term, what the *system* is' Kennedy [6]. He also wrote: 'but the sum total of this hundred years must surely move us to consider whether profound changes are not required if fatigue, as a subject, is to *achieve a proper cohesion and theoretical structure together with a more purposeful and imaginative technology.*' Today, more than 125 years since Wöhler, it is worth asking what can we learn from past studies.

We will describe the historical advances in the combined micro- and macro-mechanical study of SFM, along the levels II and III in Tables 2.2 and 2.3 of the previous chapter.

3.2 HISTORY OF THE STUDY OF PLASTICITY OF HETEROGENEOUS MATERIALS [7, 8]

3.2.1 Single Crystal Slip in Metals

Slip is the basic mechanism of plastic deformation within metal crystals.

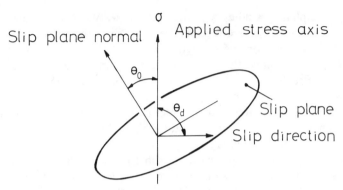

Fig. 3.1. Relation between applied stress axis, slip plane normal and slip directions.

In the last year of the 18th century, Sir James Alfred Ewing (1855–1935) and his colleagues at Cambridge University first observed slip on certain gliding planes within iron crystals when the specimen was sufficiently stressed [9]. Under an optical microscope the steps showed as dark lines which were called 'slip bands' [9].

When a single crystal is subjected to tensile stress σ, a resolved shear stress τ_s on the slip plane is given as (see Fig. 3.1)

$$\tau_s = \mu_{\text{ten}} \sigma \quad (\tau_s \leqslant \tau_c)$$
$$\mu_{\text{ten}} = \cos\theta_0 \cos\theta_d$$

(3.1)

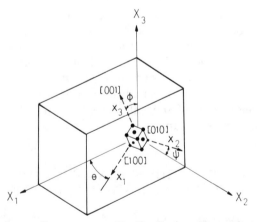

Fig. 3.2. Crystal coordinate system; X_1, X_2, X_3 are the specimen coordinates and x_1, x_2, x_3 the crystal coordinates.

where τ_c is the *critical shear stress* for crystal slip. This equation, first proposed by Erich Schmid (a professor of physics at Fribourg University, Switzerland, who was with the Metallgesellschaft A. G., Frankfurt-on-Main, in 1924), is called the *critical shear stress law* or the *Schmid law*. In eqn (3.1) μ_{ten} is called the *Schmid factor* and $M = 1/\mu_{\text{ten}}$ shows the *direction factor*.

Slip deformation of a crystal under multiaxial stresses was modeled as follows:

(i) Plastic deformation of the crystal occurs due to the summation of slips within active slip systems.
(ii) Shear stress on the active slip plane is equal to τ_c but on the non-active slip plane it does not exceed τ_c.
(iii) Slip occurs in the positive direction of shear stress and thus plastic work takes a definite positive value.

Considering crystal coordinates (Fig. 3.2) and slip in the slip system (Fig. 3.3), the following relation between stresses σ_{ij} and shear stresses τ^k is derived from equilibrium equations of stresses:

$$\left. \begin{array}{l} \tau^k = (1/2)(\mathbf{n}^k \mathbf{T} \mathbf{b}^k + \mathbf{b}^k \mathbf{T} \mathbf{n}^k) \\ \\ = \sum_{i,j=1}^{3} n_i^k b_j^k \sigma_{ij} \quad (k=1, 2, 3, \ldots, m) \\ \\ \tau^k \leqslant \tau_c^k \quad (k=1, 2, 3, \ldots, m) \end{array} \right\} \quad (3.2a)$$

On the other hand, the following relation between plastic strain

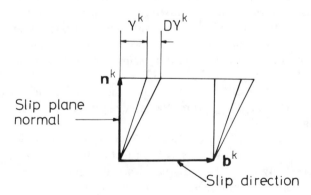

Fig. 3.3. Slip in a slip system ($k = 1, 2, \ldots, m$).

increments $D\varepsilon$ and slips $D\gamma^k$ is written as

$$D\varepsilon = (1/2)(\mathbf{n}^k\mathbf{b}^k + \mathbf{b}^k\mathbf{n}^k)D\gamma^k$$

or

$$(D\varepsilon)_{ij} = (1/2)(n_i^k b_j^k + b_j^k n_i^k)D\gamma^k \quad (3.3a)$$

and

$$D\gamma^k \geqslant 0 \quad (k = 1, 2, 3, \ldots, m)$$

Since $\tau^k = \tau_c^k$ in the active slip systems ($D\gamma^k > 0$), the following relation holds true:

$$(\tau_c^k - n_i^k b_j^k \sigma_{ij})D\gamma^k = 0 \quad (k = 1, 2, 3, \ldots, m) \quad (3.4)$$

By solving the equations from (3.2a) to (3.4) simultaneously, we can calculate the plastic deformation of the crystal under multiaxial stresses. Since eqn (3.4) is represented by five simultaneous equations involving $D\gamma^1$, $D\gamma^2$, $D\gamma^3$, $D\gamma^4$ and $D\gamma^5$, a combination of $D\gamma^m$ which produces five strain increments of $(D\varepsilon)_{ij}$ is uniquely determined when the number of slips is five. When m is larger than 5, however, many combinations exist. This is the so-called problem of *nonuniqueness in the choice of active slip systems.*

Thus the actual solution to a problem concerning plastically deformed polycrystals is that of a highly complex plastic boundary value problem for a large collection of anisotropic and continuous crystals. Specific solutions will depend on the shape of the crystallites and their orientations, both of which change with strain. At present a comprehensive study of the effects of material parameters such as strain hardening behavior, rate sensitivity, initial polycrystal textures etc., based on solutions to full boundary value problems, is not feasible.

Historically, in order to solve this problem the *principle of maximum plastic work* or the *minimum shear principle* was first proposed by G. I. Taylor, a British physicist [10], in 1938 and later by J. F. W. Bishop and Rodney Hill, British researchers into the mathematical theory of plasticity [11], in 1951, from the viewpoint of the mechanism of crystallographic slip.

Sir Geoffrey Ingram Taylor (1886–1975) is well known as a pioneer of crystal dislocation theory. He was born in London and graduated from Cambridge University. He had studied a wide field of sciences such as mathematics, meteorology, aeronautics and crystal plasticity, and was appointed to many governmental posts, such as Yallow Research Professor of the Royal Society, and received many academic awards both in Britain and abroad. He was also a member of the foreign academies of America, France, USSR, Poland, Sweden and India.

Let us take

$$\{\sigma\} = \{\sigma_{11}, \sigma_{22}, \sigma_{33}, 2\sigma_{23}, 2\sigma_{31}, 2\sigma_{12}\}$$
$$\{D\varepsilon\} = \{D\varepsilon_{11}, D\varepsilon_{22}, D\varepsilon_{33}, 2D\varepsilon_{23}, 2D\varepsilon_{31}, 2D\varepsilon_{12}\}$$
$$\{D\gamma\} = \{D\gamma^1, D\gamma^2, \ldots, D\gamma^m\}$$
$$\{\tau_c\} = \{\tau_c^1, \tau_c^2, \ldots, \tau_c^m\}$$

$$[T] = \begin{bmatrix} n_1^1 b_1^1 & n_2^1 b_2^1 & n_3^1 b_3^1 & \left(\dfrac{1}{\sqrt{2}}\right)(n_2^1 b_3^1 + n_3^1 b_2^1) & \left(\dfrac{1}{\sqrt{2}}\right)(n_3^1 b_1^1 + n_1^1 b_3^1) & \left(\dfrac{1}{\sqrt{2}}\right)(n_1^1 b_2^1 + n_2^1 b_1^1) \\ \cdots & \cdots & \cdots & \cdots & \cdots & \cdots \\ n_1^m b_1^m & \cdots & \cdots & \cdots & \cdots & \left(\dfrac{1}{\sqrt{2}}\right)(n_1^m b_2^m + n_2^m b_1^m) \end{bmatrix}$$

(3.5)

where $\{\sigma\}$ and $\{D\varepsilon\}$ are the column matrix of (6×1), $\{\tau_c\}$ and $\{D\gamma\}$ are that of $(m \times 1)$, and $[T]$ is that of $(m \times 6)$. Then eqns (3.2a) and (3.3a) are written as

$$\{\tau_c\} = [T]\{\sigma\} \qquad (3.2b)$$

and

$$\{D\varepsilon\} = [T]^T\{D\gamma\} \quad (D\gamma \geq 0) \qquad (3.3b)$$

respectively, where $[T]^T$ is the transpose matrix of $[T]$ of $(6 \times m)$.

Next let us consider the following problems:

Problem I: To maximize an objective function, $DW_I = \Sigma_{i,j=1}^{3} \sigma_{ij} D\varepsilon_{ij}$, for σ_{ij} under the restriction $\{\tau_c\} \geq [T]\{\sigma\}$.

Problem II: To minimize an objective function, $DW_{II} = \Sigma_{i,j=1}^{m} \tau_c^k D\gamma^k$, for $D\gamma^k$ under the restriction of $\{D\varepsilon\} = [T]^T\{D\gamma\}$.

When $\{\tau_c\}$, $[T]$ and $\{D\varepsilon\}$ are unknown and σ_{ij} and $D\gamma^k$ are variables, these two problems are referred to as a *dual problem* in linear programming methods. The solution has the following properties: (i) If one of the problems has the optimum solution the other also has it, and both maximum and minimum values of the objective function are equal. (ii) In the optimum solution both problems (I and II) are satisfied automatically; in other words, when the stress state has plastic work with the maximum value, τ in several slip systems reaches τ_c and it is guaranteed that the given strain increments can be realized.

Problem I corresponds to the principle of maximum plastic work as pointed out by T. Ohkubo [12]. The region of the solution for $\{\tau_c\} \geqslant [T]\{\sigma\}$ forms a polyhedron closed by five-dimensional deviatoric stress space, the so-called yield surface of a single crystal. Denoting stresses in this polyhedron as σ_{ij}^* and real stresses as σ_{ij} for given $D\varepsilon_{ij}$, the principle of maximum plastic work is written as

$$\sum_{i,j=1}^{3} (\sigma_{ij} - \sigma_{ij}^*) D\varepsilon_{ij} \geqslant 0 \tag{3.6}$$

Problem II, on the other hand, corresponds to the minimum shear principle which was proposed by Bishop and Hill, and it also corresponds to the *principle of minimum shear summation*, $\sum_{k=1}^{m} D\gamma^k = \min.$, proposed by Taylor, when the critical shear stresses in all slip systems are equal to each other.

Thus in the 1950s the *Taylor–Bishop–Hill theory* of crystal plasticity was established.

3.2.2 Crystal Plasticity of Polycrystalline Metals

A model of the yielding of polycrystals from the viewpoint of crystal slip was first proposed by Gorge Sachs (1896–1961), an American metallurgist of Russian birth, in 1928 [13]. He studied the numerical relation between the Schmid factor μ and the yield condition for fcc metals using the *constant deformation model* shown in Fig. 3.4. Assuming that (i) the crystals having various values of M have a fixed value of τ_c, (ii) τ_c under simple tension is equal to that under simple shear, and (iii) the crystal deformation is determined from the average Schmid factor $\tilde{\mu}$, we obtain an equation similar to eqn (3.1) given as

or
$$\left. \begin{array}{c} \tau_s = \tilde{\mu}_{\text{ten}} \sigma, \quad \tau_s = \tilde{\mu}_{\text{tor}} \tau \\ \\ \tau/\sigma = \tilde{\mu}_{\text{ten}}/\tilde{\mu}_{\text{tor}} \end{array} \right\} \tag{3.7}$$

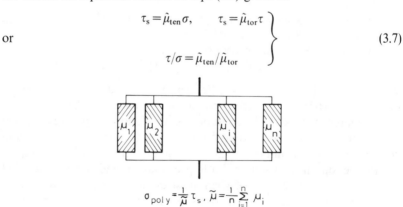

Fig. 3.4. A model of deformation in a polycrystalline body.

Table 3.2
Relation between Inhomogeneity of Distribution in μ and τ/σ for a bcc Metal (α-Iron) Reprinted from ref. 14 by permission of Japan Soc. Mech. Engrs).

	Homogeneous distribution	Case of preferential direction					
		$\{110\}, \{110\}$ $[100], [111]$		$\{100\}$ $[111]$			
		4	2	4	2	1/2	1/4
$\tilde{\mu}_{\text{ten}}$	0·472	0·478	0·475	0·478	0·476	0·470	0·467
$\tilde{\mu}_{\text{tor}}$	0·824	0·823	0·823	0·798	0·810	0·837	0·847
τ/σ	0·573	0·580	0·577	0·599	0·587	0·561	0·551

Thus the ratio of tensile yield stress σ to torsional yield stress τ is represented by these average Schmid factors.

Table 3.2 [14] shows the relation between the inhomogeneity of distribution in μ and τ/σ for a bcc metal (α-iron). In the case of homogeneous distribution $\tau/\sigma = \tilde{\mu}_{\text{ter}}/\tilde{\mu}_{\text{tor}} = 0.573$ which is very close to 0·577 in the so-called Mises yield condition. On the other hand, in the case of the preferential direction [110] we find a tendency for τ/σ to have a slightly larger value when compared to homogeneous material. In hcp polycrystals (isotropy) the relation $\tau/\sigma = \tilde{\mu}_{\text{ten}}/\tilde{\mu}_{\text{tor}} = 0.318/0.538 = 0.591$ is obtained [14]. Furthermore, Taira (1920–1978; see Section 3.4) and Abe calculated the numerical relation between number of slip planes and slip directions, and τ/σ was given as shown in Table 3.3 [15]. As is seen from the table, the values of $\tilde{\mu}_{\text{ten}}$ and $\tilde{\mu}_{\text{tor}}$ vary with respective conditions, but have a constant valye of 0·590 in the case of one slip plane or one slip direction.

Taylor [10] proposed a model for fcc polycrystalline metal in which both the multi-slips and the equality of strains in all grains were taken into consideration. Bishop and Hill developed this model to encompass multiaxial stresses as described below.

Table 3.3
Numerical Relation between Number of Slip Planes and Slip Directions, and τ/σ (Reprinted from ref. 15 by permission of Japan Soc. Mech. Engrs)

Number of slip planes	1	1[a]	1	∞	∞[b]
Number of slip directions	1	3	∞	1	∞
$\tilde{\mu}_{\text{ten}}$	0·212	0·318	0·333	0·333	0·50
$\tilde{\mu}_{\text{tor}}$	0·360	0·539	0·565	0·565	1·00
τ/σ	0·590	0·590	0·590	0·590	0·50

[a] Corresponds to hcp metals.
[b] Corresponds to polycrystalline metals.

We can see that the work done externally, i.e. $\hat{\sigma}_{ij}\tilde{D\varepsilon}_{ij}$, is equal to $\tilde{D\varepsilon}_{ij} = (1/V)\int_v \sigma^*_{ij} dV$, where the local strain increment of respective grains is assumed to be equal to the macroscopic strain increment $\tilde{D\varepsilon}_{ij}$. Here σ^*_{ij} is the stress under which $D\varepsilon_{ij}$ occurs equally in all grains and V is the volume of the polycrystalline body. Therefore we can note that the stress which produces $\tilde{D\varepsilon}_{ij}$ is σ^*_{ij}.

Considering six-dimensional stress space σ_{ij}, the distance p from the origin to the tangent plane in contact with the yield surface, as shown in Fig. 3.5 [11], is written as

$$p = (\hat{\sigma}_{ij}\tilde{D\varepsilon}_{ij})|\tilde{D\varepsilon}_{ij}| = \tilde{D\varepsilon}_{ij}(1/V)\int_V \sigma^*_{ij} dV/(\tilde{D\varepsilon}_{ij}\tilde{D\varepsilon}_{ij})^{1/2} \qquad (3.8)$$

When $\tilde{D\varepsilon}_{ij}$ is given, the stress which is produced equally in respective grains is σ^*_{ij} wherein $\hat{\sigma}_{ij}\tilde{D\varepsilon}_{ij}$ has a maximum value. Thus the distance p in Fig. 3.5 is calculated by eqn (3.8) and the yield surface is constructed by the envelope surface of the tangent spaces. Bishop and Hill calculated the yield surface of an isotropic fcc polycrystalline metal for stress states ranging from simple tension to simple torsion, and obtained 0·540 for τ/σ [11]. Taylor's theory was extended to the form which included elastic strain components by T. H. Lin [16], and also J. W. Hutchinson [17] calculated the yield condition of bcc metals. Table 3.4 [14] shows the summarized results.

The Sachs model was the best method of simplifying the physical picture, and this model effectively illustrates small strain and fatigue deformation. Since the Sachs model has the defect in which strain compatibility among the grains is not maintained due to the equality of

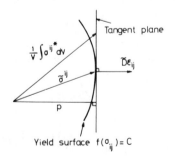

Fig. 3.5. Constructing the yield surface of a polycrystalline body.

Table 3.4
Comparison between Values of τ/σ for Several Isotropic Polycrystal Models
(Reprinted from ref. 14 by permission of Japan Soc. Mech. Engrs)

		τ/σ	Researcher
von Mises		0·577	
Single slip	fcc	0·578	Sachs [13]
(Sachs model)	bcc (α-iron)[a]	0·573	Taira et al. [14]
($\tilde{\mu}_{ten}/\tilde{\mu}_{tor}$)	bcc (Si-iron)[b]	0·578	Taira et al. [14]
	hcp	0·590	Taira and Abe [15]
Multi-slip	fcc	0·540	Bishop and Hill [11]
(Taylor model)	bcc	0·556	Hutchinson [17]

[a]Forty-two of the slip planes $\{110\}$, $\{112\}$ and $\{123\}$.
[b]Six of the slip plane $\{110\}$.

stress in respective grains, the Taylor model is preferable for large strain applications. In the Taylor model the strain continuity at the grain boundary is maintained, but fails with stress equilibrium. This is a result of the strict requirement that all grains must deform arbitrarily and thus stress continuity does not hold true. In practice grain deformation can occur at individual deformation points along the grain boundary. Also, it is not always necessarily true that each grain will deform in slip systems larger than five. For instance, in polycrystalline zinc there are only three slip systems and deformation occurs without any gaps in the grain boundary. Therefore the Taylor yielding model is not applicable to hcp metals, and the correspondence of the calculated τ/σ to the experimental value is better with the Sachs model than with the Taylor model [18]. This will be described in detail in Chapter 4.

Hill [19] established the basis of the study of *macro elastic constants of polycrystals due to the energy upper and lower bounds* method in 1952, and the study was developed by Z. Hashin, a professor of solid mechanics at Tel Aviv University, Israel, and his colleagues [20] in the 1960s. This was also developed into a self-consistent study due to the so-called *inclusion problems* in a uniform infinite body by A. V. Hershey in America [21], J. D. Eshelby in Britain [22] and Ekehart Kröner in the Federal Republic of Germany [23] in the 1950s. The *self-consistent model* was extended to the elastic-plastic problem of polycrystals by E. Kröner, B. Budianski and T. T. Wu, T. H. Lin and Marvin Ito in America, and R. Hill in the 1960s. This will be covered in the following section.

3.2.3 The Curve of Polycrystalline Metals

Over the course of history many attempts have been made to estimate the polycrystal work-hardening curve from that of single crystals.

First, in the *Sachs* [13] and *Kochendorfer* [24] *models*, it was assumed that only an active slip plane in the respective grain of polycrystals should be considered, and the work-hardening curve of crystals with various crystal directions is written $\tau = \tau(\gamma)$, where γ is the shear strain. Sachs [13] calculated the average of all possible direction factors for fcc polycrystalline metal and obtained $D\gamma = 2\cdot238 D\varepsilon$.

Second, the *Taylor model* [25] and that of *Oyane* (a Japanese researcher into metal plasticity) [26], differ from the Sachs model, because multi-slips due to the grain boundary were considered. However, the assumption $\tau = \tau(\gamma)$ was still used, and these models were only applicable to metals having small crystal direction dependence such as aluminum.

Taylor [25] assumed that shear stress had the same value in all slip directions and calculated the solutions for 44 different crystal directions in which $\Sigma_{k=1}^{m} |D\gamma^k|$ had the minimum value, arriving at the relation $M = \sigma/\tau = \Sigma_{k=1}^{m} |D\gamma^k|D\varepsilon = 3\cdot06$. In calculations performed by Bishop and Hill [11] a value of $M = 3\cdot06$ was found, where instead of strain they applied conditions of maximum plastic work within 56 stress states in the same five slip systems. This value was in agreement with the experimental value for metals having small crystal dependence on work-hardening. Table 3.5 [27–30] shows the M values in an axisymmetrical deformation of cubic crystalline metal, where the M value in a slip system with twin deformation was analyzed by applying the following equation:

$$M = (1/D\varepsilon)\left[\sum_{k=1}^{m} |D\gamma_s^k| + \alpha \sum_{k=1}^{m} |D\gamma_t^k|\right] = \sigma/\tau_s \qquad (3.9)$$

where suffixes s and t denote slip and twin, respectively, and $\alpha = \tau_t/\tau_s$. We can see that in this table M values for [110] and [111] are 50% larger than for [100].

Many attemps have been made to correct the defects in the Sachs and Taylor models, but none has completely succeeded. Studies in which both the stress and strain continuities at the grain boundary were considered have been made by Kröner (1961) [31], Budianski and Wu (1962) [32] (so-called *KBW model*), Hutchinson [33], Lin and Ito [34], Takahashi [35] and others. We will describe the KBW model briefly here (see Fig. 3.2).

Table 3.5
Value of M in Axisymmetric Deformation of Cubic Lattice Metals with Several Deformation Modes (Reprinted from ref. 27 by permission of Amer. Soc. Metals)

Deformation mode	Axial direction				Minimum	Researcher
	[100]	[110]	[111]	Random		
$\{110\}\langle111\rangle$ slip	2·449	3·674	3·674	3·067	3·28	Chin and Mammel [28]
$\{112\}\langle111\rangle$ slip	2·121	3·182	3·182	2·954	2·06	
$\{123\}\langle111\rangle$ slip	2·160	3·240	3·240	2·803	2·05	
$\{110,112,123\}\langle111\rangle$ mixed slip	2·121	3·182	3·182	2·754	2·04	
$\langle111\rangle$ pencil glide	2·121	3·182	3·182	2·733	2·03	Rosenberg and Piehler [29]
Mixture of $\{111\}\langle110\rangle$ slip and $\{111\}\langle112\rangle$ twin						Chin et al. [30]
Tension	4·243	4·243	3·182	3·66	2·94	
Compression	2·121	5·303	6·364	4·63	2·121	

The relation between $D\sigma_{ij}$ and $D\varepsilon_{ij}$ in crystal coordinates and DS_{ij} and DE_{ij} in specimen coordinates is written as

$$D\sigma_{ij} = DS_{ij} - 2G(1-\beta)(D\varepsilon_{ij} - DE_{ij}) \qquad (3.10)$$

where $\beta = (4-5\nu)/[15(1-\nu)]$, ν is Poisson's ratio and G the shear modulus. The volume average $\widetilde{D\varepsilon}_{ij}$ in the grain, in other words DE_{ij}, is given as

$$DE_{ij} = \widetilde{D\varepsilon}_{ij} = (1/8\pi^2) \int_0^{2\pi} d\theta \int_0^{\pi} \sin\phi \, d\phi \int_0^{2\pi} D\varepsilon_{ij} d\psi \qquad (3.11)$$

where θ, ϕ and ψ are Euler angles (see Fig. 3.2). Putting

$$L_{ij}^{(k)} = (\tfrac{1}{2})(n_i^{(k)} b_j^{(k)} + n_j^{(k)} b_i^{(k)})$$

in eqn (3.3a), it is written as

$$D\varepsilon_{ij} = \sum_{k=1}^{m} L_{ij}^{(k)} D\gamma^{(k)}$$

Since an incremental form of eqn (3.2a) is written as $D\tau^{(k)} = L_{ij}^{(k)} D\sigma_{ij}$, substituting eqn (3.10) and $D\varepsilon_{ij} = \sum_{k=1}^{m} L_{ij}^{(k)} D\gamma^{(k)}$ into $D\tau^{(k)} = L_{ij}^{(k)} D\sigma_{ij}$, the following equation is obtained:

$$D\tau^{(k)} = Dt^{(k)} - 2G(1-\beta) \sum_{s=1}^{m} M^{(k,s)} D\gamma^{(s)} \qquad (3.12)$$

where $Dt^{(k)} = L_{ij}^{(k)} DQ_{ij}$, $DQ_{ij} = DS_{ij} + 2G(1-\beta)DE_{ij}$, $M^{(k,s)} = L_{ij}^{(k)} L_{ij}^{(s)}$. When DS_{ij} is given, DE_{ij} is calculated as follows. (i) First, DQ_{ij} is given; (ii) calculate $D\varepsilon_{ij}$ for grain with respective crystal orientation; (iii) calculate DE_{ij} by using eqn (3.11) and, if $DS_{ij} = DQ_{ij} - 2G(1-\beta)DE_{ij}$ is not equal to the value which is expected, return to (i) and modify it.

We will briefly describe work done on the nonuniqueness in the choice of active slip systems and therefore in the predicted lattice rotations. Hill [36] first provided a sufficient condition for the uniqueness of the stress rate/strain rate relation for single crystals. Experimentation, on the other hand, has indicated that latent hardening rates are generally somewhat larger than the self-hardening rate, which means that the strain-hardening matrix is indefinite. In order to overcome this difficulty some proposals have been made for partially mitigating the uniqueness condition. One of these was the grain rotation, or 'relaxed constraints' in the interaction between crystal grains. But the method does not resolve the problem of nonuniqueness in the choice of active systems. A new *rate-dependent constitutive model* has been developed for polycrystals subjected to large strains [37].

The rate-dependent boundary value approach effectively solves the long-standing problem of nonuniqueness in the choice of an active slip system which is inherent in the rate-independent theory. Because the slipping rates on all slip systems within each grain are unique in the rate-dependent theory, the lattice rotations and anisotropy of polycrystalline metals are determined uniquely. In this model there is no division of slip systems into 'active' or 'inactive' sets. Instead all slip systems always slip at a rate which depends on the current stress and hardening properties. Therefore, once the stress state is known, slipping rates on all possible slip systems and the resulting lattice rotations which form the hardening matrix are uniquely determined. Lattice rotation will be discussed in Chapter 4 in connection with the uniqueness in the choice of active slip systems in polycrystalline metals subjected to static and pulsating cyclic stresses.

3.2.4 Crystal Grain Interaction

In the history of crystal plasticity the least understood area concerns the properties of the grain boundary and the interaction between grains. These are perhaps the most important problems in the development of new materials such as alloys, ceramics, composite materials and so on. The study of the physical properties of the grain boundary was reviewed in Section 2.3 of the previous chapter, but for crystal plasticity a more simplified approach has been used. In this approach a grain boundary is defined as an infinitely thin interface, and the mechanical behavior in each grain is calculated by the FEM method; the interaction between grains has been studied by Japanese researchers [38–40] in the 1970s.

H. Miyamoto, a Japanese researcher into solid mechanics, and his colleagues [38, 39] analyzed the mechanical behavior of a plastically deformed coarse-grained pure aluminum plate, using the *flow stress versus strain matrix* $[D_p]$ in crystal plasticity theory and the τ versus γ curve of that single crystal, and they compared it with the experimental result. The matrix $[D_p]$ was derived as follows.

The incremental form of eqn (3.2b) and the plastic components in eqn (3.3b) are expressed by

$$\{D\tau\} = [T]\{D\sigma\} \qquad (3.2c)$$

$$\{D\varepsilon^p\} = [T]^T\{D\gamma\} \qquad (3.3c)$$

respectively, where $D\varepsilon^p$ is the plastic strain increment. When the $D\tau$ versus $D\gamma$ relation is given as

$$\{D\gamma\} = [S^p]\{D\tau\} \qquad (3.13)$$

the following equation is obtained from eqns (3.2c), (3.3c) and (3.13):

$$\{D\varepsilon^p\} = [T]^T[S^p][T]\{D\sigma\} \qquad (3.14)$$

Adding the elastic strain increment $D\varepsilon^e$ to eqn (3.14), we obtain

$$\{D\varepsilon\} = ([T]^T[S^p][T] + [S^e])\{D\sigma\} \qquad (3.15)$$

where $D\varepsilon = D\varepsilon^p + D\varepsilon^e$ and $[S^e]^{-1} = [C^e]$. Transforming x_i coordinates to X_i coordinates, $\{DS\}$ is written as

$$\begin{aligned}\{DS\} &= [\phi]^T([T]^T[S^p][T] + [S^e])^{-1}[\phi]\{DE\} \\ &= [\phi]^T[C^p][\phi]\{DE\} = [D^p]\{DE\} \end{aligned} \qquad (3.16)$$

where DS and DE are stress and strain increments in the specimen coordinates, respectively, and $[D^p]$ is the flow stress/strain matrix of the crystal plasticity [40]. The FEM analysis made by using the $[D^p]$ matrix is the same as that by O. C. Zienkiewicz [41] and Y. Yamada [42].

Miyamoto and his colleagues [40] reported that the analytical results of the initiation of slip lines and of crystal deformation are in good agreement with the experimental ones. T. Jimma and his colleagues [43] pointed out that the continuity of stress at the grain boundary of coarse-grained pure aluminum was maintained and that the sequence of yielding is in agreement with the Schmid factor. We will describe the details of the experimental crystal plasticity in Chapter 4.

3.3 HISTORY OF THE THEORY OF DISTRIBUTED DISLOCATIONS AND THE CONTINUUM APPROXIMATION APPROACH

3.3.1 Theory of Distributed Dislocations

The *theory of distributed dislocations* (abbreviated as TDD) had made remarkable progress in the 1950s and 1960s, and this was related to the fact that research into a fundamental part of the crystal dislocation theory for single dislocations was almost complete. Many international conferences during 1948–1952 had summed up early advances in crystal dislocation theory. The start of TDD goes back to the 1940s.

R. E. Peierls (1940) and F. R. N. Nabarro (1947), British physicists, were the first to study the nearest neighborhood of crystals on a slip plane as the 'elastic continuum'. They proposed that the shearing stress acting on the slip plane is a harmonic function of the relative displacement between the lattice planes by taking *nonlinearity in the neighborhood which is displaced by edge dislocations* into consideration. This is known as the *Peierls–Nabarro theory*. The Peierls–Nabarro theory was originally a result of a discovery by Egon Orowan, an American physicist of Hungarian birth and one of the pioneers in crystal dislocation, as pointed out by Peierls [44]. A study of screw dislocations was also made by J. D. Eshelby (1960). Eshelby (1916–1981; see Section 3.4) also studied the screw dislocation mechanism. His research efforts proved quite successful for establishing the early theories of crystal dislocation. He effectively illustrated that the difference between the ideal shearing strength of crystals as opposed to the real strength could be interpreted from the existence of crystal dislocation. But, as we shall see in the following pages, these theories had some substantial defects.

The defects are: (i) Application of the mathematical theory of elasticity to the *dislocation core* is contradictory to prerequisites in the *Peierls–Nabarro theory*. (ii) Equality between the atomic interaction force and the elastic force disregards quantum mechanical considerations. (iii) The theory is not regarded as the model wherein the maximum stress in the dislocation core is equivalent to the critical shear stress of the crystal. To solve the problem of dislocation the core has been left as a black area in the elasticity theory of dislocation. However, the property of the dislocation core plays an important role in studies not only of the Peierls–Nabarro stress and interaction between dislocation and solute atoms, but also of the electric properties of modern semiconductors. In order to solve the problem, computer physics of the dislocation core in the 1950s and 1960s by using crystal lattice modeling and the study of simulation modeling continues today in connection with direct observation by use of a very high resolution electron microscope. It is worthy of note that the Peierls–Nabarro theory historically plays an important role in studying the theory of dislocation, distributed dislocations and the continuum approximation approach, as described in the following.

After the 1950s J. F. Nye, a British physicist [45], first proposed an interpretation of distributed crystal dislocations by means of a *geometrical* concept which had an impact on the study of materials with lattice defects. B. A. Bilby, a British physicist, and his colleagues (1955) [46], E. Kröner (1956, 1960) [47, 48], a German physicist, and K. Kondo

(1952, 1955) [49–52], a Japanese mathematical and physical engineer and a founder of the Research Association of Applied Geometry (known as RAAG), had studied the geometrical interpretation of materials with lattice defects on the basis of differential geometry of distant parallelism and non-Riemannian space. The differential geometry of imperfect crystals with lattice defects that had been developed by the RAAG in Japan was a very important contribution. The notion of the torsion of a space was introduced by the distinguished French mathematician Eli Cartan (1869–1951) in 1922, and 30 years later this theory was commonly recognized as the isomorphic notion of crystal dislocation.

The following two equations are important not only in the sense of TDD but also of SFM:

$$\alpha_{kl} = -\varepsilon_{kmn}\partial_m\beta^p_{nl} \quad (k, l, m, n = 1, 2, 3) \tag{3.17}$$

$$\mu_{kl} = -(\tfrac{1}{2})(\varepsilon_{kpq}\partial_p\alpha_{lq} + \varepsilon_{lpq}\partial_p\alpha_{kq}) \quad (k, l, p, q = 1, 2, 3) \tag{3.18}$$

where α_{kl} is a *dislocation density tensor* of the second order, μ_{kl} is an *incompatibility tensor* of the second order, β^p_{nl} is a *plastic deformation gradient* and ε_{kmn} is Eddington's tensor. Equation (3.17) shows that there are dislocations in the portions where plastic deformation varies with position. The equation was discovered by Kondo [52], Bilby *et al.* [46] and Kröner [53] independently. We see from eqn (3.18) that the compatibility condition for ε^p_{kl}, in general, does not hold true except for the uniformity in α_{kl}, and μ_{kl} is an important parameter in the continuum approximation which represents an *inhomogeneity of plastic strain*. After that the relation between strain space and the non-Euclidean quantities was studied [54, 55] and that between *misorientation*, which represents continuous change in the orientation of the subgrain of plastically deformed metal obtained by means of Laue X-ray diffraction, and the non-Euclidean quantities was also studied [56, 57]. We will review these theories in Chapter 5.

Since the 1950s the study of dislocations which interact with inhomogeneities had been motivated by the need to gain a better understanding of certain strengthening and hardening mechanisms in materials [58]. The first studies examined problems such as the interaction of dislocations with a phase boundary [59, 60], the interaction of dislocations with circular inclusions [61], the interaction of dislocations with an extremely thin oxide film or other surface layers [58, 62, 63], the interaction of dislocations with lamellar inclusions [64], the deformation of plastically heterogeneous materials [65], both elastic and plastic deformations at the crack tip [66], grain boundaries [66] etc. Such problems are studied by means of the elasticity theory. The so-called *elastic constant effect* arises from the fact that the total potential energy of both the body and the applied stresses depends on the position of dislocations with respect to the inhomogeneities.

We will review the *theory of dislocation pile-up* briefly. Pile-up plays an important role in a study of the mechanical behavior of crystalline solids. Historically, the effect of grain size on the yield and flow stresses of polycrystalline materials is well known. The *Hall–Petch relation* (1953) could be rationalized based on the pile-up model [67]. This theory was recently developed further by Yokobori and his colleagues [68] who introduced the effect of grain size on the dynamic yield stress from the *dynamic pile-up of dislocations emitted from a dislocation source*. The pile-up was used by A. N. Stroh (1953) [69] to calculate the *amount of stored energy in a work-hardened material*, and by Bilby, Cottrell and Swindeman (1963) [70] to estimate the *spread of plastic zones ahead of a crack* (known as the *BCS model*).

The analytical and numerical results of dislocation distribution, as well as the energy and stress field in the pile-up, can be determined by using two different geometrical arrangements. In the first approach the dislocations are studied as discrete singularities, each with a finite Burgers vector. This approach is exact, but the theoretical solutions are available for only a few studies. In the second approach the discrete dislocation arrangement is replaced by a continuous distribution of dislocations with infinitesimal Burgers vectors. This approach, although the approximate solution for a finite number of discrete dislocations, is more tractable mathematically and provides a solution for the limiting cases of discretely distributed dislocations. Historically, regarding methods which satisfy the free boundary condition, the *surface dislocation method* by N. Louat (1962) and M. J. Marcinkowski (1974) is superior compared with the *image dislocation method* by Head (1953), particularly because the discrete dislocations distribution is extensively applicable.

Furthermore, W. G. Johnston and J. J. Gilman, American physicists [71], proposed the relation between average moving velocity v and macroscopic plastic strain rate $\dot{\varepsilon}^p$, known as the *Johnston–Gilman equation*, in 1959:

$$\dot{\varepsilon}^p = bNv \quad (3.19)$$

where b is the magnitude of the Burgers vector and N is the average dislocation density. Equation (3.19) became a fundamental equation regarding the plastic strain rate or the creep rate of metals.

Toshio Mura, a researcher into solid mechanics in America (1963) [72], first defined a *dislocation velocity tensor* as

$$V_{lji}(\mathbf{x}) = \sum_{n,v,b,V} nV_l v_j b_i \qquad (3.20)$$

from the viewpoint of the dynamics of a continuous distribution of dislocations. In eqn (3.20) Σ denotes the summation of n, v_j and b_i provided that n dislocations having both the direction v_j and the Burgers vector b_i go through the unit cross-sectional area perpendicular to v_j, and V_l is the velocity vector of the dislocations group $(\mathbf{n}, \mathbf{v}, \mathbf{b})$. It was clarified that $V_{lji}(\mathbf{x})$ was connected with both α_{ji} of eqn (3.17) and the plastic deformation gradient rate $\dot{\beta}^p_{ji}$ by [72]

$$\dot{\alpha}_{ji} = \varepsilon_{jlk} \partial_l (\varepsilon_{mnk} V_{mni}), \quad \dot{\beta}^p_{ji} = -\varepsilon_{jkl} V_{kli} \qquad (3.21)$$

where $\dot{\beta}^p_{ji}$ shows the ji component of the plastic strain rate. If we take the vector product of the dislocation velocity tensor and the unit vector of the direction along the dislocation line and multiply it by the slip direction vector, we get $\dot{\beta}^p_{ji}$ which is the ji component of the plastic strain rate. H. Minagawa, a Japanese researcher into solid mechanics, and his colleagues [73] also showed that the stress produced by the continuous distribution of moving dislocations was obtained from $\alpha_{ji}(\mathbf{x})$ and $V_{lji}(\mathbf{x}, t)$.

3.3.2 Continuum Approximation Approach

In 1967 the International Union of Theoretical and Applied Mechanics (IUTAM) symposium on the 'generalized Cosserat continuum and the continuum theory of dislocations with applications' was held. The symposium was an attempt to bridge the gap between microscopic (or atomic) and phenomenological (or continuum mechanical) studies on the mechanical behavior of materials and to apply the theory of the continuum to the microscopic research field [74]. Studies of the relationship between bodies such as Cosserat, multipolar, micromorphic or lattice theory, nonlocality and materials with lattice defects were presented. Until this time studies of materials with microstructure had not been integrated. It was only after this that the symposium directed its efforts towards integrated research.

In the symposium observations on *crystal disclination* also had an impact on the research into the theory of generalized continua.

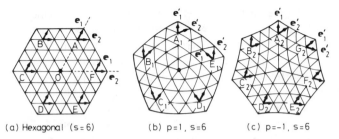

Fig. 3.6. Schematic illustration of a disclination: (a) hexagonal lattice; (b) five-fold wedge disclination; (c) seven-fold wedge disclination. (Reprinted from ref. 75 by permission of OHM-Sha)

Dislocation motion produces shearing strain and disclination denotes rotational strain and is one of the lattice defects. Figure 3.6 [75] shows two examples of wedge disclinations with positive and negative signs and which correspond to *Volterra's dislocation of the sixth kind* and rotation axes which are perpendicular to the plane. These opposite signs may be generated by adding (+) or removing (−) a wedge-shaped piece of material and rewelding the material. The wedge angle is equal to the rotational angle; its minimum angle in the hexagonal lattice is 60°. The

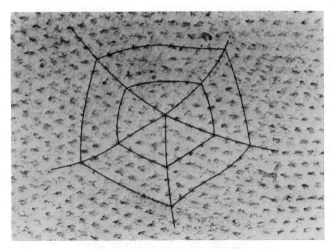

Fig. 3.7. A negative disclination in the two-dimensional lattice formed by vortex lines in type II superconductors in a mixed state. (Reprinted from ref. 76 by permission of Springer Verlag)

associated lattice structures with characteristic five- and seven-fold symmetries are shown in Figs. 3.6 (b) and (c), respectively. The wedge disclinations have been discovered through experimentation. Figure 3.7 [76] shows the negative disclination in the two-dimensional lattice formed by vortex lines in the mixed state of a type II superconductor which was first discovered by H. Trauble and U. Essmann, German physicists, in 1967–8. These days both the *disclination density tensor* and the dislocation density tensor are defined, and since both dislocation and disclination are interpreted as an arrangement of dislocations, it is quite certain that both are not conceptually independent.

The majority of technical materials such as polycrystals, dispersed materials, sintering materials and composite materials (mixed phase materials) are so-called heterogeneous materials. In general, the material has boundaries in which the physical property changes when the subdivision goes over the boundary, and the concept 'continuum' is not used directly. It is, however, very important to use the continuum mechanical approach which provides the macroscopic study of mechanical behavior of materials by extending it. This is the so-called *continuum approximation approach*. In the concept 'continuity' a possibility of infinitesimal division is included, but the possibility of deformation is maintained even though the division is infinitesimally small. In the continuum approximation subdivision is permissible unless the possibility of deformation is lost, and the applicable limit is regarded as a deformable minute element. The recognition of such a limit is a problem of the so-called *hierarchy cognition* in Chapter 2. Therefore it is important to elucidate both the submacroscopic mechanical concept, which bridges the macro properties of the material, with the micro properties and the *measure* which represents them concisely. Table 3.6 [77] shows the fundamental concepts regarding the continuum approximation, and the above measures will be found in the table.

The Cosserat mechanics in item (i) of the table was applied to, and led to progress in, the plasticity field by Sawczuk, a Polish researcher into solid mechanics, and by H. Lippman, a German researcher into solid mechanics. The theory of continuous distribution of dislocations in item (ii) also was responsible for advances by elucidating the relationship between strain/stress space and physical quantities. The *dual relation*

$$\begin{aligned}&(\text{Strain}){:}(\text{Stress function})\\&=(\text{Incompatibility}){:}(\text{Stress})\\&=(\text{Torsion}){:}(\text{Couple stress})\end{aligned} \qquad (3.22)$$

Table 3.6
Fundamental Concepts in the Continuum Approximation Approach (Reprinted from ref. 77 by permission of OHM-Sha)

Concept	Definition and characteristics
(i) Polarity and non-locality	(1) *Polarity* means the capability of rigidly local rotation in addition to the deformation of the constitutive element (particle) in heterogeneous material. *Non-locality* means that physical property of the constitutive particle is influenced by the distant particle. (2) The former is the revival and development of the Cosserat mechanics proposed by the Cosserat brothers (see Section 3.3), and the latter is that of the action-through-medium. 'Non-local elasticity' was named by Erhart Kröner in 1963. (3) In heterogeneous material with polarity, it is known that a *couple stress vector* is produced in order to allow a rigid rotation of the particle in addition to the usual stress vector. It is also known that a *scale parameter* appears in the constitutive equation between the couple stress and the rotational strain. The physical meaning of the scale parameter has not been clarified completely, but it will at least represent a stratificational concept in studying the mechanical behavior of heterogeneous elasticity.
(ii) Complexity of configuration	(1) *Complexity of configuration* means the extension of continuous media such as an elastic body to the material space with defects, in order to allow the existence of crystal dislocation and a plastic deformation. (2) That above is a geometrical approach to homogeneous deformation, and the theory of the continuum approximation of perfect crystals with lattice defects is also the theory of the continuous distribution of dislocations.
(iii) Statistical continuum mechanics	(1) *Statistical continuum mechanics* means the mechanics in which the macroscopic mechanical behavior of continuous media has a statistical constitutive equation, i.e. *random media*. (2) Through statistical examination of a heterogeneous material, the macroscopic property of a polycrystalline body, e.g. an *effective elastic constant*, has been studied, and the research after Hill (1952) [19] marked a turning-point.

was discovered by S. Amari and H. Minagawa, Japanese researchers into solid mechanics (1968), from the viewpoint of the statics of distributed dislocations.

The statistical continuum mechanics in item (iii) of the table was first studied in heterogeneous linear elastic material by B. J. Beran, an American researcher into solid mechanics, and his colleagues [78, 79] and E. Kröner [80–82] in the 1970s. As we can see, the mechanical property of the material is a random variable with respect to the position of x. The controlled equation in statistical continuum mechanics is lastly attributed to the probability differential equation. Historically two methods have been used to solve the equation. One is the method where the moment equations are solved and the associated *correlation function* is successively obtained, and the other is the method to obtain the characteristic function for the probability density function. To date it remains difficult to obtain a useful result by solving for these functions.

Heterogeneous polycrystalline materials, in general, have the following characteristics:

1. Material properties, shape and size of the constituent grains (crystallites) vary from crystal to crystal.
2. Distribution of these quantities is statistical and there is a correlation between crystals.
3. There is some difference in material characteristics between the portion of the grain boundary and that of the interior of the crystal.

In order to formulate theories of material behavior it is necessary to simplify the problem by using an assumption. Here it is quite practical to make the assumption that 'an elemental constituent is a material point'. Adopting a *multi-point correlation function* with respect to the material constant, we can then consider item 2 above. The study of heterogeneous material performed by Beran or Kröner was similar to that above. On the other hand, P. H. Dederichs and R. Zeller, German physicists [83, 84], studied the behavior of polycrystalline bodies with the provision that the material properties remained constant in the interior of single crystals. In this model both items 1 and 2 are taken into consideration.

Then, how has generalized continuum mechanics progressed? Kröner [85] pointed out the historical development by using the following three concepts:

1. Number of kinematic micro degrees of freedom, i.e. the complexity of the particle or the *polarity*.

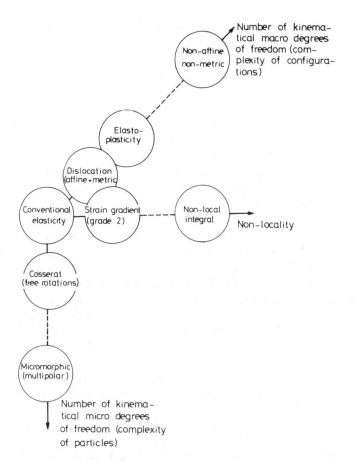

Fig. 3.8. Geometric classification of continuum mechanics. (Reprinted from ref. 85 by permission of Springer Verlag)

2. Number of kinematic macro degrees of freedom, i.e. the *complexity of the configuration*
3. *Nonlocality*.

Figure 3.8 [85] illustrates the above. Item 1 is the extension of the concept to include polar materials and item 2 is the extension of the concept to include materials with lattice defects such as crystal dislocations. Item 3 is not strictly geometrical but it provides the metric quantity with respect to a finite distance between particles from the viewpoint of an *action at a distance*. In this sense item 3 is included in both 1 and 2.

The three concepts are interrelated. For instance, the *Cosserat material with constrained rotation* in item 1 corresponds to the grade 2 material in item 3, and since the torsion tensor in the stress-space is equivalent to the *couple stress*, there is a close relation between items 2 and 1 or 3. Generalized continuum mechanics has thus progressed through three concepts but will not succeed in the theoretical connection between these three items. Therefore it is necessary to elucidate the correlation between these three concepts in connection with the physical picture.

This does not mean that generalized continuum mechanics has provided an effective explanation of the plastic behavior of technical materials. It is necessary to consider that the continuous distribution of dislocations is represented not only by the smooth dislocation density but also by the arrangement and the direction. From this point of view we notice that Kondo and his colleagues had elucidated the geometrical representation of the global lattice defect density and the direction by means of the Riemann–Christoffel curvature tensor, and we proved it by means of the *misorientation* obtained from Laue X-ray spots [56, 57]. Table 3.7 [86] shows the correspondence of both torsion and Riemann–Christoffel curvature tensors to crystal lattice defects.

It is well known that the one-point correlation function $\{\alpha_{kl}\}$ (tensor of 2nd order) of the dislocation density tensor α_{kl}, which represents the number of dislocations flowing across the area ds_i at point **x**, is equiva-

Table 3.7
Correspondence of Both the Torsion Tensor and the Riemann–Christoffel Curvature Tensor to Lattice Defects (Reprinted from ref. 86 by permission of BAIFUKAN)

	Kind of lattice defect	*Region of defect*
Torsion tensor	Dislocation with the same sign, stacking fault and so on	Local dislocation density
Riemann–Christoffel curvature tensor	Dislocation dipole with different signs, point defect; non-conservative motion of dislocations such as climbing	Global lattice defect densities containing dislocation orientation and density of vacancy in addition to dislocation density

lent to the dislocation density tensor discovered by Nye [45]. Here we find a contradiction where $\{\alpha_{kl}(\mathbf{x})\} = 0$ is realized when an equal number of positive and negative dislocations are produced when the material is subjected to simple tension. Therefore the multi-point correlation functions including the two-point correlation function $\{\alpha_{ij}(\mathbf{x})\alpha_{kl}(\mathbf{x}_1)\}$ ($\mathbf{x}=\mathbf{x}_1$) is an accurate description of the continuous distribution of dislocations. The *two-point dislocation correlation function* for points $\mathbf{x}_1$, $\mathbf{x}_2$:

$$a_{i1j1i2j2}(\mathbf{x}_1, \mathbf{x}_2) = \{t_{i1}^{(b)}(\mathbf{x}_1)b_{j1} t_{i2}^{(b)}(\mathbf{x}_2)b_{j2}\}$$

$$= (1/V)\sum_b \sum_c t_{i1}^{(b)}(\mathbf{x}_1)b_{j1} t_{i2}^{(b)}(\mathbf{x}_2)b_{j2} \quad (3.23)$$

has the following characteristics [87]. (1) When $\mathbf{x}_1=\mathbf{x}_2=0$, eqn (3.23) shows the dislocation length per unit volume. (2) The *dislocation network* with straight lines shows the two-point correlation tensor (see Fig. 3.9(a) [87]). (3) The *dislocation dipole* (see Fig. 3.9(b) [87]) is represented by the

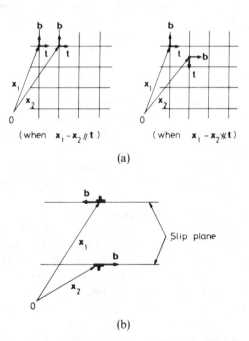

Fig. 3.9. Application of two-point dislocation correlation function to dislocation structure: (a) dislocation network; (b) dislocation dipole.

two-point correlation function. (4) The dislocation pile-up and the small-angle grain boundary are also represented by the two-point correlation function.

Thus we can see that the three approaches in Fig. 3.8 are related to each other and they are connected through the correlation function in statistical continuum mechanics, as shown in Fig. 2.6 of the previous chapter.

3.3.3 The Life and Work of Bošković, the Cosserats, and Volterra in the Continuum Approximation Approach

Rudjer Josif Bošković (1711–1787)
Hideki Yukawa (1907–1981) and Takeshi Inoue, Japanese physicists who studied the theory of elementary particles [88], evaluated the 'Theoria Philosophiae Naturalis' [89] by Bošković as follows.

On the premise that the material world consists of a perfect number of particle couples, and that interaction takes place among all the couples, Bošković interpreted all physical phenomena as the transition of the relative positions of the particles. In his thoery of *action at a distance* of particles, Newtonian mechanics plays a central role and it seems that the theory is far from the concept '*action through a medium*'. Since each particle acts on another particle, we are able to regard the individual particle point as the center of the action. Therefore we should note that *Bošković's theory of material particles* has not only a general application in the 18th century based on the concept 'action at a distance' but also the element which connects with the concept 'action through a medium' in the 20th century.

Bošković critically extended both *Leibnitz's law of continuity of matter* based on the sophisticated concept '*monade*' and the study of attractive and repulsive forces in the famous thirty-first query of 'Elective Attractions' which is included in the second edition (1717) of Newton's 'Optics as a Treatise of the Reflections and Colours of Light'. Bošković wrote: 'My theory was established as the system which connected Leibnitz's theory of monade with the Newtonian mechanics'.

Željko Marković, the author of Bošković's biography, also wrote as follows [90]: 'It was because of its consequences for the constitution of matter that the law of forces was particularly important. In Bošković's natural philosophy the *first elements* of matter became mere points-real, homogeneous, simple, indivisible, without extension, and distinguished from geometrical points only by their possession of inertia and their mutual interactions. Extended matter becomes the dynamic configuration of a finite number of centers of interaction. Many historians have seen in Bošković's derivation of matter from forces an anticipation of the concept of the field, an anticipation still more clearly formulated very much later by Michael Faraday in 1844. Matter, then, is not a continuum but a discontinuum. Mass is the number of points in the volume, and drops out of consideration as an independent entity. In the special case of high-speed particles, Bošković even envisaged the penetrability of matter'.

Simeon-Dennis Poisson (1781–1840), a distinguished French mathematician and physicist and well known for *Poisson's ratio*, was one of the first researchers who applied Bošković's theory to the bending of elastic plate, in 1812.

Bošković was born in Dubrovnik, Yugoslavia, on 18 May 1711, the son of Nikola Bošković, a merchant of Dubrovnik. He began his education in the Jesuit college of Dubrovnik and continued it in Rome. He learned science in a way characteristic of his later career, through independent study of mathematics, physics, astronomy, and geodesy. In 1735 he began studying Newton's Optics and the Principia at the Collegium Romanum, and he made himself an enthusiastic propagator of the new natural philosophy.

In 1759 Bošković began his travels to European countries for four years and was appointed a member of the Royal Society, Britain. Returning to Italy, he was appointed a professor of mathematics, University of Pavia, and played a leading role in the organization of the Jesuit observatory at Brera near Milan in 1764.

Marković wrote [90]: 'Bošković was perhaps the last polymath to figure in an important way in the history of science, and his career was in consequence something of an anachronism and presents something of an enigma.... Sharp in thought, bold in spirit, independent in judgement, zealous to be exact, Bošković was a man of eighteenth-century European science in some respects and far ahead of his time in others. Among his works are writings that still repay study, and not only from a historical point of view'.

Eugène Maurice Pierre Cosserat (1866–1931) and François Cosserat (?–1914)

The Cosserat brothers, Eugène Maurice Pierre and François, were born and raised in France. The especially notable younger brother E. Cosserat, professor of differential calculus and astronomy and one of the moving forces at the University of Toulouse, is well known today as the founder of the *theory of finite elastic deformation* in 1896 [91], and he proposed the mechanics of continuous media which was composed of particles with direction vectors in 1909 [92]. His elder brother François, chief engineer of the Service des Ponts et Chaussées, collaborated with Eugène and he was the main participant in tests on synthesis and philosophical concepts, the mathematical framework of the research being furnished by Eugène. It was important that the book entitled 'Théorie des Corps Déformables' (Theory of Deformable Bodies) [92] gave the prototype of the *Cosserat continuum mechanics* of a field of the so-called theory of continuum approximation or generalized continuum mechanics, which clarified the mechanical behavior of materials with microstructure.

H. Minagawa (1972) wrote in a review [93]: 'they intended to complete a theory of 'ether' which involved many contradictions and exposed its defects in those days. In the book (p. 115) both all the *mechanical theories of ether* from James MacCullaugh (1809–1847), an Irish physicist, to Lord Kelvin (1824–1907), the distinguished British physicist, and Maxwell's *theory of electron-magnetic field* (1873) were united and integrated in their theory'. MacCullaugh (1839) thought that bodies were composed of elements which resisted only rotational torsion, and explained the optical phenomenon by the mechanical law of the ether. Afterwards Kelvin made the mechanical model of the ether. The model did resist rotational force but did not resist translational motion. Nineteenth century scientists defended the mechanical model of the ether (aether) with magnificent tenacity. In this the history of the theory of the ether suffered a tortuous course. After the ether had been ignored in the *relativity theory* by Albert Einstein (1879–1955), the history of the ether was forgotten rapidly and the study by the Cosserat brothers was also buried. After the Second World War, since the relationship between the Cosserat theory and the crystal dislocation theory was pointed out, attention focused on the work done by the Cosserat brothers, and the theory progressed and rapidly became popularized [94].

The Cosserat brothers described a combination of points M and the associated rectangular three unit vectors, e_1, e_2, e_3, as *trièdre*, and an assembly of the trièdre as *milieu déformable*. The trièdre was the so-called particle point of new continuous media and deformation of the body was regarded as both the translation and the rotation of the trièdre (Fig. 3.10 [93]). Is the microstructure of a material expressed precisely by the Cosserat theory? Through history this point has received great attention.

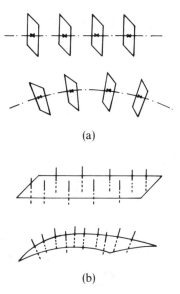

Fig. 3.10. Cosserat continuum concept in one- and two-dimensions. (a) One-dimensional Cosserat continuum composed of plane elements. (b) Two-dimensional Cosserat continuum composed of line elements. (Reprinted from ref. 93 by permission of Mater. Sci. Soc., Japan)

The study of Cosserat continuum mechanics was part of the revival of the contradictory concepts 'action at a distance' and 'action through a medium'.

Vito Volterra (1860–1940)
We will refer to the biography written by E. Volterra, University of Texas [95], to describe the life and work of Vito Volterra, a pioneer in *elastic dislocation and the disclination in multiconnected bodies*. He was born in Ancona, Italy, the only child of Abramo Volterra, a cloth merchant, and his wife Angelica Almagia, on 3 May 1860. His ancestors had moved to Volterra, a small city in Tuscany, Italy, from Bologna at the beginning of the fifteenth century. Volterra lost his father when he was two years old and he was taken into the home of his mother's brother, Alfonso Almagia. Later he lived in Turin and in Florence and spent the greater part of his youth in Florence.

Volterra was a precocious child and graduated with a doctorate in physics from the University of Pisa in 1882, and succeeded Engnio

Beltrami in the chair of mathematical physics at the University of Rome in 1900. Volterra's scientific work covers the period from 1882 to 1940. His most important contributions were in higher analysis, mathematical physics, celestial mechanics, the mathematical theory of elasticity, and mathematical biometrics. The paper 'Sur l'équilibre des corps élastiques multiplement connexes' (On the equilibrium of multiconnected elastic bodies) (116 pages) in 1907 [96] was published when he was 47 years old. He first studied the discontinuity of multiconnected elastic bodies with hollow cylinder shaped specimens and analyzed both stress and deformation in elastic dislocations. Afterwards this was developed by Carlo Somigiliana, an Italian mathematical physicist, in the 1910s [97]. Somigiliana was one of the researchers who appreciated Volterra's paper greatly. J. M. Burgers (1939, 1940), a Dutch researcher into fluid and solid mechanics, known as the originator of the screw dislocation, 'Burgers vector' and the 'Burgers circuit' in crystal dislocation theory, studied Volterra's dislocation field. The theory of elastic dislocation has also contributed to the theory of the continuous distribution of dislocations.

Volterra received numerous honors from nearly every major scientific academy, but scientific research did not occupy all his activity. He devoted the last quarter of his life to the Resistance and fought against fascism in Italy. Volterra was an intimate friend of many well known scientific, political, literary, and artistic men of his time. His biographer wrote: 'He has been compared to a typical man of the Italian Renaissance for the variety of his interests and knowledge, great scientific curiosity, and his sensitivity to art, literature, and music'.

3.4 ADVANCES IN FRACTOLOGY

3.4.1 The Formation of Fractology

In 1955, Takeo Yokobori [2] attempted to integrate research into the fracture of materials, in which *both micro and macro research are connected nonlinearly*, through the following:

1. An interdisciplinary approach.
2. The study of all solid materials.
3. All kinds of failure phenomena (yielding, brittle fracture, ductile fracture, environment assisted cracking, fatigue, creep, and so on).

He called it *fractology*, and pointed out [2] the necessity for research into such areas as continuum mechanics of notch and crack, phenomenological fracture mechanics, criteria of fracture, combined micro- and macro-fracture mechanics (physical fracture mechanics, abbreviated as CMMFM), stochastic process theory of fracture, theory of transfer on fracture, theory of rate process in fracture, kinetic theory of fracture, and, in special cases, nucleation theory of fracture, and he discussed the relationship between these theories.

For example, in the crack propagation behavior found in metallic fatigue and metallic creep, it is well known that a discontinuous phenomenon is associated with the incubation period, but the empirical formulation rules of macroscopic crack propagation rate can also be applied (see Fig. 2.4 in the preceding chapter). The main reason why we can observe this discontinuous phenomenon in crack propagation is that the following interactions occur so as to connect one fracture process with another:

1. Interaction between macrocracks and visco-plasticity (slip dislocations) ahead of the crack tip.
2. Interaction between macrocracks and microcracks or microvoids ahead of the crack tip.
3. Interaction between the crack and the environment at the crack tip.

Therefore the *interaction* in various fracture processes plays an important role in the variety of the fracture modes of the materials.

The recognition of the variety in the fracture of materials and that the discontinuity in material space and time is due to the interaction in various fracture processes brought about the formation of the field of study we call fractology. Fractology has been remarkably fruitful in studying yielding and various fracture properties of materials having technical uses. We can find many typical examples in research areas such as fracture mechanics, material fatigue, environment assisted cracking, high-temperature fracture in the creep region, and so on, in the 1960s and 1970s.

3.4.2 Advances in Fracture Mechanics

These days so-called *fracture mechanics* (abbreviated as FM), in the narrow sense, is research concerning the '*singular field of the crack tip*' and the application to the *criterion* of certain fracture properties of materials by using the continuum mechanical approach, which includes linear elastic fracture mechanics (LFM) and nonlinear fracture mechanics (NLFM). Historically, LFM was started by George R. Irwin, an American researcher into solid mechanics, and his colleagues in the US Navy Research Laboratory during 1948–54 [98, 99]. The study of FM forms an important part of fractology but, as a matter of course, it does not include all the mechanical approaches to fracture; it has greatly contributed to progress in the new technological fields of fractology from a macroscopic viewpoint. This also stimulated study of thermodynamics of the fracture of materials on the microscopic level.

In Griffith's theory [100], the energy G of an elastic body containing a crack in a thermodynamic equilibrium state is released when the crack grows by a unit interface area, and the fracture strength of a cracked body was derived by calculating the elastic energies at the first and the last loading, respectively, without any discussion of the fracture process at the crack tip. This brilliant theory was published by A. A. Griffith (1893–1963), a British aeronautical engineer, in 1920 (see Section 3.4.4). For the crack to grow it is necessary for the *energy release rate G* to become larger than the work G_c required to create a crack interface in the unit area and equivalent to the so-called surface energy $2\gamma_s$. Based on this idea, Irwin and Kies proposed the fundamental relation $G > G_c$ as the determining basis of whether the crack grows or not. They called G_c the *fracture toughness* of cracked materials and it afforded an important opportunity for studying the fracture toughness of materials. This lucid idea was presented orally, entitled 'Fracture dynamics', at the ASM in October 1947, and thus LFM was founded. Irwin was also the first to define the *stress intensity factor*, which represents the singular stress field at the crack tip, in 1957–8 [101, 102].

Thus, for the singular field where elastic stress and strain become infinitely large at the crack tip, the K value has been useful in calculating stress and strain at the crack tip while avoiding the inconvenience of the singular point, and has practical applications such as in *damage tolerance design* [103, 104] from the viewpoint of *fail-safe*. The concept of the 'singular stress field' had been put forward by T. Terada (1878–1935), a Japanese physicist [105], in 1931; although K had not been formulated, it had been discovered previously by Sir Nevill F. Mott, a British physicist [106]. Mott found that the stress ahead of the crack tip could be written as $\sigma \simeq E(a/r)^{1/2}$ (where E is Young's modulus, a is the atomic distance, and r is the radial distance from the crack tip), and the crack opening displacement y (COD) is written as $y \simeq (ar)^{1/2}$. Irwin's research, however, has greatly contributed to LFM in elucidating the meaning of the strength and fracture of materials. In these studies of FM the following considerations were important.

Stratificational Cognition in the Continuum Mechanical Approach
First let us consider the characteristics of the continuum mechanical approach in FM. One is a fracture problem of structural materials for technical use which includes technological assessment for the prevention of fracture [107], and another is a fracture problem concerning heterogeneous materials with microstructure [108, 109]. In these, an *energy momentum tensor*, P_{jm}, had a special meaning; it was first proposed by John Douglas Eshelby (1916–1981), a British researcher into solid mechanics [110], in 1951 (see Section 3.4.4).

P_{jm} was constructed from the following resultant force vector, F_m, which is imposed on a singular point or an inhomogeneity point within the loop which it is encircled by Σ as

$$F_m = \int_\Sigma P_{jm}\,d\Sigma_j, \qquad P_{jm} = (1/2)\sigma_{ik}\varepsilon_{ik}\delta_{jm} - \sigma_{ji}(\partial u_i/\partial x_m) \qquad (3.24)$$

The energy momentum tensor had played an important role not only in his studies such as the analysis of force in stationary and moving cracks in an elastic body in 1968, but also in NLFM. In 1968 James R. Rice [111] and G. P. Cherepanov [112], researchers into solid mechanics in America and the USSR respectively, proposed the *J-integral*, which was first given for two-dimensional linear and nonlinear elastic bodies as (see Fig. 3.11)

$$J = \int_\Gamma [W\,dy - T_i(\partial u_i/\partial x)\,ds] \qquad (3.25)$$

where $W = \int_0^{\varepsilon_{ij}} \sigma_{ij}\,d\varepsilon_{ij}$, T_i is the surface traction vector on loop Γ, u_i is the displacement vector and ds is the line element of Γ. Afterwards it was discovered that the J-integral was equivalent to the resultant force

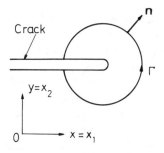

Fig. 3.11. Path in the J-integral.

around the crack tip, i.e. the energy release rate, G, when the crack grows. They also observed that the J-integral demonstrated *path independence* with respect to Γ. Thus the following conclusions were made:

1. In LFM the J-integral is equivalent to the energy release rate, G.
2. The function in the J-integral is a kind of energy momentum tensor, P_{jm}, and the J-integral is the application of F_m to the crack tip.

Considering the edge dislocation pile-up as one of the singular points shown in Fig. 3.12(a) [113], the resultant force around the loop which encloses only the fixed dislocation is obtained by eqn (3.24). Let us replace Σ by the loop Σ_1 which includes moving dislocations; since the resultant force in moving dislocations is zero, the J-integral for Σ_1 must be equal to that for Σ. Therefore the force in continuously distributed dislocations is given by the J-integral for the loop which surrounds only the front of pile-up dislocations. Furthermore, assuming that the singular point at the crack tip is equivalent to that due to the pile-up of crack dislocations as shown in Fig. 3.12(b) [114], the *extension force*, i.e. the energy release rate, is equivalent to the resultant force due to crack dislocation-pile-up, and is also equivalent to the J-integral around Γ in Fig. 3.12(b).

Next let us consider whether the J-integral for a heterogeneous material really has an independent value with respect to the integration path. Since we presume that there is an uneven distribution of dislocations ahead of the crack tip, the J-integral value will vary with the variation in path from Γ to Γ_1 shown in Fig. 3.12(c) [114]. Therefore we can stipulate that the J-integral for path Γ_1 is represented by the sum of the resultant force due to the singularity at the crack tip and that due to the singularity of the fixed pile-up dislocation as

$$J_{\Gamma 1} = J_e + J_d \tag{3.26}$$

where J_e and J_d are the values for the paths which surround only the crack tip (the so-called J-integral in LFM) and the fixed dislocations, respectively. Thus the J_d-integral varies with the path, and it should be noted that the path-independence does not always hold true in the case of a heterogeneous material with a dislocation cloud ahead of the crack tip. In other words, it is notable that the resultant force in the dislocation cloud ahead of the crack tip varies with the change of dislocation density, and even if the dislocation density has the same force it is changeable with the distribution. In connection with the differential

HISTORY OF THE MICRO- AND MACRO-MECHANICAL APPROACH

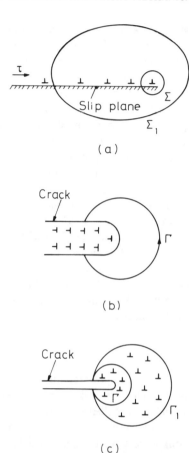

Fig. 3.12. Force on dislocations and the correspondence to the J-integral. (a) Edge dislocation pile-up on a slip plane. (b) Crack dislocation pile-up within the crack tip. (c) Dislocation cloud ahead of the crack tip. (Reprinted from refs 113 and 114 by permission of Academic Press and BAIFUKAN, respectively)

geometry as shown in Table 3.7, we will see that the J-integral is related to the Riemann–Christoffel curvature tensor (see Chapter 7).

Recognition of the Irreversible Process and Non-equilibrium in the Fracture of Materials
Historically the contradiction inherent in the thermodynamic reversible process in SFM has been formally avoided by using the so-called 'rate

process theory with stress dependence', and the importance of the 'kinetic theory of fracture' was emphasized from the viewpoint of the combined nucleation theory along with the dynamic theory of distributed dislocations and the stochastic theory [115].

Regarding thermodynamic non-equilibrium in the fracture process of solids, the *principle of local equilibrium* has been assumed, by which we are able to consider the state function at an individual point of the system in the case of a low-degree non-equilibrium state if the system is in the irreversible process. Since the propriety of the principle has not yet been proven, it is necessary to obtain agreement between the theoretical and experimental results. The so-called *critical J-integral criterion*, $J_c = 2\gamma_p$ (eqn (2.5) in Chapter 2) is applicable strictly to only the initial condition for crack growth in the cracked material [114], where γ_p is the plastic surface energy. For instance, ASTM E813, BS, and JSME S001 are the standard test methods for measuring the J_c factor of *post-yield fracture toughness*. In this case, the evaluation of γ_p for technical use must be made by taking various stratificational factors into consideration (see Section 2.3 in Chapter 2). It should be noted here that the work done by the spread local plasticity zone ahead of the crack tip is not included in γ_p [116].

Stratificational Recognition in CMMFM
The continuum approximation approach in which the interaction between macro and micro defects is taken into consideration has been productive in the field of the so-called micromechanics of the fracture of solids. For example, the theory of fracture in which both the necessary and sufficient conditions were satisfied has succeeded in the formulation, shown in Table 3.8 [117], due to the validity of the 'hierarchy' in both continuum mechanical and continuum approximation approaches. In the nucleation theory of fracture, however, it is unknown whether both processes, before and after the critical size of the embryo, are interpreted at the same microscopic level or not (see Section 2.4 in Chapter 2). We can note, at least, that the irreversible process exists in quantum mechanics, and both processes may be interpreted from the uncertainty principle.

3.4.3 Advances in Studies of Metallic Fatigue and High-temperature Strength of Materials

Another typical example of fruitful progress in fractology is the study of metallic fatigue. Table 3.9 [118] shows the characteristics of research done in the 1960s and 1970s.

Table 3.8
Comparison between Combined Micro and Macro Fracture Mechanics (CMMFM) and So-called Fracture Mechanics (FM) (Reprinted from ref. 117 by permission of IWANAMI-Shoten)

	CMMFM	FM
Kind of defect	Contains both crack type and dislocation array type	Only crack type is considered
Number Arrangement	Plural Contains mutual interference	In LEFM the number is one and there is no mutual interference
Kind of crack	Contains not only a crack ($\rho = 0$) but also a notch	In LEFM only a crack ($\rho = 0$) is considered
Fracture criterion	(a) In so-called brittle fracture Necessary condition I: energy condition Necessary condition II: critical local stress condition	In LEFM only an energy criterion, i.e. K_c = constant (critical value), is considered
	(b) In time-dependent fracture, only a condition including thermal activation process is taken into consideration	In NLFM a critical COD condition, i.e. COD = constant (critical value), is considered

LEFM, Linear elasticity fracture mechanics
NLFM, Nonlinear fracture mechanics
ρ, Radius of crack tip

We will examine especially the mechanics of crack initiation and propagation, the study of the relation between *threshold stress intensity range* ΔK_{th} and *fatigue limit* σ_w. For example, the *Paris–Erdogan rule* (1963) [119] for the fatigue crack propagation rate as a function of ΔK has greatly contributed to technical application, but some difficulties remain. The phenomenon of *fatigue crack opening and closure* was discovered by E. Elber, an American researcher into FM [120], and it has contributed toward overcoming these difficulties. Later a highly accurate technique to measure both the crack opening and the closure was devised by M. Kikukawa, an emeritus professor at Osaka University, and his colleagues. It was found that the crack propagation

Table 3.9
Characteristics of Study of Metallic Fatigue in the 1960s and 1970s (Reprinted from ref. 118 by permission of OHM-Sha)

Research field	Characteristics
Study of microscopic mechanism of crack initiation and propagation	(1) Study of model of fatigue crack initiation and propagation in the 1960s (2) Establishing electron fractography in the 1960s and X-ray fractography in the 1970s (3) Advances in information concerning the fatigue mechanism by means of TEM and EMR in the 1960s (4) Great progress in information processing techniques by means of computer, and advances in the fracture mechanical approach such as *fatigue fracture toughness*, stop of microcrack, *threshold stress intensity range ΔK_{th}*, in the 1970s (5) Recognition that ΔK_{th} or fatigue limit σ_w is not always an *unchangeable property of the material* (6) ASTM Symposium on Fatigue Mechanism (1978); ASM Materials Science Seminar on Fatigue and Microstructure (1978)
Reliability engineering of crack initiation and propagation	(1) Advances in studies of the stochastic model of fatigue crack propagation, distribution characteristics of fatigue strength and fatigue life, statistical treatment of data, in the 1960s (2) Combined micro and macro stochastic study of fatigue crack propagation and study of fatigue fracture under random loading from viewpoint of two-step transition probability (3) Publication of many data bases
Study of low-cycle fatigue	(1) Nonlinear fracture mechanical approach to fatigue crack propagation (2) Studies of the cumulative damage rule on the basis of strain partitioning (3) Studies of low-cycle fatigue under multiaxial stresses; ASTM International Symposium on Fatigue under Biaxial/Multiaxial Stresses (1982, 1985) (4) Structural design codes, such as ASME Boiler and Pressure Vessel Code Section III (1968)
Study of high-speed effect and of fatigue under impact	(1) Study of strain rate effect on yielding and plastic flow (2) Study of ultrasonic fatigue (3) Study of fatigue under impact load (4) Study of relationship between those above

TEM, Transmission electron microscope
EMR, Electron microscope replica techniques

rate was represented uniquely by ΔK_{eff} which corresponds to ΔK during crack-opening [121]. It was, however, also found that ΔK_{th} and an even $(\Delta K_{\text{eff}})_{\text{th}}$ value were influenced by loading histories such as the K-increasing test or the K-decreasing test, and the environment and structure of the material at the crack tip. Thus it has been clarified that ΔK_{th} and $(\Delta K_{\text{eff}})_{\text{th}}$ are not always invariable properties of materials for technical use as well as σ_w.

The interdisciplinary study of the high-temperature strength of materials made great progress in the 1970s. In the 1960s collaborative systems for creep rupture testing were arranged around the world. Up until the 1960s these collaborative systems were organized as shown in Table 3.10 [122]. Figure 3.13 is a photograph showing the creep testing apparatus with nearly a thousand creep rupture testing machines in the National Research Institute of Metals (NRIM), Japan. The NRIM has provided much long-term creep rupture data (during 100 000 h; nearly 11 years) concerning heat-resistant alloys for technical use, and has supplied this information to the world. The studies in the 1970s were characterized as follows.

Above all, in this period researchers recognized that the strength properties of materials, at high temperatures or in a corrosive environment, such as creep, corrosion cracking and thermal fatigue, do not occur alone but are 'compound' phenomena with various interactions. Therefore it has been very useful in studying the variety of the phenomenal aspects of this 'interaction' so that they can be accurately understood. A characteristic of the high-temperature strength of materials during this 20 year period was that stratificational interaction was recognized as the single most important aspect.

As an example of researchers who discovered the importance of the interactive effect on the high-temperature strength properties of materials, two scientists will be discussed, Shuji Taira (1920–1978), a former professor at Kyoto University, and Robert M. Goldhoff (1920–1978), a metallurgy researcher working for the General Electric Company in Schenectady.

Taira was noted for his excellence as a researcher and was a leader of workers in high-temperature studies and X-ray diffraction studies of the mechanical behavior of materials. In 1971 he received the Academy Award from the Japanese Academy of Sciences, an annual award for outstanding achievement and contribution to science, for his and Yokobori's research on the strength and fracture of metallic materials. After his death, A. J. McEvily, professor of metallurgy at the University

Table 3.10
Organization of Creep Research Collaboration in the World until the 1960s
(Reprinted from ref. 122 by permission of OHM-Sha)

Country	Situation of research system
Switzerland (about 60 creep testing machines)	(1) Laboratories in the industries, such as Sulzer Brothers Ltd. (founded 1834) and Escher Wyss Ltd. (founded 1966) are the leading researchers in the field of creep, and EPMA (Eidgenössische Materialprüfungs und Versuchsanstalt) does not perform research as the main theme. (2) Creep research in industry focuses on the study of the properties of the material for technical uses, and materials testing for structural design, rather than the development of heat-resisting materials.
West Germany (about 230 creep testing machines)	(1) Under the leadership of VDEh (Verein Deutscher Eisenhüttenleute; founded 1869), the Achterausschauss was established in collaboration with universities and industries in 1950, and creep data of heat-resisting metals were obtained at temperatures below 750°C. In 1962, the Zwelferausschauss was established, and data above 750°C were obtained for the purposes of chemical plants and power generators. (2) MPA (Staatliche Materialprüfungsanstalt), TH Darmstadt, Max Planck Institut für Metallforschung, Max Planck Institut für Eisenforschung, and Mannesmann Aktiengesellschaft are the best-known research laboratories in West Germany.
France (about 40 creep testing machines)	IRSID (Institut de Recherches de la Siderurgie Française) was founded as the research center for iron and steel in France in 1952, and has performed creep tests as commissioned by industry.
Britain (about 650 creep testing machines)	(1) Creep research system is well organized and there are some large-scale collaboration systems: ERA (Electrical Research Association); CERL (Central Electricity Research Laboratory); United Steel Co. Ltd., Swindon Laboratories; NPL (National Physical Laboratory; founded 1900); NEL (National Engineering Laboratory; separated from NPL in 1947). (2) BSCC (British Steelmakers Creep Committee) was founded in 1956 by BISRA (British Iron and Steel Research Association; founded 1850). Since that foundation, respective creep data have been collected and long-term tests for 100 000 h performed.

United States (about 200 creep testing machines for item (2))	(1) The fruits of creep research in industry, especially manufacturers, have been great, and the collaborative system in this country is quite different from that in Britain and West Germany. The research has been performed individually, and the forum has been held in ASME, ASTM, ASM and so on. These have been published in a series of books from ASTM (STP) and ASME-MPC. (2) Central Research Laboratory in Westinghouse Electric Co., United States Steel Research Center, Battle Memorial Institute (Battle Columbus Institute), NBS (US Department of Commerce, National Bureau of Standards), and EPRI (Electric Power Research Institute) are typical of creep research laboratories.
USSR	(1) Research laboratories are well organized in this country. A. A. Baikov Metals Institute, Academy of Sciences of USSR in Moscow, Institute for Problems of Strength, Academy of Sciences of UkrSSR in Kiev, Leningrad University, and Institute of Mechanical Engineering in Moscow are the best-known laboratories. (2) Institute for Problems of Strength in composed of 12 divisions, and the following research is studied: strength and fracture criteria; low-temperature strength of materials; long-term creep strength; strength and physical properties; centrifugal strength of disk; impact strength; strength in gas flow for gas turbines; composite materials; fatigue and thermal fatigue; ultrasonic fatigue; mathematical modeling; vibration and reliability.
Japan (about 1500 single-type and 300 multi-type creep testing machines)	(1) After the establishment of JIS Z 2271 (1956) and JIS Z 2272 (1956) in creep testing methods, the importance of creep has been recognized since the 1960s through the activity of Japanese industry. (2) Creep collaboration has taken place at the following places: 123rd Committee (1957) and 129th Committee (1960) in JSPS (Japan Society for the Promotion of Science; founded 1932); Creep Committee (1960) in ISIJ (Iron and Steel Institute of Japan; founded 1914); National Creep Laboratory (1968) in NRIM (National Research Institute of Metals; founded 1956). Creep research laboratories in steelmaking, fabrication facilities and universities in this country have been greatly improved. (3) A creep data base has been established in JSPS. The creep data, especially NRIM long-term creep data for various heat-resisting steels and alloys, have been greatly appreciated in the world. NRIM installed 880 single-type and 144 multi-type creep testing machines in 1980.

(a)

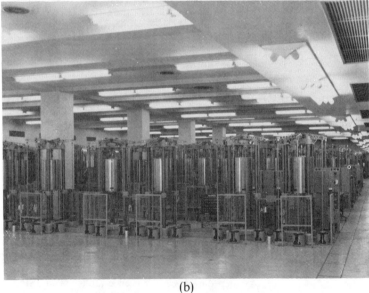

(b)

Fig. 3.13. Photographs showing the creep testing apparatus in the National Research Institute of Metals (NRIM), Japan: (a) single-type tensile creep testing machines; (b) multi-type tensile creep testing machines. (Courtesy NRIM)

of Connecticut, wrote: 'the international community has lost one of its most capable and dedicated members'.

L. F. Coffin, a distinguished materials researcher working for the General Electric Company, wrote in the obituary book which contains letters and notes in Taira's memory sent from fifty-eight friends in eleven countries abroad [123]: 'I first met Professor Taira when he visited our laboratory in 1963 at the time of the Joint International Conference on Creep in New York. He had presented a paper there with Professor M. Ohnami on 'Fracture and deformation of metals subjected to thermal cycling combined with mechanical stress' [124]. This work supported Professor Taira's earlier findings that a linear cumulative damage rule could be used to predict the interactive effects of cyclic thermal stresses and tensile means stress. Because of my interest in thermal stress fatigue, we developed a common bond of professional association and personal friendship. Professor Taira was highly regarded, not only in Japan but through the world, for his many contributions to his professional field. His many visits to the United States and to other countries attest to that. His role in founding the International Conference on Mechanical Behavior of Materials (the 1st ICM, 1971, Kyoto) will be a lasting memory to him'.

The ASME transaction journal had a special and unique issue focusing on the life and work of R. M. Goldhoff, nicknamed Bob, in the Journal of Engineering Materials and Technology (Vol. 191, No. 4, 1979); it was the first time the journal had dedicated an entire issue to a single individual. The issue opens with the following obituary contributed by L. F. Coffin and others: 'At General Electric Company, Bob's work involved conducting and later managing applied research and development studies concerned with the metallurgy and properties of materials used in power generation equipment such as steam and gas turbines and electric generators. He gained international recognition for his contributions to and expertise in material behavior under complex regimes of time, temperature and stress. His work focused on the interdisciplinary area between physical metallurgy, material properties and engineering design. He was an active member of the ASME-ASTM-MPC (The Metals Properties Council Inc.) Join Committee on the Effect of Temperature on the Properties of Metals. The task group of this Committee concerned with *Time–Temperature Parameters* (*TTP*) (Bob was the chairman) did an extensive study of metals for the extrapolation method of short-time rupture data which might serve as a basis for a standard practice' [125].

The long-term creep rupture data in the NRIM, Japan, had also served as the data base which helped establish the extrapolation method. Thus, in the 1970s the following studies of the high-temperature strength of materials will be examined:

1. The structural design method for high-temperature applications and the estimation method of the *remaining life* from the viewpoint of safety integrity.
2. The research into *creep/fatigue interaction* and the *influence of oxide environment* from the viewpoint of *crack initiation and propagation*.
3. A marked advance in the creation of *new alloys* having heat resistance, high-temperature corrosion resistance and a high-temperature melting point.

Table 3.11
Characteristics of Mechanical and Physical Approach to High-temperature Strength of Materials in the 1960s and 1970s (Reprinted from ref. 122 by permission of OHM-Sha)

---

(1) FEM analysis of plasticity/creep, and thermal ratcheting of large-scale structural elements having an arbitrary temperature gradient
(2) Similitude law in nonlinear elasticity, approximate solution in creep problems by means of reference-stress method, frame method, energy method, exact solution of large deflection of axisymmetric shell, and creep buckling
(3) Plasticity/creep constitutive equation, including experimental examination of the effects of the 1st and 2nd invariants of stresses and strain history
(4) FEM analysis of creep fracture of thick-walled tubes, notched and cracked materials, and criterion of creep fracture. Dissipated theory of creep fracture in USSR and development into damage mechanics
(5) High-temperature fatigue crack propagation equation by Coffin, in ASTM International Symposium on Fatigue at Elevated Temperatures (1972), elucidating the meaning of 'Manson–Coffin's equation' (1953, 1954). Suggestion of an interactive relation between frequency effect and high-temperature oxide environmental effect
(6) 'Strain-partitioning method' by Manson (1971). Frequency effect, load sequence effect, effect of axisymmetric strain wave and oxide environmental effect in high-temperature fatigue
(7) Nonlinear fracture mechanics of crack propagation behavior in creep and fatigue at elevated temperatures
(8) Thermal fatigue research in Japan (collaboration and recommendation of thermal fatigue testing method in Soc. Mater. Sci., Japan, in 1974)
(9) High-temperature biaxial fatigue and notch effect
(10) Material damage evaluation and remaining life assessment of crept materials

---

FEM, Finite element method

Table 3.11 [122] shows the characteristics of the research done in fields 1 and 2 above. This research has contributed to technological progress in, for example, ASME B & PV Code Case N-47 in the United States [126] and the design code of a high-temperature heat exchanger in the Japanese R & D efforts in nuclear steelmaking in 1973–80 [127, 128]. Especially in the latter design code, progress in the design procedure requiring a temperature of 1273K drew the attention of the world, because at that high temperature in helium gas-flow even a small increase in the design temperature caused significant difficulties. Combined micro- and macro-mechanical studies of creep/fatigue, thermal ratcheting in the creep region, and the environmental effect of decarburizing and carburizing helium and so on were undertaken in order to establish the design code (see Chapter 8).

3.4.4 The Life and Work of Ewing, Griffith, and Eshelby in Fractology

Sir James Alfred Ewing (1855–1935) and Men Associated with Ewing's Work
Two fundamental concepts, 'crack initiation and propagation' and 'inelastic hysteresis', in the fatigue mechanism of structural metals go back 80 years to the time of Sir James Ewing and Sir Leonard Bairstow (1880–1965) in Britain.

Ewing and Humfrey in Cambridge University [129] are of the opinion that the specimen surfaces on which slipping occurs continue to be planes of weakness and that the effect of repeated slipping and grinding results in the production of a rough and jagged irregular edge, suggesting the *accumulation of debris*. This repeated grinding tends to destroy the cohesion of the crystal on the slip surface, and in certain cases it actually develops into a crack. Once a crack occurs, it grows rapidly because of the orientation of stress at the crack tip. The test showed how a crack may grow to failure under alternating stresses. Even though, in the paper by Ewing and Humfrey, there are some defects in terminology or the representation of the fatigue mechanism, the experiment greatly contributed by providing a basic explanation as to why fatigue fracture shows no sign of local elongation, and also why the specimen which had been subjected to many stress reversals showed no deterioration detectable by the usual tensile tests. Thus Ewing first cleared the old theory of the 1840s in which material deterioration due to fatigue was thought to be the result of 'recrystallization' and elucidated that the fatigue damage of metals might be the result of the *'accumulation of repeated slipping'*.

The fundamental concept that the fracture of materials might be the result of an initiation of a microcrack and subsequent growth was proposed by A. A. Griffith in 1920 [100] and Japanese physicists [130, 131, 105] in 1929–30, but the theory had its origin in the paper by Ewing and his colleagues.

Leonard Bairstow, a scientific researcher in the National Physical Laboratory (founded in 1900, abbreviated NPL) and later a professor of aeronautical engineering at London University, made a great many experiments concerning the inelastic hysteresis loop of iron and steel under repeated stresses in 1910 [132].

When a specimen of axle steel was subjected to a repeated stress of 57% of the static yield point σ_y, the material response at first was represented by the straight line shown in Fig. 3.14 [132]. As the number of cycles of stress increased, the straight line changed into a loop, and after 18 750 cycles the width of the loop was about 11% of the original elastic extension as shown in loop ABCD. Bairstow was of the opinion that at a slightly smaller stress of about 52% of the yield point the specimen would be perfectly elastic and that no number of cycles would cause the development of a loop. This range of elastic limit corresponds to Bauschinger's 'natürliche Elastizitätsgrenze' and Bairstow supported this theory. After 29280 cycles of stress at 81% of the yield point, the loop became very wide and had the shape shown in EFGH in Fig. 3.14. In all

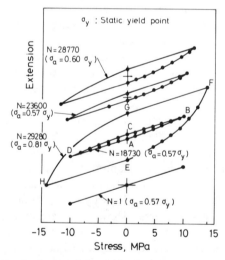

Fig. 3.14. Change in the hysteresis loop of a Swedish iron bar specimen subjected to push-pull. (Reprinted from ref. 132 by permission of The Royal Society)

these loops the lines FG and HE were found to be parallel to the original elastic line. He found that, at a stress lower than the yield point, the width of the loop increased with increased number of cycles of stress and then tended to become constant, and at higher stress the width of the loop actually decreased with increased number of cycles. The former today corresponds to the phenomenon wherein after *cyclic softening* the loop saturates, and the latter corresponds to the case of *cyclic hardening*. Bairstow concluded that iron and steel were capable of 'adjusting themselves' to cyclic stressing after a sufficient number of stress cycles. Phenomena of both cyclic softening and cyclic hardening of iron and steel may suggest that the local cyclic stress/strain fields in the materials adjust themselves with the number of cycles of stress. Nowadays 'adjusting itself' means the 'interaction' between the many damage processes described in Chapter 2.

Thus Bairstow presumed that the dissipated energy which corresponded to the area of the hysteresis loop was related to the formation of 'slip bands' which corresponds to the theory of fatigue crack initiation proposed by Ewing. When Bairstow's adjustment is complete, the specimen is perfectly elastic throughout the stress cycles and fatigue failure does not occur, even if slips occur in the adjusting process. The exact connection between fatigue crack initiation and slip bands is not clear today, but Bairstow's concepts had stimulated progress in research into metal fatigue mechanisms such as 'extrusion' and 'intrusion' proposed by P. J. E. Forsyth and his colleagues at the Royal Aircraft Establishment (RAE) in Britain [133] and the 'persistent slip band' (PSB) theory proposed by N. Thompson in America [134] in the 1950s.

Ewing was born in Dundee, Scotland, on 27 March 1855, the son of a church minister. He studied at the University of Edinburgh on an engineering scholarship. He was under the influence of Peter Tait and Fleming Jenkins; through Jenkins he came into contact with William Thomson (Lord Kelvin) and participated in laying three transatlantic telegraph cables. Following Jenkins's recommendation, Ewing went to Japan in 1878 as a 23 year old professor of mechanical engineering at the University of Tokyo and lectured on steam engines, applied mechanics and mechanisms. The university provided Ewing with the means to establish a seismological observatory, and in 1879 he installed the instruments and recorded earthquakes. He founded the Society of Seismology in Tokyo in 1880 in collaboration with John Milton (1850–1913), a professor of the KOHBU-DAIGAKU (later joined to the University of Tokyo). In 1889 he discovered that the thermoelectric effect lags behind the applied stress and introduced the term '*hysteresis*' (from the Greek word which means 'to be late') to describe it. He then turned to the study of hysteresis in magnetization. In 1882 he observed that the area enclosed by the hysteresis loop is proportional to the work done during a complete cycle of magnetization and demagnetization. His study of magnetism in Japan was continued by the Japanese physicists Hantaro Nagaoka (1865–1950) and Kohtaro Honda (1870–1954). Ewing was awarded the Japanese Order of Precious Treasure in 1881. He returned to Britain in 1886, after 5 years in Japan, and in 1890 he was appointed professor of mechanism and applied mechanics at Cambridge University. He received honorary degrees from Oxford, Cambridge, Durham and St. Andrews universities. He was appointed a Fellow of the Royal Society in 1887 and received the Royal Medal for his research on magnetism in 1885. He passed away on 7 January 1935 at the age of 79.

In a paper presented in BAAS in 1889 [135], Ewing wrote the following: 'These experiments make it clear that in conditions of loading such as those they deal with, there is a decided departure from Hooke's law, one effect of which is that work is done upon the material when it is put through a cycle of stress change. If the relation of strain to stress were represented graphically, the curves would form a loop enclosing a certain area. This result has a sufficiently obvious bearing on the conclusions made by Wöhler with regard to the deteriorating effect of repeated vibrations of stress'. It was only nine years after W. Thomson and about twenty years before Bairstow.

Alan Arnold Griffith (1893–1963)
Griffith was a 'man of vision' [136–139]. He was born in London on 13 June 1893. Subsequent to this graduation from the Department of Mechanical Engineering at the University of Liverpool, he joined the Royal Aircraft Factory (later the RAE) in July 1915 and received short general workshop training until November 1916. During the next 4 years he successively occupied positions as draftsman, technical assistant and senior technical assistant in the Physics and Instrument Department at RAE, and was appointed senior scientific officer of the department in April 1920 when he was 27 years old.

In December 1917, a paper by A. A. Griffith and G. I. Taylor [140] was awarded the Thomas Hawksley Gold Medal. This paper dealt with novel methods for estimating torsional stresses in sections of complicated shapes. The need for the determination of stresses in the aerofoil sections of aircraft blading and other problems had originally led to the above work, and the methods proposed by Griffith proved to be exceedingly accurate and convenient. Professor A. H. Gibson remarked that it was virtually the only real advance in our knowledge of the general problem of torsion since the study by Adhemar-J.-C. Barrede Saint-Venant (1797–1886), and Sir Leonardo Bairstow also praised it. He was nicknamed 'Bubble Griffith'. In 1920 he produced what is perhaps one of his best known contributions to the science of the mechanical behavior of materials. This paper, 'The phenomenon of rupture and flow in solids', was introduced in Europe by Adolf Gustav Smekal in 1922 [141] and Karl Wolf in 1923 [142]. In Japan it was first referred to in a paper by Morizo Hirata, a founder of the stochastic theory of fracture, in 1929 [130], and was later introduced in detail by Takeo Yokobori in 1955 [3]. G. I. Taylor commended Griffith's paper as follows [136]: 'Griffith's great contribution to ideas about the strength of materials was that he realized that the weakening of material by a crack could be treated as an equilibrium problem in which the reduction in strain energy of a material containing a crack, when the crack extends, could be equated to the increase in surface energy due to the increase in surface area'. Sir Alan H. Cottrell, a British researcher into the fracture of solids, reviewed the career of Griffith, who has been called the 'father of the science of fracture', and the research that has been done since Griffith in his interview at the 4th International Conference on Fracture [143].

Griffith had been interested in the aircraft gas turbine at RAE. He progressed by designing high-efficiency gas turbines and reported this in 1926 in RAE Report H1111, 'An aerodynamic theory of turbine design'. In this famous classic paper, he demonstrated that the axial gas turbine was feasible as a power unit for aircraft propulsion and determined the basis for a complete design. The tests in 1928 demonstrated that blading designed following the aerodynamic theory in Griffith's 1926 paper yielded efficiencies that accorded well with his prediction, reaching 91% of expected values, including the blade, casing and rotor friction losses but neglecting bearing friction losses, at speeds of around 15 000 rev/min. This was the first time that high efficiency was obtained from an axial type compressor, the type that now predominates in all high-output high-efficiency gas turbines. Griffith may be said to be the true originator of the multi-stage axial engine. Early in 1945, while development of the Whittle type engine continued, Griffith commenced a detailed study of a simple jet engine with an axial compressor. He realized the need for an engine incorporating the latest ideas on axial compressor blading, and the design would reduce fuel consumption and considerably reduce specific weight. The success of the Conway engine in the Boeing 707-420 and the Douglas DC-8/40, and the fact that the by-pass principles are now being followed by engine manufacturers throughout the world, is a complete vindication of the soundness of his proposals. He also studied vertical take-off and landing (VTOL), developed the 'flying bedstead' (1953) and the Short SC1 VTOL (1954), and made the VTOL SST proposal which ultimately led to Concorde. Griffith retired from his position as chief scientist at Rolls Royce in June 1960. His health was declining but he continued his consulting work for the next 2 years. He died on 11 October 1963.

Griffith's brilliant work earned him Fellowship of the Royal Society in 1941, a C.B.E. in 1948, the silver medal of the Royal Aeronautical Society in 1955, the Bleriot medal in 1962, and the Brequet trophy in 1963. A. A. Rubbra, Griffith's biographer [136] wrote: 'Griffith, a man of vision, tall, slim and generally of serious countenance, was always calm and quite, reticent and somewhat aloof; yet those closely associated with him found a very charming and friendly personality always with a ready and puckish wit'. The obituary [138] states: 'He has always avoided any sort of public recognition of his work. He was of a naturally retiring disposition and he talked about his work very little. There were very few people able to keep up with his reasoning, and nearly all his work has been secret, precluding public discussion. This is why he gave no lectures or papers'. Biographical Memoirs of the Royal Society [136] reported that papers of his were published including 10 papers under a single name, but 107 papers had not been published. Despite his avoidance of publicity, his work has been recognized by many people in the world. His life was full of intelligent curiosity and quietness.

John Douglas Eshelby (1916–1981)
'The sudden death of Jock Eshelby, professor of the theory of materials at the University of Sheffield, robs us of a great scholar', wrote B. A. Bilby, a British researcher into solid mechanics, in Eshelby's obituary [144]. We refer to that obituary as follows.

Eshelby was born on 21 December 1916. From the age of 13 he missed his formal schooling and learned instead from tutors at home. Having 'to work things out for himself' perhaps helped to make him the very original and creative thinker that he was. He obtained a First in physics at Bristol in 1937 and began experimental research which was interrupted by the war. Returning to Bristol at an exciting time for solid state physics, when rapid advance was being made in understanding the deformation of crystals, he took up theoretical work and made his first mark in dislocation theory, obtaining a PhD in 1950. Eshelby was elected a Fellow of the Royal Society in 1974 for his distinguished theoretical studies of the micromechanics of crystalline imperfections and material inhomogeneities. He made a major contribution to the theory of moving dislocations, including a discussion of the equation of motion and an elegant use of quantum mechanics to discuss the interaction of kinks with elastic waves. He solved a number of important problems in dislocation statics and made numerous contributions to the theory of point defects. By an elegant use of the theory of potential he obtained some remarkable results on *ellipsoidal inclusion and inhomogeneities, applicable to elastic and viscous materials.*

In 1951 [110] he introduced, in analogy with the Maxwell tensor, the elastic energy momentum tensor, and he was much concerned with the development of this concept during his latter years. In 1968 he published an account of its application to the calculation of forces on static and moving cracks in elastic media, followed by a discussion of the equation of motion of a crack, including the proper energy balance near the moving crack tip. A related study on the Maxwell tensor was also made for elastic-plastic materials and published independently by J. R. Rice in 1968. This has progressed in a widespread application to FM, sometimes in a way to which Eshelby did not always subscribe. The ASME awarded him the Timoshenko medal in 1977. He died on 10 December 1981. He was an original and creative thinker, and contributed greatly to theoretical studies of solid micromechanics and fracture mechanics.

REFERENCES

1. M. Ohnami, in *Science and Technology on Strength and Fracture of Materials*, Vol. 1 in A Series of Comprehensive Lectures on Strength and Fracture of Materials (ed. M. Ohnami), OHM-Sha, 1984, p. 72 (in Japanese).
2. T. Yokobori, *An Interdisciplinary Approach to Fracture and Strength of Solids*, 2nd Edn, IWANAMI-Shoten, 1974, p. 3 (in Japanese).
3. T. Yokobori, *Strength and Fracture of Materials*, GIHODO, 1955, pp. 1–300 (in Japanese) (English Edn, *Strength, Fracture, and Fatigue*, P. Noordhoff, The Netherlands, 1965, pp. 1–372).
4. J. J. Burke, N. L. Reed and V. Weiss (Eds), *Fatigue: An Interdisciplinary Approach* (Proc. 10th Sagamore Army Materials Research Conf., 1964), pp. 1–404, Syracuse University Press.
5. M. Ohnami, What can we learn from the history on and present research on fracture of structural materials? (Review), *Journal of the Japanese Society for Strength and Fracture of Materials*, **20** (1986), 41 (in Japanese).

6. A. J. Kennedy, Fatigue since Wöhler–a century of research, *Engineering*, **186** (1958), 781, Design Council, London.
7. M. Ohnami and K. Shiozawa, *Strength and Fracture of Polycrystalline Body*, BAIFUKAN, 1976, pp. 13–38 (in Japanese).
8. Ref. 1, pp. 131–3, 145–50, 163–5, 183–6.
9. J. A. Ewing and W. Rosenhain, Experiments in micro-metallurgy: Effects of strain (Preliminary notice), *Proceedings of the Royal Society*, **67** (1899), 887, The Royal Society, London.
10. G. I. Taylor, Plastic strain in metals, *Journal of the Institute of Metals*, **62** (1938), 307, Institute of Metals, London.
11. J. F. Bishop and R. Hill, A theory of the plastic distortion of a polycrystalline aggregate under combined stresses, *Philosophical Magazine*, **42** (1951), 414, Taylor and Francis, London.
12. T. Ohkubo, Crystallography, 2-I, II (Problem of mechanical anisotropy), *Journal of Japan Society for Technology of Plasticity*, **11** (1970), 152, 223, CORONA-Sha.
13. G. Sachs, Zur Ableitung einer Fliessbedingung, *Zeitschrift Verein Deut. Ing.*, **72** (1928), 734.
14. S. Taira, T. Abe and N. Nagao, Crystallographic interpretation of yield condition of polycrystalline metals (1st rep., bcc metal; 2nd rep., hcp metal), *Transactions of the Japan Society of Mechanical Engineers*, **33** (1967), 199, 1535 (in Japanese).
15. S. Taira and T. Abe, Yield condition of polycrystalline metals (3rd rep., Relation between slip and yield condition), *Transactions of the Japan Society of Mechanical Engineers*, **34** (1968), 1839 (in Japanese).
16. T. H. Lin, Analysis of elastic and plastic strains of face-centered cubic crystals, *Journal of Mechanics and Physics of Solids*, **5** (1957), 143, Pergamon Press.
17. J. W. Hutchinson, Plastic deformation of b.c.c. polycrystals, *Journal of Mechanics and Physics of Solids*, **12** (1964), 25.
18. J. L. M. Morrison, The yield of mild steel with particular reference to the effect of size of specimen, *Proceedings of the Institute of Mechanical Engineers*, **142** (1940), 193.
19. R. Hill, The elastic behavior of a crystalline aggregate, *Proceedings of the Physical Society*, **A65** (1952), 349, Physical Society, London.
20. Z. Hashin and S. Shtrikman, A variational approach to the theory of the elastic behavior of polycrystals, *Journal of Mechanics and Physics of Solids*, **10** (1962), 343.
21. A. V. Hershey, The elasticity of an isotropic aggregate of anisotropic cubic crystals, *Journal of Applied Mechanics*, **21** (1954), 236, Amer. Soc. Mech. Engrs.
22. J. D. Eshelby, The determination of the elastic field of an ellipsoidal inclusion and related problems, *Proceedings of the Royal Society*, **A241** (1957), 376.
23. E. Kröner, Berechnung der Elastischen Konstanten des Vielkristalls aus Konstanten des Einkristalls, *Zeitschrift für Physik*, **4** (1960), 273, Deutsche physikalische Gesellschaft.

24. A. Kochendorfer, *Plastische Eigenschaften von Metallen und Metallischen Werkstoffen*, Springer Verlag, Berlin, 1941.
25. G. I. Taylor, Strains in crystalline aggregates, in *Deformation and Flow of Solids* (ed. R. Grammel) (IUTAM Colloquium, 1956), Springer Verlag, Berlin, p. 3.
26. M. Oyane and K. Kojima, Plastic deformation of fcc polycrystalline metals in consideration of the deformation and rotation of grains (Part I, II), *Transactions of the Japan Society of Mechanical Engineers*, **21** (1955), 817, 823 (in Japanese).
27. G. Y. Chin, The inhomogeneity of plastic deformation (Papers presented at a seminar of ASM, Oct. 1971) (1973), p. 83, Amer. Soc. Metals.
28. G. Y. Chin and W. L. Mammel, Computer solutions of the Taylor analysis for axisymmetric flow, *Transactions of Metallurgical Society of AIME*, **239** (1967), 1400, Amer. Soc. Metals.
29. J. M. Rosenberg and H. R. Piehler, Calculation of the Taylor factor and lattice rotations for bbc deforming by pencil glide, *Metallurgical Transactions*, **2**, (1971), 257, Amer. Soc. Metals.
30. G. Y. Chin, W. L. Mammel and M. T. Dolan, Taylor analysis for $\{111\}$ $\langle 112 \rangle$ twinning and $\{111\}$ $\langle 110 \rangle$ slip under conditions of axisymmetric flow, *Transactions of Metallurgical Society of AIME*, **245** (1969), 383.
31. E. Kröner, Zur plastische Verformung des Vielkristalls, *Acta Metallurgica*, **9** (1961), 155, Pergamon Press.
32. B. Budianski and T. T. Wu, Theoretical prediction of plasticity of polycrystals, *Proceedings of 4th US National Congress of Applied Mechanics* (1962), p. 1175, Amer. Soc. Mech. Engrs.
33. J. W. Hutchinson, Elastic-plastic behavior of polycrystalline metals and composites, *Proceedings of the Royal Society*, **A319** (1970), 247.
34. T. H. Lin and M. Ito, Latent elastic strain energy due to the residual stress in a plastically deformed polycrystal, *Journal of Applied Mechanics*, **34** (1967), 606, Amer. Soc. Mech. Engrs.
35. H. Takahashi, Stress-strain relation of polycrystalline metals: Case of hexagonal crystalline metal (Paper of the Japan Society of Mechanical Engineers, No. 730–2, 1973, 239, in Japanese).
36. R. Hill, Generalized constitutive relations for incremental deformation of metal crystals by multislips, *Journal of Mechanics and Physics of Solids*, **14** (1966), 95.
37. D. Peirce, R. J. Asaro and A. Needleman, Overview No. 32—Material rate dependence and localized deformation in crystalline solids, *Acta Metallurgica*, **31** (1983), 1951.
38. H. Miyamoto et al., Fracture mechanics and numerical analysis (Review), *Journal of the Japan Society of Mechanical Engineers*, **74** (1971), 663 (in Japanese).
39. H. Miyamoto et al., Mechanical property from viewpoint of constitutive elements of material (Review), *Journal of the Japan Society of Mechanical Engineers*, **75** (1972), 575 (in Japanese).
40. H. Miyamoto, M. Sumikawa and T. Miyoshi, Interpretation of mechanical behavior of pure aluminum in terms of microstructure, *Proceedings of the*

International Conference on Mechanical Behavior of Materials (1972), Vol. 1, p. 140, Soc. Mater. Sci., Japan.
41. O. C. Zienkiewicz and Y. K. Chenung, The finite element method in structural and continuum mechanics, McGraw-Hill, New York, 1967.
42. Y. Yamada, N. Yoshimura and T. Sakurai, Plastic stress-strain matrix and its application for the solution of elastic-plastic problem by the finite element method, International Journal of Mechanical Science, 10 (1968), 343, Pergamon Press.
43. T. Jimma, T. Murota and T. Ichiyanagi, Stress-strain matrix of metal crystal and its application, Journal of the Japan Society of Mechanical Engineers, 75 (1972), 602 (in Japanese).
44. R. E. Peierls, Commentary on the 'Peierls–Nabarro force', in Dislocation dynamics (ed. A. R. Rosenfield, G. T. Hahn, A. L. Bement Jr and R. I. Jaffee) McGraw-Hill, New York, 1968, pp. xiii–xiv.
45. N. J. Nye, Some geometrical relations in dislocated crystals, Acta Metallurgica, 1 (1953), 153.
46. B. A. Bilby, R. Bullough and E. Smith, Continuous distributions of dislocations: a new application of the methods of non-Riemannian geometry, Proceedings of the Royal Society, A231 (1955), 263.
47. E. Kröner, Die Versetzung als elementare Eigenspannungsquelle, Zeitschrift für Naturforschung, 11a (1956), 969, Verlag der Zeitschrift für Naturforschung, Tübingen.
48. E. Kröner, Allgemeine Kontinuumstheorie der Versetzungen und Eigenspannungen, Archives of Rational Mechanics and Analysis, 4 (1960), 273, Springer Verlag.
49. A. C. Egingen (Ed.), Dedication to Prof. K. Kondo on the occasion of his seventieth birthday, International Journal of Engineering Science, 19–12 (1981), Pergamon Press.
50. K. Kondo, On the geometrical foundation of the theory of yielding, Proceedings of 2nd Japan Congress of Applied Mechanics (1952), p. 41 Gakujitsubunken-Fukyukai.
51. K. Kondo, Non-Riemannian geometry of imperfect crystals from a macroscopic viewpoint, RAAG Memoirs, 1, D-I (1955), 458.
52. K. Kondo, Non-holonomic geometry of plasticity and yielding, RAAG Memoirs, 1, D (1955), pp. 453–72.
53. E. Kröner, Der fundamentale Zusammenhang zwischen Versetzungsdichte und Spannungs Funktionnen, Zeitschrift fur Physik, 142 (1955), 463, Deutsche physikalische Gesellschaft.
54. K. Kondo, Geometry of elastic deformation and incompatibility, RAAG Memoirs, 1, C–I (1955), 361.
55. S. Amari, On some primary structure of non-Riemannian plasticity theory, RAAG Memoirs, III, D–IX (1962), 99.
56. M. Ohnami and K. Shiozawa, Definition of misorientation and the application to evaluate strength of metallic materials, Transactions of the Japan Society of Mechanical Engineers, 39 (1973), 769 (in Japanese)
57. K. Shiozawa and M. Ohnami, Study of flow stress of metal by means of the geometrical aspect of the continuously dislocated continuum, Mechanical Behavior of Materials (1974), 93, Soc. Mater. Sci., Japan.

58. J. Dundurs, Elastic interactions of dislocations with inhomogeneities, in *Mathematical Theory of Dislocations*, Amer. Soc. Mech. Engrs, 1969, pp. 70–115.
59. A. K. Head, The interaction of dislocations and boundaries, *Philosophical Magazine*, **44** (1953), 92.
60. A. K. Head, Edge dislocations in inhomogeneous media, *Proceedings of the Physical Society*, **1366** (1953), 793.
61. J. Dundurs, On the interaction of screw dislocation with inhomogeneities, in *Recent Advances in Engineering Science* (ed. A. C. Eringen), Vol. 2, Gordon and Breach, New York, 1967, pp. 223–33.
62. R. Week, J. Dundurs and M. Stippe, Exact analysis of an edge dislocation near a surface layer, *International Journal of Engineering Science*, **6** (1968), 365.
63. M. Ohnami and R. Imamura, Effect of vacuum levels on creep-rupture properties of polycrystalline metals at elevated temperature (based on the relation between oxide film and substructure of metal), *Proceedings of 22nd Japan Congress on Materials Research* (1979), p. 41, Soc. Mater. Sci., Japan.
64. Y. T. Chou, Screw dislocations in and near lamellar inclusions, *Physica Status Solidi*, **17** (1966), 509, Akademie Verlag.
65. B. F. Ashby, The deformation of plastically non-homogeneous materials, *Philosophical Magazine*, **21** (1970), 399.
66. M. J. Marcinkowski, *Unified Theory of the Mechanical Behavior of Matter*, Wiley–Interscience, New York, 1979, Chapter 13, pp. 231–52.
67. J. C. M. Li and Y. T. Chou, The role of dislocations in the flow stress-grain size relationships, *Metallurgical Transactions*, **1** (1970), 1145.
68. A. T. Yokobori, Jr, T. Yokobori and H. Nishi, Stress rate and grain size dependence of dynamic stress intensity factor by dynamical piling-up of dislocations (Paper of the Japan Society of Mechanical Engineers, 84-0376A, March 1985, in Japanese).
69. A. N. Stroh, A theoretical calculation of stored energy in a work-hardened material, *Proceedings of the Royal Society*, **A272** (1953), 391.
70. B. A. Bilby, A. H. Cottrell and K. H. Swindeman, The spread of plastic yield from a notch, *Proceedings of the Royal Society*, **A272** (1963), 304.
71. W. G. Johnston and J. J. Gilman, Dislocation velocities, dislocation densities and plastic flow in lithium fluoride crystals, *Journal of Applied Physics*, **30** (1959), 129, Amer. Institute of Physics.
72. T. Mura, Continuous distribution of moving dislocations, *Philosophical Magazine*, **8** (1963), 843.
73. S. Minagawa and T. Nishi, Stresses produced by a continuous distribution of moving dislocations in an isotropic continuum, *Physical Review Letters*, **28** (1972), 353, Amer. Physical Society.
74. E. Kröner (Ed.), *Mechanics of Generalized Continua* (IUTAM symposium on the generalized Cosserat continuum and the continuum theory of dislocations with applications, 1968), Springer Verlag, pp. 1–358.
75. M. Ohnami (Ed.), *Introduction to Micromechanics*, OHM-Sha, 1980, pp. 221–3 (in Japanese).
76. K. Anthony, U. Essmann, A. Seeger and H. Trauble, Disclinations and the Cosserat continuum with incompatible rotations, Ref. 74, p. 355.

77. Ref. 1, p. 184.
78. M. J. Beran and J. J. McCoy, Mean field variations in the statistical sample of heterogeneous linearly elastic solids, *International Journal of Solids and Structures*, **6** (1970), 1035, Pergamon Press.
79. M. J. Beran and J. J. McCoy, The use of strain gradient theory for analysis of random media, *International Journal of Solids and Structures*, **6** (1970), 1267.
80. E. Kröner, Elastic moduli of perfectly disordered composite materials, *Journal of Mechanics and Physics of Solids*, **15**, (1967), 319.
81. E. Kröner and H. Koch, Effective properties of disordered materials, *Solids Mechanics Archives*, **1** (1976), 183, Sijthoff and Noordhoff International Publishers.
82. E. Kröner, Bounds for effective elastic moduli of disordered materials, *Journal of Mechanics and Physics of Solids*, **25** (1977), 137.
83. P. H. Dederichs and R. Zeller, Elastische Konstanten von Vielkristallen, *Berichte der Kernforschungsanlage Jülich* (1972), Juli-877-FF.
84. R. Zeller and P. H. Dederichs, Elastic constants of polycrystals, *Physica Status Solidi*(b), **55** (1973), 831.
85. E. Kröner, Ref. 74, p. 330.
86. M. Ohnami, Ref. 7, p. 163.
87. E. Kröner, Initial studies of a plastic theory based upon statistical mechanics, in *Inelastic Behavior of Solids* (ed. M. F. Kanninen et al.), McGraw-Hill, New York, 1970, p. 137.
88. H. Yukawa and K. Inoue, Science thought in the 19th century, in *Modern Science*, Vol. 1, CHUOKOHRON-Sha, 1973, pp. 12–13 (in Japanese).
89. R. J. Bošković, *Theoria Philosophiae Naturalis*, 1st Edn, Venice, 1763 (English Edn, trans. J. M. Child, *Theory of Natural Philosophy*, Open Court, Chicago, London, 1922).
90. Charles Coulston Gillispie (Ed. in chief), *Dictionary of Scientific Biography*, Vol. 2, Charles Scribner Sons, New York, 1981, pp. 326–32.
91. E. Cosserat and F. Cosserat, Sur la théorie de l'élasticité, *Anales de la Faculté des Sciences de Toulouse*, **1–10** (1896), 1.
92. E. Cosserat and F. Cosserat, *Theorie des Corps Deformables*, Librairie Scientifique A. Hermann, Paris, 1909, pp. 1–221 (English Edn, *Theory of Deformable Bodies*, NASA TT F-11, 561 (1967), pp. 1–220, National Aeronautics and Space Administration, Washington, DC).
93. H. Minagawa, Cosserat continuum theory and related problems, *Journal of the Materials Science Society of Japan*, **9**–5.6 (1972), 289 (in Japanese).
94. Ref. 75, Chapters 1, 2 and 4.
95. C. C. Gillispie (Ed. in chief), *Dictionary of Scientific Biography*, Vol. 14, Scribner, New York, 1981, pp. 85–8.
96. V. Volterra, Sur l'équilibre des corps élastiques multiplement connexes. *Annales Scientifiques de l'Ecole Normale Supérieure*, **3–24** (1907), 401–517, Paris.
97. C. Somigiliana, Sulla teoria delle distorsioni elastiche, Nota I, *Rendiconti delle Sedute della Reale Accademia dei Lincei, Classe di Scienze fisiche, matematiche e naturali*, Memorie e Note di Soci o Presentate da Soci (1914), 463–90.

98. G. R. Irwin, Fracture dynamics, in *Fracturing of Metals*, American Soc. Metals, 1948, p. 152.
99. G. R. Irwin and J. A. Kies, Fracturing and fracture dynamics, *Welding Journal*, **31** Research Suppl. (1952), 95s, Amer. Welding Society.
100. A. A. Griffith, The phenomena of rupture and flow in solids, *Philosophical Transactions of the Royal Society*, **A221** (1920), 163, The Royal Society, London.
101. G. R. Irwin, Analysis of stresses and strains near the end of a crack travelling in a plate, *Journal of Applied Mechanics*, **24** (1957), 361.
102. G. R. Irwin, Fracture, in *Handbuch der Physik* (ed. S. Flügge), **IV** (1958), 551.
103. MIL-STD-1530(A), Aircraft Structural Integrity Program. Airplane Requirements (1975).
104. Amer. Soc. Mech. Engrs, ASME Boiler and Pressure Vessel Code, Section III, Rules for Construction of Nuclear Power Plant Components (1972).
105. T. Terada, Analogy of crack and electron, *Proceedings of Imperial Academy*, **7** (1931), 215, The Japan Imperial Academy.
106. N. F. Mott, Fracture of metals: some theoretical considerations, *Engineering*, **165** (1948), 16.
107. Y. Asada and K. Koibuchi (Eds), *Strength and Fracture of Mechanical Structure*, in Comprehensive Lectures on Strength and Fracture of Materials, Vol. 8, OHM-Sha, 1984, pp. 1–205 (in Japanese).
108. Ref. 75, pp. 185–201.
109. T. Mura, *Micromechanics of Defects in Solids*, Martinus Nijhoff Publishers, 1982, pp. 1–480.
110. J. D. Eshelby, The force on an elastic singularity, *Philosophical Transactions of the Royal Society*, **A244** (1951), 87.
111. J. R. Rice, A path-independent integral and the approximate analysis of strain concentration by notches and cracks, *Transactions of the American Society of Mechanical Engineers*, **E.35** (1968), 379.
112. G. P. Cherepanov, Cracks in solids, *International Journal of Solids and Structures*, **4** (1968), 811.
113. B. A. Bilby and J. D. Eshelby, Dislocations and the theory of fracture, in *Fracture* (ed. H. Liebowitz), Vol. 1, Academic Press, 1968, p. 99.
114. Ref. 7, pp. 161–2.
115. Ref. 2, pp. 256–71.
116. V. Vitek and G. G. Chell, An assessment of some post-yield fracture criterion, *Materials Science and Engineering*, **27** (1977), 209, Elsevier Sequoia S.A.
117. Ref. 2, p. 159.
118. Ref. 1, p. 181.
119. P. Paris and F. Erdogan, A critical analysis of crack propagation laws, *Transactions of the American Society of Mechanical Engineers*, **D.85** (1963), 528.
120. W. Elber, The significance of fatigue crack closure; damage tolerance in aircraft structures, *ASTM Special Technical Publication*, **486** (1974), 230, Ameri. Soc. Test. Mater.
121. M. Kikukawa, H. Johno, K. Tanaka and M. Takatani, Measurement of

fatigue crack propagation and crack closure at low stress intensity level by unloading elastic compliance method, *Journal of the Society of Materials Science, Japan*, **25** (1976), 899 (in Japanese).
122. Ref. 1, pp. 172–9.
123. R. Ohtani and K. Tanaka (Eds), *In Memory to Shuji Taira*, July 1979, pp. 7–8.
124. S. Taira and M. Ohnami, Fracture and deformation of metals subjected to thermal cycling combined with mechanical stress, *Proceedings of Joint International Conference on Creep* (1963), p. 3–57, Inst. Mech. Engrs.
125. R. M. Goldhoff, Towards the standardization of time-temperature parameters usage in elevated temperature data analysis, *Journal of Testing and Evaluation*, **2** (1974), 387, Amer. Soc. Test. Mater.
126. Amer. Soc. Mech. Engrs, ASME Boiler and Pressure Vessel Code Case N-47, Class 1 Components in Elevated Temperature Service, Section III, Division 1 (1978).
127. M. Kitagawa, K. Mino, H. Hattori, A. Ohtomo, M. Fukagawa and Y. Saiga, Some problems in developing the high temperature design code for 1.5 MWt helium heat exchanger, *Elevated Temperature Design Symposium* (1976), p. 33, Amer. Soc. Mech. Engrs.
128. M. Kitagawa, H. Hattori, A. Ohtomo, T. Teramae and J. Hamanaka, Lifetime test of a partial model of high temperature gas-cooled reactor helium-helium heat exchanger, *Nuclear Technology*, **66** (1984), 675.
129. J. A. Ewing and J. C. W. Humfrey, The fracture of metals under repeated alternations of stress, *Philosophical Transactions of the Royal Society*, **200** (1903), 241.
130. M. Hirata, On Fracture of Glass (Report of the Science and Physics Institute), **8–1** (1929), 52 (in Japanese).
131. K. Taguchi, Propagation of Failure (Report of the Science and Physics Institute), **10–2** (1931), 110 (in Japanese).
132. L. Bairstow, The elastic limit of iron and steel under cyclical variations of stress, *Philosophical Transactions of The Royal Society*, **A210** (1910), 35.
133. P. J. E. Forsyth and C. A. Stubbington, The slip-band extrusion effect observed in some aluminum alloys subjected to cyclic stresses, *Journal of the Institute of Metals*, **83** (1955), 395.
134. N. Thompson, N. J. Wodsworth and N. Louat, The origin of fatigue fracture in copper, *Philosophical Magazine*, **1** (1956), 113.
135. J. A. Ewing, On hysteresis in the relation of strain to stress, *British Association Report* (1889), p. 502, British Association for the Advancement of Science.
136. A. A. Rubbra, *Biography Memoirs of Fellows of The Royal Society*, Vol. 10, The Royal Society, 1964, pp. 117–36.
137. F. W. Armstrong, The aero engine and its progress — fifty years after Griffith, *Aeronautical Journal*, **80** (1976), 499, Royal Society of Aeronautics.
138. Obituary, *The Engineer*, 25 October 1963.
139. Private letter from D. W. Gopode, Chief librarian, Royal Aircraft Establishment, 14 March 1983.
140. A. A. Griffith and G. I. Taylor, Use of soap film solving torsion problems, *Proceedings of the Institute of Mechanical Engineers*, **93** (1917), 755.

141. A. G. Smekal, Technische Festigkeit und molekulare Festigkeit, *Die Naturwissenschaften*, **10** (1922), 799.
142. K. Wolf, Zur Buchtheorie von A. Griffith, *Zeitschrift für angewandte Mathematik und Mechanik*, **A221** (1920), 163, Akademie Verlag.
143. A. H. Cottrell, Fracture and society, in *Fracture*, ICF 4, Vol. 4, 1977, pp. 177–200.
144. B. A. Bilby, Obituary of Professor John Douglas Eshelby, *Journal of Mechanics and Physics of Solids*, **30**–3 (1982), 193–4.

Chapter 4

Plasticity Laws of Polycrystalline Bodies

A series of studies of the crystal plasticity of polycrystalline bodies have been made by using a combined micro- and macro-mechanical approach to the study of the strength and fracture of materials. This chapter describes mainly the study of the crystal plasticity of polycrystalline metals constructed from inhomogeneous crystals. We will examine crystal plasticity under simple tension and cyclic stressing, and microstructural studies of the cyclic strain hardening behavior in a biaxial stress state at an elevated temperature. This chapter also describes continuum mechanical studies of the influence of strain history on cyclic deformation and creep at room and elevated temperatures and the effect of hydrostatic stress on flow stress and creep at room and elevated temperatures. In addition we will also examine the current topic of the cyclic constitutive equation in a biaxial stress state at an elevated temperature.

4.1 CRYSTAL PLASTICITY UNDER SIMPLE TENSION

When a polycrystalline metal is subjected to simple tension, the plastic deformation consists mainly of deformation in the interior of the crystal grain and grain boundary sliding. The former occurs due to slip deformation in the definite slip system (Fig. 4.1(a)) and also by twin deformation in the twin system (Fig. 4.1(b)). Grain boundary sliding is frequently observed in high-temperature creep of metals (Fig. 4.1(c)). Plastic deformation of crystals also occurs due to combined external force and phase transformation (Fig. 4.1(d)), and can be caused by the self-diffusion of atoms (Fig. 4.1(e)). In this section the characteristics of inhomogeneity in crystal plasticity due to slip deformation will be discussed.

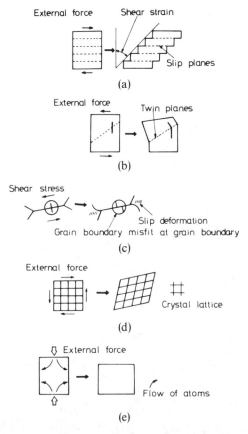

Fig. 4.1. Mode of typical plastic deformation of polycrystalline metal. (a) Slip deformation; (b) twin deformation; (c) grain boundary sliding; (d) deformation due to both external force and phase transformation; (e) deformation due to self-diffusion of atoms. (Reprinted from J. Kihara, *J. Japan Soc. Mech. Engrs*, **75** (1972), 563)

4.1.1 Inhomogeneity in Crystal Plasticity

Plastic deformation of polycrystalline metals does not occur uniformly and thus inhomogeneity is observed. Jimma *et al.* [1] analyzed the spreading of plastic deformation occurring on a sheet specimen (area 30×31 mm, thickness 1 mm) of coarse-grained pure commercial aluminum under static tension by means of FEM as described in Section 3.2.4. The specimen was divided into 512 triangular elements and the

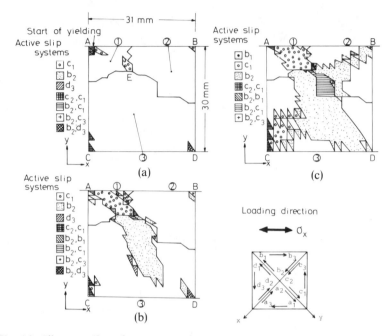

Fig. 4.2. Slip spreading of a coarse-grained pure aluminum sheet specimen under static tension by means of FEM; the active slip systems are shown in Table 4.1. (a) $\sigma_x = 11 \cdot 1$ MPa; (b) $\sigma_x = 12 \cdot 1$ MPa; (c) $\sigma_x = 12 \cdot 5$ MPa. (Reprinted from ref. 1 by permission of Japan Soc. Mech. Engrs)

displacement increment was designed so that the number of simultaneously yielding slip systems would be less than ten. Lin and Ito [2] also analyzed this problem.

Figure 4.2 [1] shows the propagation of yield regions at individual stages. Table 4.1 [1] shows both the Schmid factor μ_G and the active slip systems for three grains in Fig. 4.2. As seen from Fig. 4.2(a), the metal first begins yielding at corner point A, next the elements yield near corner points B, C, D and triple point E. As elongation advances, the region from A to E yields first and the plastic deformation spreads to grain 3 over the grain boundary between 1 and 3, and hence Fig. 4.2(a) is transformed into Fig. 4.2(b). As elongation advances, the central yield region also spreads to points B and C, and final deformation is represented as shown in Fig. 4.2(c). Here we find that in the interior of the grain single slip occurs in the slip system (Table 4.1), but double slip also occurs at the strain-concentrated portion of the grain boundary and

Table 4.1
Schmid Factor μ_G and Active Slip Systems (Reprinted from ref. 1 by permission of Japan Soc. Mech. Engrs)

Grain no.	Schmid factor μ_G	Active slip systems (see Fig. 4.2)
1	0·481	c_1
2	0·418	b_2
3	0·454	b_2

specimen corners. Comparing stage (b) with stage (c), we can see that stress is released in the slip systems b_2 and d_3 near points A and D. This suggests that the stress in the strain-concentrated portions varies relative to the advance in plastic deformation.

It is well known that crystals embedded in polycrystalline metal behave like a mechanical anisotropy due to the following factors: (i) the influences from their orientation on the elastic constant and critical shearing stress [3]; (ii) the variety in multi-slip and latent hardening [4]; (iii) the degree of restriction by neighbors. These three main factors play an important role not only in elucidating the plasticity law but also in clarifying the inelastic behavior of metals.

Figure 4.3 [5] shows the experimental relation between crystal rotation and strain of the coarse-graind 0·03% carbon steel sheet specimen (width 25 mm, gauge length 65 mm, thickness 1·3 mm) under simple tension. Figure 4.3 (a) and (b) show the results for both the polycrystalline specimen and the specimen with a single crystal zone. Rotation was measured by using the Laue back-reflection X-ray method. Aston [6] measured the rotation in a coarse-grained aluminum specimen in 1927 and reported that the rotation near the grain boundary was smaller than that in the interior. We can conclude from Fig. 4.3 that the crystal rotation increases in the polycrystalline structural specimen relative to elongation but remains nearly constant in the monocrystalline or bamboo-structural specimen.

Figure 4.4 [7] shows the strain concentration factor ψ in the above specimens, where ψ is defined as the ratio of axial strain of the individual grain to axial strain of the specimen. After comparing both specimens, we find that the numerical value of ψ varies from grain to grain after comparatively small elongation of the specimens, but for the polycrystalline structural specimen the value of ψ comes up to unity with advanced elongation of the specimen. In addition, the bamboo-structural specimen demonstrates no asymptotic behavior of ψ like that found in the

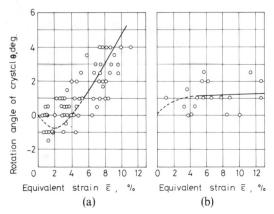

Fig. 4.3. Experimental relation between the rotation angle, θ, of grain and the Mises equivalent strain, $\bar{\varepsilon}$, in two types of grains of coarse-grained 0.03% carbon steel sheet specimen. (a) Polycrystalline specimen; (b) specimen with single crystal zone. (Reprinted from ref. 5 by permission of Soc. Mater. Sci., Japan)

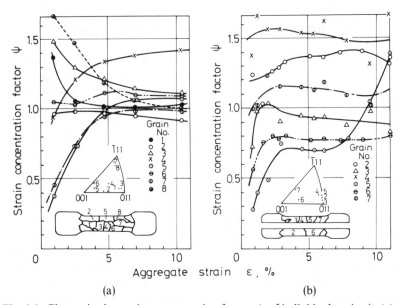

Fig. 4.4. Change in the strain concentration factor, ψ, of individual grains in (a) a polycrystalline specimen of 0.03% carbon steel and (b) a specimen of 0.03% carbon steel with a single-crystal zone, with increased simple tensile aggregate deformation (0.5 mm/min). (Reprinted from ref. 7 by permission of Soc. Mater. Sci., Japan)

polycrystalline structural specimen. We have interpreted these variations in crystal deformation as follows.

Both the influence of the orientation and restriction by neighbors are the most important factors determining the yield conditions of the crystals embedded in the aggregate. It is therefore suggested, for the polycrystalline specimen, that the former is a predominant factor compared with the latter for comparatively small elongation of the specimen but the latter is an influential factor as elongation advances. Therefore a large elongation of the specimen causes the influence of orientation to vanish. For the monocrystalline zone specimen, on the other hand, the variation of plastic deformation from crystal to crystal is controlled mainly by the dependence on the orientation and hence there is no remarkable change in ψ as the specimen is elongated.

Let us consider the directions of principal strain in the interior of the grain. In the following section we will examine, with the intention of bridging a gap between crystal plasticity and continuum mechanics, non-proportional loading, and find the discrepancy between both principal stress and strain directions which is both large and remarkable under multiaxial stresses. Let us consider this problem from the viewpoint of crystal plasticity.

Figure 4.5 shows the experimental distribution of the direction of principal strain in a sheet specimen (width 25 mm, thickness 2·7 mm) of coarse-grained pure aluminum subjected to 5% elongation by Saga et al. [8]. Measurement of two-dimensional strain and the direction angle of principal strain was made by using computer-aided Rosset analysis.

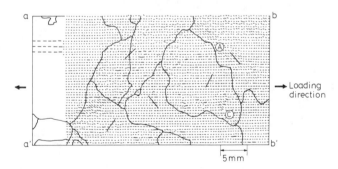

Fig. 4.5. Distribution of both directions of principal strain (fine lines) and maximum shear stress (thick lines) in 5% plastically deformed coarse-grained 99·99% pure aluminum. (Reprinted from ref. 8 by permission of Japan Soc. Mech. Engrs)

Most of the direction of principal strain coincides with the tensile direction of the specimen but the misfit is observed locally in the portion of localized slip bands such as point A and in the local portion with inhomogeneous deformation near the grain boundary such as point C. In the following experiments we will see clear evidence supporting these facts.

4.1.2 Relation between Principal Strain Direction and Misorientation

Figure 4.6 [5] shows an example of the experimental relation between

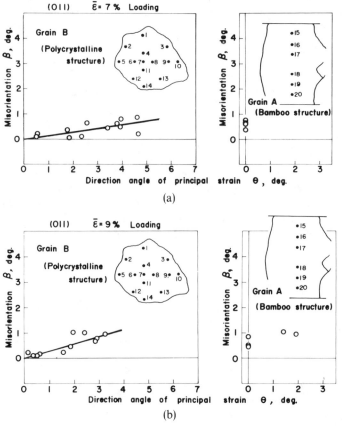

Fig. 4.6. Experimental relation between misorientation, β, and direction angle of principal strain, θ, in two types of grains of coarse-grained 0·03% carbon steel sheet specimen under static tension. (Reprinted from ref. 5 by permission of Soc. Mater. Sci., Japan)

the direction angle of principal strain, θ, and misorientation, β, in the same specimen as Figs 4.3 and 4.4 under static tension. The physical meaning of β will be analyzed in Chapter 5, but we must note here that β is significant not only with regard to dislocation density but also continuous change in the orientation of the diffraction plane which results from the redistribution of dislocations and interaction between impurities and moving dislocations. The misorientation measured while directing an X-ray beam perpendicularly to the specimen surface is calculated as follows. Distortions δ and γ in the Laue back-reflection spot can be read by using the Greninger chart, and the misorientation β is calculated using the equation $\beta = (\delta^2 + \gamma^2)^{1/2}$. The misorientation of the deformed metal is determined by subtracting the misorientation before deformation from that after deformation. The local strain and the direction angle of principal strain were measured by using a Shadowgraph with a precision of 1 μm for the printed mesh on the specimen.

In the polycrystalline specimen the principal strain direction varies from location to location in the interior of a grain, and the principal strain axes show a nearly proportional relation with the misorientation as shown in Fig. 4.6. In contrast, the direction angle of principal strain is nearly zero in the monocrystalline structural specimen as shown in the figure. The discrepancy between the tensile axis of the polycrystalline metal specimen and the principal strain axes in the individual grains denotes inhomogeneous distribution of the principal stresses applied. We can thus conclude that the change of structure of grain embedded in the polycrystalline metal increases as the multiaxiality of stresses increases.

4.1.3 Variation in X-Ray Quantities within a Grain and between Neighboring Grains

Figure 4.7 [5] shows the variation in the misorientation within the interior of both polycrystalline and monocrystalline structural grains of 0.03% carbon steel. In these figures β_{center} denotes the misorientation in the central portion 7 for the polycrystalline structural grain and the average at two points 17 and 18 for the monocrystalline structural grain, and β_L denotes the misorientation at the individual points. No variation in $|\beta_L - \beta_{center}|$ is observed for the monocrystalline specimen. A large inhomogeneity in the misorientation occurs near the grain boundary, but does not occur for the monocrystalline structural specimen.

Figure 4.8 [9] shows the change of the discrepancy between excess dislocation densities, ρ, at two points of the neighboring grain over the

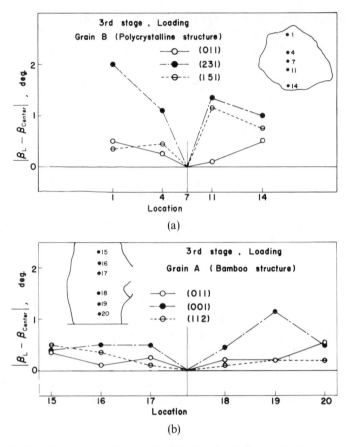

Fig. 4.7. Experimental result showing the distribution of misorientation, β, in two types of grains of coarse-grained 0·03% carbon steel sheet specimen under static tension. (Reprinted from ref. 5 by permission of Soc. Mater. Sci., Japan)

grain boundary, and that in the interior of the grain having increased axial strain on the specimen. In this case ρ was measured by using the X-ray microbeam technique. Note that near the grain boundary the discrepancy increases as the specimen elongates but does not do so in the interior of the grain. This phenomenon was also observed in the discrepancies found in lattice strain, $(\Delta d/d)$, which was measured by using the X-ray microbeam technique [9].

From the above experimental comparison between crystal plasticity in polycrystalline structural grains and monocrystalline structural grains under static tension, the following can be concluded.

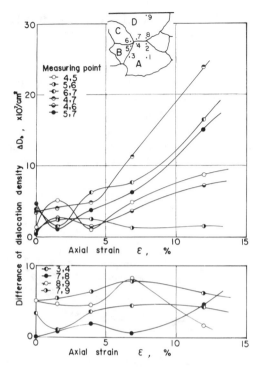

Fig. 4.8. Difference between excess dislocation densities, ρ, in two points of the neighboring grain over the grain boundary, and that in the interior of the grain with increased axial strain (for 0·03% carbon steel). (Reprinted from ref. 9)

(1) The stress state in the interior of the monocrystalline structural grain was uniaxial but it was multiaxial in that of the polycrystalline structural grain. This resulted in the difference between both types of structural grains in monotonic plastic deformation.

(2) The misorientation measured by the Laue back-reflection X-ray diffraction method increased as Mises type local equivalent strain in the interior of the grain increased, but the change in the polycrystalline structural grain was larger than that in the monocrystalline structural grain. The misorientation in this case was remarkably dependent on the local direction of principal strain, and the misorientation increased as the change in the local direction angle of principal strain increased.

(3) The inhomogeneity in misorientation in the interior of the polycrystalline structural grain was larger compared with that of the monocrystalline structural grain, and was especially pronounced in the neighborhood of the grain boundary.

4.2 CRYSTAL PLASTICITY UNDER CYCLIC STRESSING

Let us consider the crystal plasticity of polycrystalline metals subjected to cyclic stressing from the viewpoint of the inhomogeneity in the interior of the grain and the interaction between neighboring grains. Figure 4.9(a) [7] shows an example of cycle-dependent deformation of individual grains of a polycrystalline structural 0·03% carbon steel sheet specimen subjected to low-cycle pulsating stress. In this figure, solid curves represent the deformation for individual grains and the dashed curve represents the aggregate. The test was performed under a maxi-

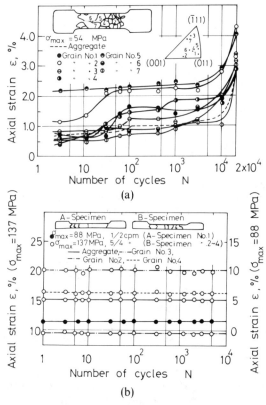

Fig. 4.9. Variation of the cycle-dependent axial deformation from grain to grain of two types of coarse-grained 0·03% carbon steel sheet specimens during pulsating tensile stress cycling. (a) Polycrystalline structure; (b) monocrystalline structure. (Reprinted from ref. 7 by permission of Soc. Mater. Sci., Japan)

mum stress of 54 MPa, a level slightly higher than that of the static yield strength of the polycrystalline structural aggregate, and a speed of 5 cpm. In the figure the orientation of grains is also presented on the basic triangle. We see that the deformations of individual grains with repeated stressing are different from one another. The deformations of grains 2 and 5 were notably larger at an early stage of the cycles than those of other grains. This can be attributed mainly to the variation in orientation from grain to grain. On the other hand, no cycle-dependent deformation was observed in the monocrystalline structural specimens as shown in Fig. 4.9(b) [7]. In order to examine these phenomena, the behavior of the crystals during deformation and their interaction will be discussed in the following pages.

4.2.1 Changes in the Structure of Materials Subjected to Cyclic Stressing
It is possible that changes in the structure of metals under cyclic stressing will be produced mainly by the difference in the direction of mobility in each dislocation and the variation in the arrangement of

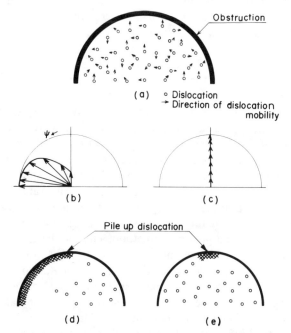

Fig. 4.10. Model interpreting the structural change of plastically deformed materials subjected to cyclic stressing.

dislocations. In the present section, the following dislocation model is presented in order to examine the difference between change in the structure of metals under changing principal axes of stress and that under fixed axes [10].

Imagine a model in which dislocations distribute in an arbitrary obstruction as illustrated in Fig. 4.10(a). In the present model we assume that each dislocation has some direction of mobility determined by geometrical and physical properties. First, let us consider the case of cyclic stressing under changing principal axes of stress as shown in Fig. (b). In this figure arrows show the change of the stress vector during a half cycle of pulsation. Assuming that the direction of mobility of dislocation is close to that of stress, the majority of dislocations move easily under these stress conditions. Figure (d) shows the pile-up of dislocations around an obstacle. As in the case of fixed axes of principal stresses shown in Figs (c) and (e), the number of movable dislocations and that of pile-up dislocations during a half cycle of pulsating stress are less than those in Fig. (d). We may deduce from this model that the difference between the change in the structure of the material during cyclic stressing under changing principal axes of stress and that under the fixed axes can be attributed to the variation of the mobility of dislocations and the distribution of dislocation density. It is conceivable that this is a result of the difference in back-stress due to pile-up dislocations around obstacles during cyclic stressing applied while changing the principal axes of stresses from fixed axes.

In order to examine the correctness of the above model, the following experiments were performed. Figure 4.11 [10] shows the experimental results of change in misorientation of grains of both polycrystalline and monocrystalline structures under low-cycle pulsating stress. The test was performed under a maximum stress of 60 MPa and a speed of 5 cpm. In this figure experimental plots were adjusted to the same orientation factor of the diffraction plane μ_P of the polycrystalline structural grain as that of the bamboo-structural grain, where μ_p was determined graphically by using the stereo-projection method. This can be attributed to the experimental relation of misorientation β of each diffraction plane to the number of cycles N. We observed that the misorientation is obviously dependent on the crystal lattice plane in both structural grains. From this figure we find that the misorientation in the polycrystalline structural grain is larger than that in the bamboo-structural grain. Considering the influence of structural anisotropy on the cycle-dependent plastic deformation of both types of structural grains, we can

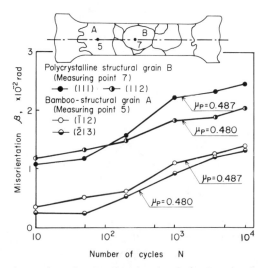

Fig. 4.11. Experimental results showing the change in the misorientation of both types of grains within the polycrystalline and bamboo-structural specimens composed of 0·03% carbon steel under low-cycle pulsating stress. (Reprinted from ref. 10 by permission of Soc. Mater. Sci., Japan)

strongly suspect that this influence exists in polycrystalline structural grains in which the mobility of dislocations is easily facilitated. These considerations are important in the sense that we recognize that most of the structural metals employed for technical use are subjected to repeated loadings.

4.2.2 Influence of Fluctuations of Microscopic Stress and Strain in a Material Subjected to Cyclic Stressing on the Yield Condition

In order to discuss the change in the structure of polycrystalline metals from the viewpoint of the global property of material space, we must in essence consider a macroscopic overview of the microscopic structure of the material.

Let us consider a case where macroscopic uniform stresses are distributed on an arbitrary cross-section of a material with microstructure, i.e. a heterogeneous material. Although our observation is concerned with a macroscopically uniform substance, it is necessary to assume that the intensity of stress is distributed microscopically. In this case the microscopic incremental strain $(D\varepsilon)_{ij}$ of an individual material point in the space will be decomposed into the mean incremental strain $\overline{(D\varepsilon)}_{ij}$ and the

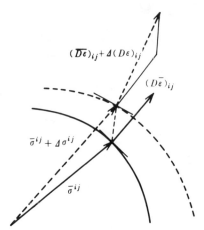

Fig. 4.12. Schematic representation of the translation of location and change in form of the yield curve.

deviation $\Delta(D\varepsilon)_{ij}$. In the same manner, the microscopic stress σ_{ij} will be decomposed into both $\bar{\sigma}_{ij}$ and $\Delta\sigma_{ij}$. Therefore the existence of $\Delta(D\varepsilon)_{ij}$ and $\Delta\sigma_{ij}$ results in the translation of location and the change in form of the yield curve, i.e. development of anisotropy of the material.

Figure 4.12 [10] shows an interpretation of this phenomenon. The yield curve represented by a solid line demonstrates the shape of the yield surface in which only $\bar{\sigma}_{ij}$ is taken into consideration and a dashed line shows the shape of the yield surface under the condition $(\bar{\sigma}_{ij}+\Delta\sigma_{ij})$. Applying the law of maximum plastic work, the exterior normal direction on the yield curve is given by the sum of the direction of the mean incremental strain $\overline{(D\varepsilon)}_{ij}$ and that of the fluctuation $\Delta(D\varepsilon)_{ij}$. This will give a result which illustrates the origin of the variation in the shape of the individual yield curve in material space.

In connection with this, Fig. 4.13 [10, 11] shows the change in distribution of the misorientation measured along the longitudinal direction of the specimen under low-cycle pulsating stress in the same manner as seen in Fig. 4.11. Similar measurements were also made along the area perpendicular to the specimen axis of the polycrystalline and monocrystalline structures, respectively. In the case of the diffraction plane (101) the variation in distribution of the misorientation in the polycrystalline structural grain is very small at an early stage of the stress cycle, but becomes pronounced at advanced cycles. Note that the contrary variation was observed in the case of the diffraction plane (111). On the

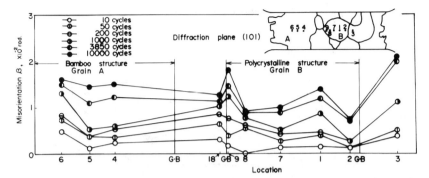

Fig. 4.13. Experimental results showing the change in the distribution of misorientation measured along the longitudinal direction of coarse-grained 0·03% carbon steel sheet specimen under low-cycle stressing. (Reprinted from ref. 10 by permission of Soc. Mater. Sci., Japan)

other hand, the bamboo-structural grain showed no pronounced variation in distribution of the misorientation.

4.2.3 Interaction between Neighboring Grains and the Mechanism of Dependent Crystal Plasticity

Figure 4.14 [11] shows the change in misorientation of a coarse-grained 0·03% carbon steel sheet specimen after repeated stress cycles in connection with the Schmid factor μ_G of the respective grains. The data were plotted with an orientation factor of the diffraction plane $\mu_P = 0.472$. We can see that misorientation β increases as μ_G increases, and this is especially remarkable in the region of μ_G from 0·48 to 0·50. At an early stage of the stress cycle there is a similar relation between β and μ_G as that under monotonic stressing but the dependence on μ_G vanishes during the advanced stages of the cycling, and the rate of increase in β for grains having small μ_G is larger than that for large μ_G. This fact suggests that the structure at early stages of the stress cycle is interpreted by the so-called constant stress model in which the dependence of crystal orientation is dominant as well as monotonic deformation. But during advanced stages of the cycle the so-called constant strain model is proper where the interaction between neighboring grains is taken into consideration.

From the present analytical and experimental studies of the crystal plasticity of polycrystalline metals subjected to cyclic stressing, we arrived at the following conclusions.

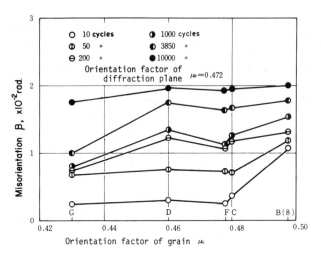

Fig. 4.14. Experimental relation between misorientation, β, and the Schmid factor, μ_G, of grain with an orientation factor of diffraction plane, $\mu_P = 0.472$, in a coarse-grained 0·03% carbon steel sheet, specimen under pulsating tensile stressing. (Reprinted from ref. 11 by permission of Soc. Mater. Sci., Japan)

(1) The measured value of misorientation in the coarse-grained polycrystalline metal increased with an advanced number of cycles of low-cycle pulsating stress. Moreover, the change of misorientation in the polycrystalline structural grain under cyclic pulsating stress was larger than that in the monocrystalline structural grain.

(2) The misorientation in polycrystalline structural grains was dependent not only on the orientation factor μ_P of each diffraction plane but also on the Schmid factor μ_G of the grain at an early stage of cyclic stressing. However, both of these dependencies decreased at an advanced stage of the cycles.

(3) The polycrystalline structural grain had a remarkable degree of variation in misorientation in its favor when compared with that of the monocrystalline structural grain.

(4) These facts implicitly supported the validity of the model in which the difference between the mobility of dislocations under changing principal axes of stress and that under the fixed axes was taken into consideration.

(5) Evidence also suggested that changes in the structure of the material under varying principal stress axes showed a correlation between the fluctuation of microscopic yield stress σ_{ij} in the material and

the microscopic strain increment $(D\varepsilon)_{ij}$. Furthermore, we came to the conclusion that the existence of $\Delta\sigma_{ij}$ and $\Delta(D\varepsilon)_{ij}$ would result in the translation of location and the change in form of the yield curve, i.e. development of anisotropy of the material. This will be discussed in the following Sections 4.3 and 4.4.

4.3 INFLUENCE OF STRAIN HISTORY ON PLASTIC DEFORMATION OF POLYCRYSTALLINE METALS SUBJECTED TO CYCLIC STRESSING

In this section the strain history of the plastic deformation of polycrystalline metals subjected to low-cycle pulsating stress is investigated with the intention of bridging a gap between continuum mechanics and crystal plasticity. In the continuum mechanical approach, in which material is regarded as the aggregate constructed from inhomogeneous crystals, as a matter of course, statistical homogeneity of the structure is assumed. In order to study the inelastic behavior of this material under general loading, it is necessary to take into account the change of structure and statistical inhomogeneity of microscopic stress and strain distributions in the material.

Hence it is necessary to investigate the mechanical factors determining the yield condition of the material. For example, these factors are the first invariant of stress J_1, the second invariant of deviatoric stress tensor J'_2, time t, temperature T, and number of stress or strain cycles, N. Therefore the yield condition is written as

$$f\left(J'_2, \int d\varepsilon_{ij}\right) = F \quad \text{and} \quad F = F\left(\int d\varepsilon_{ij}, J_1, t, T, N\right) \quad (4.1)$$

In this case the need for analytical and experimental examinations of the relations between the influence of strain history and the structural change of the material is emphasized from the viewpoint of microscopic stress and strain fields of the materials.

The material used in this study is a comparatively large-scaled thin-walled tubular cylindrical 0·15% carbon steel specimen (gauge length 150 mm, outer diameter 32 mm, inner diameter 29 mm). Figure 4.15 [12] shows an example of the experimental cycle-dependent deformation curves of circumferential, axial and shearing strains on the entire gauge length of the specimen, and the localized gauge lengths of regions A, B and C, respectively, under combined loading of fixed axial tension and repeated torsion shown in the figure. The central part of the gauge length

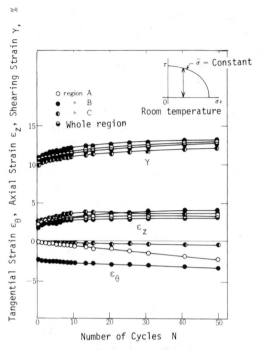

Fig. 4.15. Experimental cycle-dependent deformation curves of axial strain, ε_z, circumferential strain ε_θ, and shearing strain, γ, in each part along the length of a thin-walled tubular 0·15% carbon steel test specimen, during cycles of stress with fixed axial tension and repeated torsion. Equivalent stress $\bar{\sigma} = 382$ MPa ($\sigma_z = 118$ MPa, $\tau = 210$ MPa). (Reprinted from ref. 12 by permission of Soc. Mater. Sci., Japan)

of the test specimen is denoted as region B and the two sides as regions A and C. The test was performed under the maximum Mises equivalent stress of 382 MPa and a cycle speed of 1/4 cpm. Special care was taken to ensure precise, continuous and simultaneous measurements of the axial elongation, the displacement of the internal and external diameters, and the twisting angle along the specimen [12].

We can see from Fig. 4.15 that the axial strain, ε_z, gradually increases during repeated torsion, and this means that torsional stressing has an influence on axial strain. We can hypothesize that the variation in the cycle-dependent deformation of each part along the length of the specimen is due mainly to the variation in the mechanical properties of the individual structural element. We can interpret this as follows.

Figure 4.16 [12] shows that both shearing strain γ and axial strain ε_z

along the entire gauge length for region B increase, thus maintaining a linear relation. We can also see that regions A and C show a linear relation at the stage of smaller γ, but the relation of γ to ε_z changes relative to the progress of γ. This means that the variation in the shearing strain from structural element to element results in the redistribution of flow stress. As a continuation of this we will determine the development of anisotropy of the material and the Bauschinger effect during the stress cycle.

Figure 4.17 [12] shows a typical example of graphical representation of the yield curves of each part along the length of the specimen with progressive low-cycle pulsating stress. As a concrete form of eqn (4.1) the following equation of the so-called anisotropic hardening of the yield function was applied to this problem [13, 14]:

$$f = (1/2)C_{ijkl}S_{ij}S_{kl} - B\varepsilon_{ij}S_{ij} \qquad (4.2)$$

where

$$C_{ijkl} = G_{ijkl} + AL_{ijkl} + A^2 M_{ijkl}$$

$$G_{ijkl} = (1/2)(\delta_{ik}\delta_{jl} + \delta_{il}\delta_{jk})$$

$$L_{ijkl} = (1/2)(\delta_{ik}\varepsilon_{jl} + \delta_{jl}\varepsilon_{ik} + \delta_{il}\varepsilon_{jk} + \delta_{jk}\varepsilon_{il})$$

$$M_{ijkl} = (1/2)(\varepsilon_{ik}\varepsilon_{jl} + \varepsilon_{il}\varepsilon_{jk}) \qquad (4.3)$$

A = anisotropic scalar parameter

B = scalar parameter of Bauschinger effect

S_{ij} = deviatoric stress tensor

$\varepsilon_{ij} = \int d\varepsilon_{ij}$

Assuming both the flow rule and the constant volume law, the yield condition of the material subjected to an arbitrary strain history ε_{ij} is given as

$$P(\sigma_z/\bar{\sigma})^2 + Q(\sigma_z/\bar{\sigma})(\tau/\bar{\sigma}) + 3R(\tau/\bar{\sigma})^2$$
$$- 3B\{(\varepsilon_{33}/\bar{\sigma})(\sigma_z/\bar{\sigma}) + (2\varepsilon_{23}/\bar{\sigma})(\tau/\bar{\sigma})\} - 1 = 0 \qquad (4.4)$$

where

$$P = 1 + A\varepsilon_{33} + (3/4)A^2\varepsilon_{33}^2 - (2/3)A^2\varepsilon_{23}^2$$
$$Q = 2A\varepsilon_{23} + 5A^2\varepsilon_{33}\varepsilon_{23} \qquad (4.5)$$
$$R = 1 + (1/2)A\varepsilon_{33} - (1/2)A^2\varepsilon_{33}^2 + A^2\varepsilon_{23}^2$$

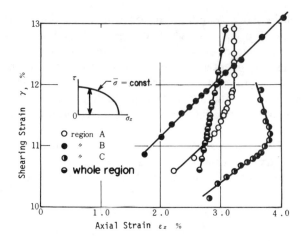

Fig. 4.16. Experimental relation between the shearing strain, γ, and axial strain, ε_z, in each part along the length of the test specimen of 0·15% carbon steel in low-cycle fatigue. Equivalent stress $\sigma = 382$ MPa ($\sigma_z = 118$ MPa, $\tau = 210$ MPa). (Reprinted from ref. 12 by permission of Soc. Mater. Sci., Japan)

In these equations, σ_z, τ and $\bar{\sigma}$ are axial, shearing and Mises equivalent stresses, respectively, and the plastic strain tensor ε_{ij} is represented by the Cartesian localized coordinate system $\mathbf{x}_i$ (see Appendix II). In this case, radial, circumferential and axial directions of the specimen are represented by $\mathbf{x}_1$, $\mathbf{x}_2$ and $\mathbf{x}_3$ respectively.

The numerical values of the anisotropic parameter A of the material, the parameter of the Bauschinger effect B, and the rotation angle of the yield ellipse θ (deg.) are also shown in Fig. 4.17. We found that the stress ellipse of each part along the length of the test specimen translates in a variety of ways with the repeat of the stress cycles. We also found that the stress ellipse is translated from the initial location to the right side of the σ_z axis and that the form of the ellipse is shortened in both directions of σ_z and τ. The fact that the yield ellipse becomes smaller than that of the initial isotropic condition during stress cycling signifies a drop in yield stress. This can be discussed from the viewpoint of the Bauschinger effect. After the 30th cycle, parameter A assumes a negative value at region C as shown in Fig. 4.17(c). In this case, rotation of the yield ellipse is in the opposite direction and the stress ellipse expands in the direction of σ_z.

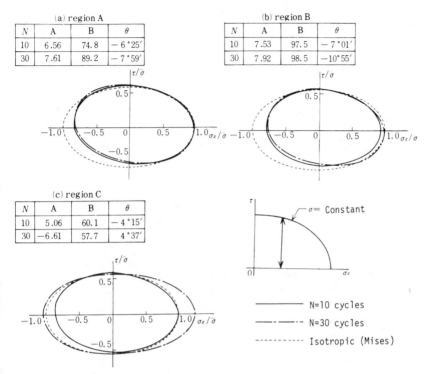

Fig. 4.17. Analytical representation of the translation of location and form of the yield curves based on combined kinematic and anisotropic hardenings, for 0·15% carbon steel ($d_0 = 32$, $d_i = 29$, $l = 150$ mm), in each part along the length of the test specimen in low-cycle fatigue. Equivalent stress $\bar{\sigma} = 382$ MPa ($\sigma_z = 118$ MPa, $\tau = 210$ MPa). Room temperature. (Reprinted from ref. 12 by permission of Soc. Mater. Sci., Japan)

Figure 4.18 [10] shows the comparison between the experimental cycle-dependent deformation curves along the entire gauge length of the same specimen as Fig. 4.15 under changing principal axes of stress and those under fixed axes, in which the deformation is represented by Mises equivalent strain. The tests were performed under a maximum equivalent stress of 382 MPa and a speed of 1/4 cpm. In this figure, pulsating stress paths (1, 2, 3, 4) are expressed as stress ellipses. We see from the figure that at an early stage of the cycling the difference between the equivalent strain curves along the entire gauge length of the specimen under each pulsating stress path is very small but becomes pronounced

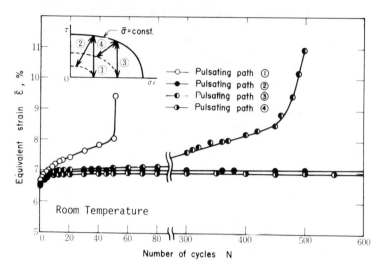

Fig. 4.18. Experimental relation between the Mises equivalent strain in each pulsating stress path and the number of cycles of combined stresses of fixed axial tension and repeated torsion. Test conditions as for Fig. 4.17. (Reprinted from ref. 10 by permission of Soc. Mater. Sci., Japan)

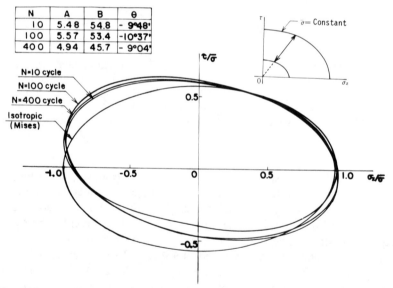

Fig. 4.19. Graphic representation of the translation of location and form of the yield curves based on combined kinematic and anisotropic hardenings of the test specimen, during the progress of low-cycle fatigue under fixed principal axes of stress. Test conditions as for Fig. 4.17. (Reprinted from ref. 10 by permission of Soc. Mater. Sci., Japan)

at the stage of advanced cycles. More concisely, the deformation under the changing principal axes of stress (pulsating stress paths 1 and 3) is considerably larger than that under the fixed axes of paths 2 and 4.

Figure 4.19 [10] shows a graphical representation of the yield curve of the test specimen under fixed principal axes of stress. The representation under changing principal axes is not shown here, having been described in Fig. 4.17. The numerical values of the anisotropic parameter A of the material, the parameter of the Bauschinger effect B and rotation angles of the yield ellipse θ (deg.) of the material with the progress of low-cycle fatigue are also shown. No change in the form of the yield ellipse is observed during cyclic pulsating stress under fixed axes of principal stress, but little rotation is observed compared with that under changing principal axes of stress. Therefore the material behaves much according to a type of kinematic hardening yield curve under fixed principal axes of stress.

Figure 4.20 [10] shows the experimental correlation between the principal direction angle of stress ϕ and that of strain ψ in each pulsating stress path. In this case axial, shearing and circumferential strains were

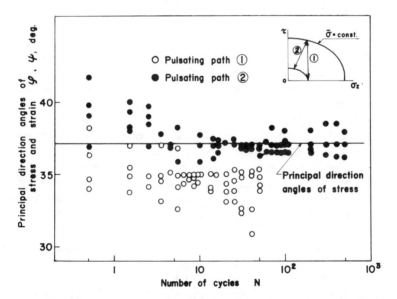

Fig. 4.20. Experimental relation between the principal direction angle of stress, ϕ, and that of strain, ψ, in each pulsating path, during low-cycle fatigue tests under combined axial tension and repeated torsion. Test conditions as for Fig. 4.17. (Reprinted from ref. 10 by permission of Soc. Mater. Sci., Japan)

measured at four points of the central region, the two sides and the entire portion of the test specimen. The gauge length of each region on the test specimen was 25 mm. From this data we find that in the case of path 1 the difference between ϕ and ψ is larger than that for path 2. In this figure, test results are shown with data points at the specified number of cycles of stress representing both the entire gauge length and individual regions of the test specimen. From these facts we can conclude that the disagreement between the principal angle of stress and that of strain results mainly from the development of inhomogeneity in microscopic stress and strain distribution in the material subjected to cyclic stressing under changing principal axes of stress.

We can make certain interpretations from the changes in the structure of material subjected to strain history in each pulsating stress path, by employing both models shown in Fig. 4.10, and the influence of fluctuations of microscopic stress and strain in the material shown in Fig. 4.12. A microstructural study of the cyclic stress/strain relationship of polycrystalline metals will also be described in Section 5.3 of Chapter 5 from a micromechanical viewpoint.

From the present analytical and experimental studies on the influence of strain history on the plastic deformation of polycrystalline metals subjected to cyclic stressing, the following conclusions can be made.

(1) The yield condition of eqns (4.2) and (4.3) was applicable to the representation of so-called anisotropic hardening in the yield surface of the polycrystalline metal under cyclic pulsating stress.

(2) An interpretation of the mechanism of low-cycle fatigue deformation of the metal was made from the redistribution of the flow stress field in material space.

(3) Cycle-dependent plastic deformation occurred in a different manner in each element of the polycrystalline metal, because of variations both in mechanical properties from structural element to structural element, and in its structural change during the stress cycle. After analysis we concluded that this resulted from the anisotropy of the material, the Bauschinger effect, and also from the redistribution of the yield stress of each structural element.

(4) Cycle-dependent deformation of the material under changing principal axes of stress was remarkably larger than that under the fixed axes. This is the result of the difference in the change of structure of the material due to different stress paths. In this case the change in the structure of material subjected to pulsating stress was interpreted by using a model in which the difference between the mobility of dislo-

cations under changing principal axes of stress and that under fixed axes was taken into consideration.

4.4 CONTINUUM MECHANICAL AND MICROMECHANICAL STUDIES OF CYCLIC STRAIN-HARDENING BEHAVIOR IN A BIAXIAL STRESS STATE AT ELEVATED TEMPERATURES

Recent progress in computer-aided numerical analysis calls for an accurate stress–strain equation for materials used at elevated temperatures. This is especially necessary for cyclic loading conditions, since any deviation in the equation from the actual material may result in a gross error in calculation. Additionally, formulation of an accurate cyclic stress–strain relationship is necessary as basic data for low-cycle fatigue. It should be noted that whether a cyclic stress–strain relationship accurately expresses the actual material behavior or not can only be proven by testing in either biaxial or multiaxial stress states and studying crystal plasticity [15–20].

The aim of this section is to study, by means of dislocation structures, crystal plasticity and continuum mechanics, the difference in cyclic stress amplitude between proportional and non-proportional loadings in strain-controlled biaxial low-cycle fatigue of polycrystalline metal at elevated temperatures.

4.4.1 Cyclic Stress–Strain Relation in Proportional and Non-proportional Loadings

Figure 4.21 [21a] shows the cyclic stress–strain curves of a thin-walled cylindrical type 304 solution heat-treated austenitic stainless steel specimen (gauge length 20 mm, I.D. 9 mm, O.D. 12 mm) with ASTM No. 3.5 grain size, at a half of the failure life in proportional loading, obtained by the 'companion specimen test method'. Here several test specimens were used for total strain range controlled tests of fully reversed loading, as shown in Fig. 4.22(a). The proportional loading of combined axial and torsional loadings was performed under a constant Mises total strain range $\Delta\bar{\varepsilon}$, as shown in Fig. 4.23(a), while using an originally developed electron-hydraulic testing machine [22] (see Fig. 4.24). Mises equivalent stress and strain have no negative sign but in the figure the sign is set so as to emphasize the cyclic loading.

We found from Fig. 4.21 that the cyclic stress–strain curve is almost independent of principal strain ratio, $\phi = \varepsilon_3/\varepsilon_1$, and that the cyclic

146 PLASTICITY AND HIGH TEMPERATURE STRENGTH OF MATERIALS

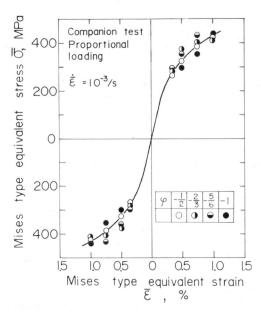

Fig. 4.21. Cyclic Mises equivalent stress/strain curves of SUS 304 stainless steel under proportional loading at 823 K in air by using the companion specimen test method. (Reprinted from ref. 21a by permission of Japan Soc. Mech. Engrs)

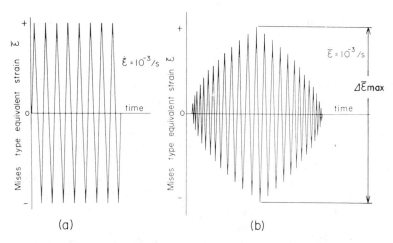

Fig. 4.22. Test methods: (a) companion specimen test; (b) incremental/decremental (ID) test.

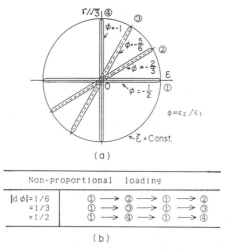

Fig. 4.23. Strain paths: (a) proportional loading; (b) non-proportional loading.

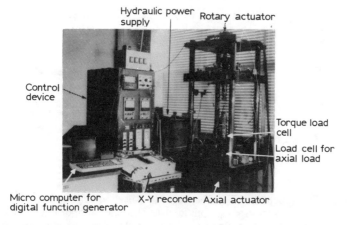

Fig. 4.24. Photograph of the high-temperature biaxial low-cycle fatigue testing apparatus.

stress–strain relation in a biaxial stress state under proportional loading is represented by a straight line only in the $\Delta\bar{\sigma}/2$ versus $\Delta\bar{\varepsilon}_\mathrm{p}/2$ relation of log-log plotting, as shown in Fig. 4.25 [21a], where $\Delta\bar{\sigma}$ and $\Delta\bar{\varepsilon}_\mathrm{p}$ are equivalent stress and plastic strain amplitudes of the Mises base, respectively. Figure 4.26 [21a] shows a similar representation of the data

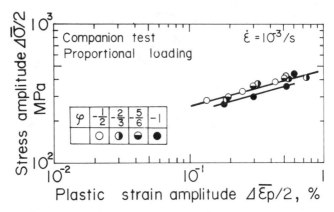

Fig. 4.25. Experimental relation between the Mises equivalent stress amplitude, $\Delta\bar{\sigma}/2$, and the plastic strain amplitude, $\Delta\bar{\varepsilon}_p/2$, under proportional loading by using the companion specimen test method. Test conditions as for Fig. 4.21. (Reprinted from ref. 21a by permission of Japan Soc. Mech. Engrs)

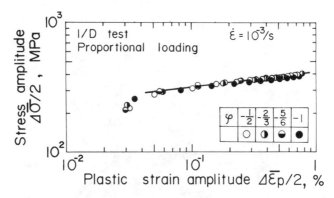

Fig. 4.26. Experimental relation between the Mises equivalent stress amplitude, $\Delta\bar{\sigma}/2$, and the plastic strain amplitude, $\Delta\bar{\varepsilon}_p/2$, under proportional loading by using the incremental/decremental test method. Test conditions as for Fig. 4.21. (Reprinted from ref. 21a by permission of Japan Soc. Mech. Engrs)

obtained by using the 'incremental/decremental (ID) test method' shown in Fig. 4.22(b). In this method the equivalent total strain range, $\Delta\bar{\varepsilon}$, varies in each cycle. One block has 25 cycles, and Fig. 4.26 shows data at one-half of the failure blocks. Note that under smaller strain levels we found lower stress yields in the companion test (Fig. 4.25) compared with the

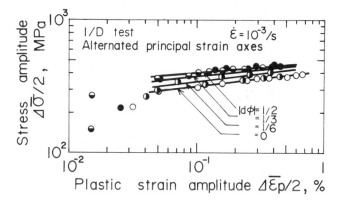

Fig. 4.27. Experimental relation between the Mises equivalent stress amplitude, $\Delta\bar{\sigma}/2$, and the plastic strain amplitude, $\Delta\bar{\varepsilon}_\mathrm{p}/2$, under non-proportional loading, by using the incremental/decremental test method. Test conditions as for Fig. 4.21. (Reprinted from ref. 21a by permission of Japan Soc. Mech. Engrs)

ID test (Fig. 4.26) but, conversely, stress in the companion specimen test becomes larger than that in the ID test as the strain increases. Also note that the cyclic stress–strain relation in the biaxial stress state obtained by using the ID test is also independent of the strain ratio, $\phi = \varepsilon_3/\varepsilon_1$.

On the other hand, Fig. 4.27 [21a] displays similar test data under non-proportional loading as shown in Fig. 4.23(b). We see that the cyclic stress–strain relation under non-proportional loading is evidently dependent on the variation in strain ratio, $|d\phi|$. Therefore it is necessary for the cyclic stress–strain in a biaxial stress state under non-proportional loading to be formulated, by taking the variation in ϕ into consideration.

Provided that the relation of $\Delta\bar{\sigma}/2$ to $\Delta\bar{\varepsilon}_\mathrm{e}/2$ and $\Delta\bar{\varepsilon}_\mathrm{p}/2$ is written as

$$\Delta\bar{\sigma}/2 = E'(\Delta\bar{\varepsilon}_\mathrm{e}/2)$$
$$\Delta\bar{\sigma}/2 = K'(\Delta\bar{\varepsilon}_\mathrm{p}/2)^{n'} \quad (4.6)$$

the following cyclic stress–strain equation is obtained:

$$\Delta\bar{\varepsilon}/2 = (\Delta\bar{\varepsilon}_\mathrm{e}/2) + (\Delta\bar{\varepsilon}_\mathrm{p}/2) = (\Delta\bar{\sigma}/2)/E' + [(\Delta\bar{\sigma}/2)/K']^{1/n'} \quad (4.7)$$

where $\Delta\bar{\varepsilon}_\mathrm{e}$ is the elastic strain range of the Mises base, E' is the tangent modulus in the $\Delta\bar{\sigma}/2$ versus $\Delta\bar{\varepsilon}_\mathrm{e}/2$ relation, and n' and K' are the cyclic strain hardening exponent and the cyclic strain hardening coefficient in the $\Delta\bar{\sigma}/2$ versus $\Delta\bar{\varepsilon}_\mathrm{p}/2$ relation, respectively.

Table 4.2 [21a, b] summarizes the material constants in eqn (4.7) for

Table 4.2
Material Constants in the Cyclic Constitutive Equation of Type 304 Stainless Steel at 824 K in Air (Reprinted from ref. 21a by permission of Japan Soc. Mech. Engrs)

Test method	Mode	Tangent modulus E' ($\times 10^5$ MPa)	K' (MPa)	n'	$\sigma_{0.2}$ (MPa)
Companion specimen test	Proportional loading	1·25	480	0·265	313
Incremental decremental test	Proportional loading	1·25	414	0·107	353
	Non-proportional loading				
	$\|d\phi\| = 1/6$	1·3	440	0·102	374
	$= 1/3$	1·4	480	0·099	409
	$= 1/2$	1·5	495	0·917	427

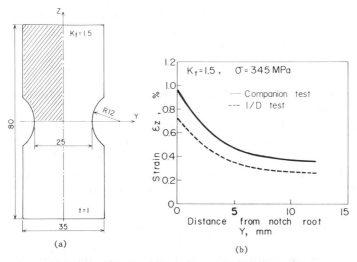

Fig. 4.28. Strain distribution along the notch section of the double-notched plate during tensile loading, calculated by using both the cyclic stress/strain relation derived from the companion specimen and the incremental/decremental test method. (S. Nishino, M. Sakane and M. Ohnami, not published)

the test data obtained by both test methods. From this table we find that there is some difference in both test methods for the values of n' and K' during proportional loading. This difference can be considered a result of different strain histories of the metals. Figure 4.28(b) shows an example of the difference in axial strain distribution along the notch section of the double notched plate specimen in Fig. 4.28(a), under tensile nominal stress of 345 MPa, calculated by using the cyclic stress–strain equation under proportional loading obtained by both test methods. The FEM elasto-plastic analysis was made by multi-linearization of the stress–strain relations in Table 4.2 using an isoparametric element with 8 nodes. It should be noted here that the observable difference in the inelastic stress/strain distribution in the structural analysis is a result of the discrepancy in the cyclic stress–strain relations obtained from both test methods, and has not been disregarded.

Besides, the difference between the cyclic stress–strain relations under proportional and non-proportional loadings brings an observable change in the magnitude of E', K' and n', and in the case of non-proportional loading these reach larger values as the variation in $\phi = \varepsilon_3/\varepsilon_1$ increases. It is important to recognize that a condition of non-proportional loading frequently arises in machine structural elements subjected to combined cyclic mechanical and thermal stresses. We can understand the large increase in cyclic strain hardening under non-proportional loading from the following microstructural study.

4.4.2 Microstructure in Proportional and Non-proportional Cyclic Loadings

Figure 4.29 [23] illustrates cyclic hardening in the Mises equivalent tensile stress amplitude of the same test specimen as Figs 4.25–4.27. The tests were performed by applying the Mises equivalent strain range of $\Delta\bar{\varepsilon} = 1.0\%$ in P, T and APT loading modes at frequency $v = 0.1$ Hz and 823 K in air. P, T and APT here denote the following strain waves, respectively:

(a) Push-pull test, abbreviated to P loading (Fig. 4.30(a)).
(b) Reversed torsion tests, abbreviated to T loading (Fig. 4.30(b)).
(c) Alternating push-pull/reversed torsion test, abbreviated to APT loading (Fig. 4.30(c)).

While principal strain axes do not change direction in modes (a) and (b) they do change $\pi/4$ radians per cycle in (c). In strain wave (c) a push-pull cycle is changed to a torsional cycle at zero strain (the cycles are counted separately). Note that APT loading can change principal strain axes while maintaining the zero mean equivalent strain. Usually, out-of-phase loading is employed as a means of non-proportional loading [24],

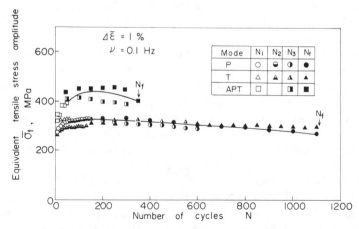

Fig. 4.29. The Mises equivalent stress amplitude, $\bar{\sigma}_t$, versus number of strain cycles, N, in SUS 304 austenitic stainless steel at 823 K in air. (Reprinted from ref. 23 by permission of Fatigue of Engineering Materials Ltd)

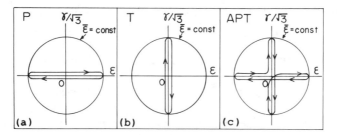

Fig. 4.30. The first test program: (a) push-pull (P); (b) reversed torsion (T); (c) alternating push-pull/reversed torsion (APT).

but mean stress or strain fluctuate in out-of-phase loading so that it is not suitable for studying the constitutive behavior of materials.

In Fig. 4.29 there is almost no difference in the stress amplitude between P and T modes. Only APT loading exhibits greater cyclic work-hardening, with a stress amplitude about 40% greater than for P and T loadings. Figure 4.31 [23] shows dislocation structures at N_1, N_2 and N_3 in loading modes at $\Delta\bar{\varepsilon} = 1\%$, where N_1 is the cycle in the early stage where larger cyclic work-hardening is in progress, N_2 the cycle wherein

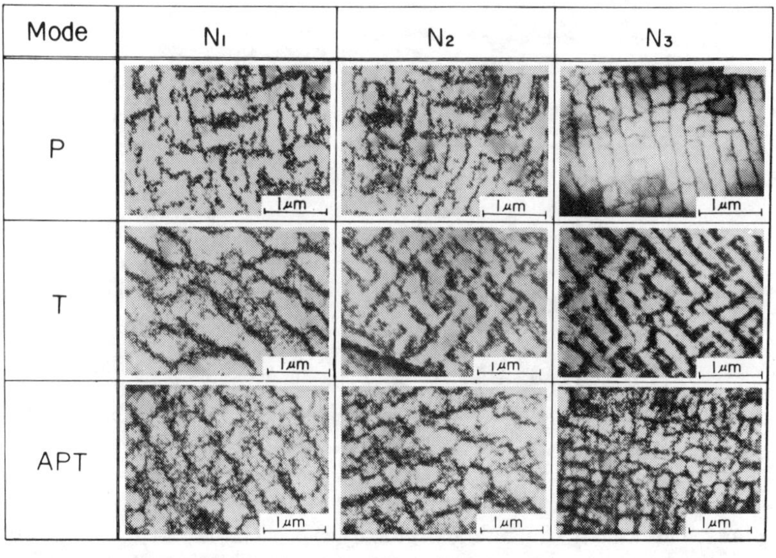

Fig. 4.31. TEM observations under push-pull, reversed torsion and alternating push-pull/reversed torsion loadings at 823 K in air. (Reprinted from ref. 23 by permission of Fatigue of Engineering Materials Ltd)

larger cyclic work-hardening almost reaches saturation, and N_3 the cycle of a half-life (N_1 was about 0.05–$0.1N_f$, N_2 0.2–$0.3N_f$ and N_3 $0.5N_f$, where N_f is the number of cycles to failure). Dislocation structures were observed by using JOEL JEM100C (100KV) TEM and the observation foil was prepared by jet polishing with reagent grade acetic acit containing 6% perchloric acid and finish polishing with reagent grade acetic acid containing 25% perchloric acid. Dislocation structure in the push-pull test is mostly ladder, partly maze and somewhat cell, while in reversed torsion it is mostly maze, partly ladder and slightly cell. APT loading yields only cell structure. The photos in column N_1 show that in all the loading modes dislocation structures form in the early stage of the strain cycles and that they are firmly established as the strain cycle progresses.

Figure 4.32 [23] shows the geometrical configuration of ladder and maze structures at N_3, where the equivalent strain range is 1%. Schematic diagrams on the right of the figure show {111} planes, i.e. the primary active slip planes of the grains on which the maximum shear

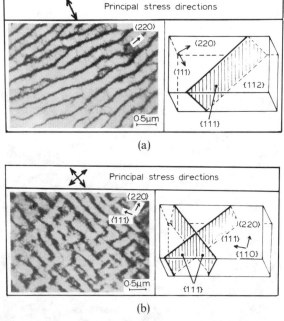

Fig. 4.32. Dislocation structure and primary slip planes ($\Delta\bar{\varepsilon}=1\%$, $N/N_f=0.5$): (a) ladder type under P loading; (b) maze type under T loading. (Reprinted from ref. 23 by permission of Fatigue of Engineering Materials Ltd)

stress occurs. They are distinguished from other slip planes according to loading directions. Since the electron beam penetrates from the top to the bottom of the schematic figure, it is clear that the dense parts of ladder or maze disolocations are on the {111} plane. This is also confirmed at N_1 and N_2.

Regarding the relationship between cyclic stress response and dislocation structure, the findings are as follows. Generally, cell structure is formed in uniaxial low-cycle fatigue tests at room temperature. At elevated temperatures, however, ladder or maze structures form because of the thermally activated process which rearranges the dislocation into a more ordered array having lower elastic strain energy. Mura et al. [25] reported that the maze and ladder structures have less strain energy than cell structures, from elasticity calculations. On the other hand, in the alternation test the maximum principal shear strain changes direction by 45° in each cycle, so that a larger interaction of slip systems occurs. In the alternation tests this larger interaction results in cell structure because it overcomes the rearrangement of the dislocations with the assistance of the thermal process. The resistance to the dislocation glide is of course greater in the cell structure which results in larger cyclic strain hardening. Figure 4.33 [23] shows microbeam X-ray diffraction patterns under the three loading modes at N_1. As total misorientation β_t corresponds to the continuous change of crystal orientation of the affected grains, a much greater intensity of plastic deformation in grains occurs in the alternation test which contributes to the increase in cyclic work-hardening.

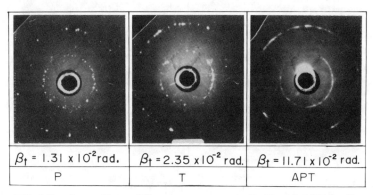

Fig. 4.33. Microbeam X-ray diffraction patterns in P, T and APT loadings at 823 K in air ($N = N_1$, $\Delta\bar{\varepsilon} = 1\%$). (Reprinted from ref. 23 by permission of Fatigue of Engineering Materials Ltd)

4.4.3 Effect of Strain History on Cyclic Stress Response

In the previous section we discussed dislocation structures which formed in three different loading modes. Now we will examine how these structures vary with changes of the loading mode.

This test was performed according to the program shown in Figs 4.34 (a) and (b), where in Fig. (a) the APT mode is changed to the P mode after straining in the former mode up to $N_2 = 110$ cycles and in Fig. (b) the loading is changed from P mode to T mode at $N_2 = 130$ cycles. Figure 4.35 [23] shows that the equivalent stress amplitude starts to decrease immediately after changing the load from APT to P mode, but does not fall to the same level as that of simple P loading. On the other hand, in changing from P to T mode, the stress amplitude increased immediately after the change, then peaked and fell to a level near that of simple P loading.

Figure 4.36(a) [23] shows TEM observations after the mode change from APT to P. The dislocation structure is seen to be mostly ladder or maze and partly cell, indicating that much of the cell structure which

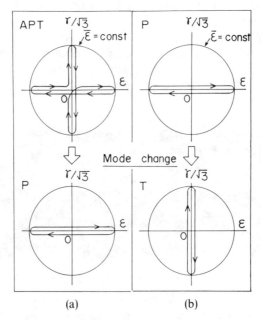

Fig. 4.34. The second test program: (a) alternating push-pull/reversed torsion → push-pull loadings; (b) push-pull → reversed torsional loadings.

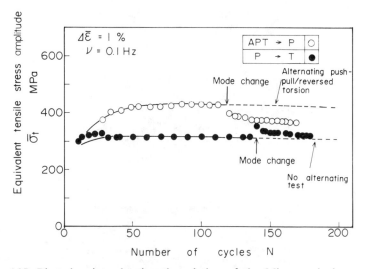

Fig. 4.35. Plotted points showing the relation of the Mises equivalent stress amplitude, $\bar{\sigma}_t$, versus the number of strain cycles, N, during the loading mode change test where $\Delta\bar{\varepsilon}=1\%$. Test conditions as for Fig. 4.21. (Reprinted from ref. 23 by permission of Fatigue of Engineering Materials Ltd)

forms in APT loading is suitable for, and is easily rearranged to, a ladder or maze structure by P loading. Rearrangement of most of the dislocation structure from cell to ladder or maze brings about a decrease in stress amplitude. However, the persistence of some residual, apparently very stable cell structure from APT loading checks this decrease and keeps the stress amplitude from attaining the same lower value as that in simple P loading.

Figure 4.36(b) [23] also shows TEM observations after the mode change from P to T loading. This figure reveals that the dislocation structure after the mode change is ladder or maze and that the dislocations are on {111} planes where maximum shear stress occurs in T mode; the dislocations formed in P mode moved to the slip planes of maximum shear stress in T mode because they were previously on the planes of the maximum shear stress in P mode. Larger interaction is to be expected when loading modes are changed, as this is the main cause of a rapid increase in stress amplitude. After a few strain cycles in T mode, the dislocations form on the maximum shear stress planes leading to a decrease in stress amplitude. Stress amplitude reaches the same value as P mode once reloading of the dislocations is completed.

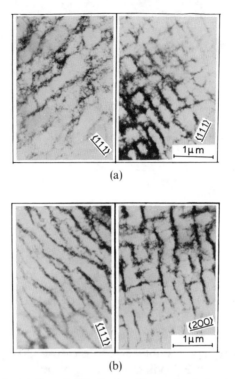

Fig. 4.36. TEM observations during the loading mode change test at 823 K in air ($\Delta\bar{\varepsilon}=1\%$): (a) APT loading → P loading; (b) P loading → T loading. (Reprinted from ref. 23 by permission of Fatigue of Engineering Materials Ltd)

4.4.4 Stress Amplitude Ellipse and Cyclic Yield Stress Ellipse

At the end of this section we will discuss the test results of the stress amplitude ellipse and cyclic yield stress ellipse according to the test program shown in Fig. 4.37 (a) and (b). In test (a), material prestrained for several cycles in the push-pull mode was subsequently strained 3–5 cycles in a direction with a principal strain ratio $\varepsilon_3/\varepsilon_1$ to obtain a data point of the cyclic yield stress ellipse of that material. This type of loading will be referred to as CPT loading. The change of mode from P to CPT was made at a level of zero stress and strain. Then the material was reloaded in the push-pull mode 'to initialize it' before the CPT test. After this a test to obtain a different point on the curve was performed. This sequence completed the cyclic yield of the stress elliptic curves of the prestrained material in the push-pull mode. As shown in Fig. 4.37(b),

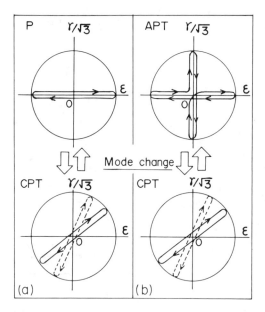

Fig. 4.37. The third test program: (a) push-pull → cyclic combined axial and torsional loadings; (b) alternating push-pull/reversed torsion → cyclic combined axial and torsional loadings.

the same procedure was carried out for the prestrained material in APT mode. Note that there is almost no efffect of CPT loading on the material behavior after the initialization in the push-pull mode. This was confirmed by the fact that after initialization the material yielded almost the same cyclic stress amplitude as the virgin material. Also, the dislocation structure after initialization, as well as in the push-pull test, was of the simple ladder type.

Figure 4.38 (a) and (b) [23] shows the equivalent stress amplitude ellipse in P and APT modes with those of the virgin material at a strain range of 1·0% and 0·7%, where the equivalent stress amplitude is the tensile and the compressive peak stress of a hysteresis loop. The figures show that the virgin material exhibits an almost isotropic stress response because the data for the material are on the Mises ellipse. However, the cyclic stress response is not isotropic for the prestrained material in P mode. The shape of the ellipse of that material is being elongated along the shear stress axis. The data points are almost on the ellipse expressed by the equation $\sigma^2 + 2\tau^2 = \bar{\sigma}_p^2$, where $\bar{\sigma}_p$ is the equivalent stress amplitude

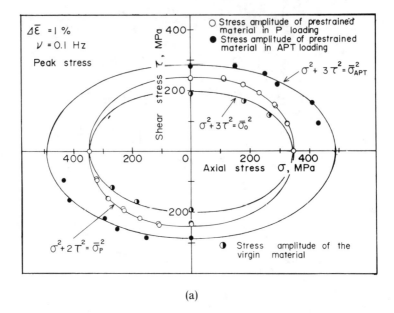

(a)

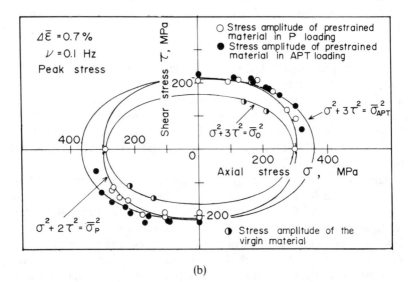

(b)

Fig. 4.38. Equivalent cyclic yield stress ellipses at peak stress levels: (a) $\Delta \bar{\varepsilon} = 1\%$; (b) $\Delta \bar{\varepsilon} = 0.7\%$. Test conditions as for Fig. 4.21. (Reprinted from ref. 23 by permission of Fatigue of Engineering Materials Ltd)

in P mode. As with the virgin material, the cyclic response for the prestrained material in APT mode is isotropic, but the ellipse is larger than that which results from the virgin material. The equivalent stress amplitude in APT mode is 1·4 times larger than that for virgin material, and the yield condition in APT mode complies with the so-called isotropic hardening rule. This is the same tendency as that found for the cyclic yield stresses ellipse [23].

Figures 4.31, 4.32 and 4.36 show that the dislocation structure is ladder or maze in P mode and that dislocations are on {111} planes where maximum shear stress is operative. The figures also show that the planes with high dislocation density are oriented in a specific direction so that the material with this microstructure exhibits an anisotropic response, because it has greater resistance to dislocation glide crossing a high-density plate. On the other hand, the material in APT mode develops cell structure which has almost the same resistance to dislocation glide in all directions. Hence it exhibits isotropic cyclic stress response, i.e. isotropic cyclic hardening.

We can summarize the present test results of cyclic strain hardening behavior of type 304 austenitic stainless steel in a biaxial stress state at elevated temperatures as follows:

(1) In the cyclic stress–strain relation in proportional loading, there was no dependence on principal strain ratio, $\varepsilon_3/\varepsilon_1$, and the Mises base was applicable to the formulation.

(2) In both cyclic stress–strain equations for proportional loading obtained by companion specimen and incremental/decremental test methods, there was some difference in the cyclic strain hardening exponent n' and the coefficient K'. There was also a case where the discrepancy in the cyclic stress–strain relation brought about a difference in the inelastic stress/strain analysis of the structure.

(3) In non-proportional loading, the values of the tangent modulus in the elastic part of the $\Delta\bar{\sigma}/2$ versus $\Delta\bar{\varepsilon}/2$ relation, E', the cyclic strain hardening exponent n', and the coefficient K' increased as the variation in $\phi = \varepsilon_3/\varepsilon_1$ increased, compared with those in proportional loading. We found that a larger increase in E', n' and K' under non-proportional loading resulted in larger inelastic stress/strain in structural analysis.

(4) There was no significant difference between the cyclic stress amplitude in the push-pull and reversed torsion tests on a Mises base, but alternation of the principal stress axes increased stress amplitude by approximately 40%.

(5) Most of the dislocation structure in the push-pull test was ladder

while in the reversed torsion test it was maze. Dislocations were on the main slip planes {111} where maximum shear stress occurred in both modes. Only the alternation test yielded cell structure because of the complex interaction of slip systems.

(6) Changing the loading mode from alternation to push-pull decreased stress amplitude but it did not fall to the same level as that for P loading. (It instead leveled off 20% greater.) After the loading mode changed, most of the dislocation structure was ladder with some cell. Changing the loading mode from push-pull to reversed torsion resulted in a rapid increase in stress amplitude, but after peaking the stress amplitude gradually decreased until it reached the same values found in simple reversed torsion loading. In this case, following the change in loading, the dislocation structure was either maze or ladder.

(7) Prestrained material in the push-pull test yielded an anisotropic stress amplitude ellipse in spite of the virgin material being isotropic. This anisotropy was related to the ladder dislocation structure. Prestrained material in the alternation test, on the other hand, yielded an isotropic stress amplitude ellipse. The size of the ellipse was larger than that for the virgin material and the yield condition of material in the alternation test complied with the so-called isotropic hardening rule. This hardening rule was related to the cell dislocation structure. This is the same tendency as that found for the cyclic yield stress ellipse.

4.5 PLASTICITY LAWS IN CREEP OF POLYCRYSTALLINE METALS AT ELEVATED TEMPERATURES

In this section we will examine the plasticity laws for creep of polycrystalline metals at room and elevated temperatures, with the intention of establishing an accurate creep constitutive equation in connection with the change of structure in the material, as a continuation of the previous section.

Historically, the discrepancy between the prediction of creep deformation of structural materials and the experimental results has been attributed to the following effects of several factors on plasticity laws in creep mechanics:

1. The strain history (e.g. the effect of the strain path, recovery of creep, structural recovery, Bauschinger effect, plastic deformation/creep interaction, and others).
2. The volume constant law.

3. The influence of hydrostatic stress.
4. The yield condition and the associated flow-rule.

We have pointed out the need for experimentation to examine the validity of the plasticity laws in metallic creep and for elucidating the limitations of those laws. Surveying the papers we referred to previously, however, we note that many physical and analytical studies on the respective aspects of the above effect have been made without clarifying the unified correlation between them. It is important to state that these effects are not independent but are closely connected. In the present section we will discuss the influences of strain history and hydrostatic stress. To arrive at total strain values instantaneous strain will be linearly added to creep strain, and we will also use logarithmic strain values. This is due to the assumption that there is no remarkable interaction between plastic deformation and creep strain because the time-independent strain is very small compared with creep strain, and the total strain is represented by Cartesian localized coordinates (see Appendix II) for the sake of simplicity in analysis.

4.5.1 Influence of the Strain Path on High-temperature Creep Deformation Under Fixed and Changing Principal Stress Axes

Simple examination of the influence of strain history was made under simple torsion. Axial creep rate $\dot{\varepsilon}_z$ and circumferential creep rate $\dot{\varepsilon}_\theta$ of a thin-walled cylindrical specimen under simple torsion are found by applying eqns (4.2), (4.3) and the flow-rule, $\dot{\varepsilon}_{ij} = \lambda(\mathrm{d}f/\mathrm{d}S_{ij})$, as

$$\dot{\varepsilon}_z = \dot{\varepsilon}_\theta = (1/3)\lambda A \tau \dot{\gamma} \qquad (4.8)$$

Another simple examination was made for comparison between the types of creep test data of simple tension and simple torsion. The ratio of uniaxial tensile creep rate $\dot{\varepsilon}_z$ to simple torsional creep rate $\dot{\gamma}$ at the same total equivalent strain $\bar{\varepsilon}$ is given as (see Appendix II)

$$r = (\dot{\gamma}/\sqrt{3})/\dot{\varepsilon}_z = \{1 - (1/4)A\bar{\varepsilon}^2 + (3/4)A^2\bar{\varepsilon}^2\}/\{1 + A\bar{\varepsilon} + (3/4)A^2\bar{\varepsilon}^2\} \qquad (4.9)$$

where the relations $\sigma_z = \sqrt{3}\tau = \bar{\sigma}$ and $\varepsilon_z = \gamma/\sqrt{3} = \bar{\varepsilon}$ are assumed on the basis of the Mises condition. In the equation we also assume $B=0$. As we can see from eqns (4.8) and (4.9), three kinds of result hold; (1) $\dot{\varepsilon}_z = \dot{\varepsilon}_\theta > 0$ and $r<1$, (2) $\dot{\varepsilon}_z = \dot{\varepsilon}_\theta = 0$ and $r=1$, and (3) $\dot{\varepsilon}_z = \dot{\varepsilon}_\theta < 0$ and $r>1$ are derived in accordance with $A>0$, $A=0$ and $A<0$, respectively.

Figure 4.39 [14] demonstrates the comparison between test data of a thin-walled tube 0·15% carbon steel specimen (gauge length 80 mm, inner dia. 18 mm, outer dia. 20·7 mm), obtained from both a uniaxial

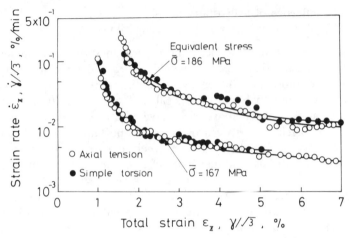

Fig. 4.39. Comparison of two types of creep tests illustrating the results of simple tension and simple torsion of 0·15% carbon steel specimen under constant Mises equivalent stresses at 723 K in air. (Reprinted from ref. 14 by permission of Soc. Mater. Sci., Japan)

tensile creep test and a simple torsional creep test at 723 K in air. We find that r is nearly constant at unity during the progress of creep but becomes slightly larger than unity. This is also coincident with the experimental results with a slight decrease in the length of the specimen during simple torsional creep. In summary, there is no remarkable development of anisotropy of the material during the creep test under fixed principal stress axes.

Figure 4.40 [14] shows an example of the creep curves under stress variation along two Mises stress ellipses of which the equivalent stress levels are $\bar{\sigma}_I = 167$ and $\bar{\sigma}_{II} = 186$ MPa with a test period of 2 h in the representation of axial, circumferential and shearing strains, respectively. The circumferential stress was imposed on the tubular specimen by internal pressure. Dashed and chained curves show the analytical creep determined from the data of simple creep tests by using time-hardening and strain-hardening hypotheses which are the most simple hardening rules, respectively. The analytical curves based on the time-hardening hypothesis are rather closer to the experimental curves than those based on the strain-hardening hypothesis. However, the data points are not in good agreement with the former analytical curves. This disagreement is particularly remarkable immediately after inverse loading of simple

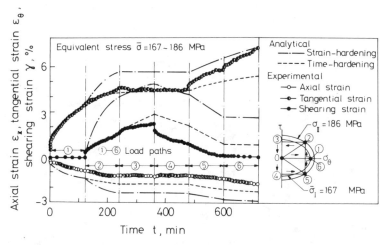

Fig. 4.40. Experimental and analytical creep curves of axial strain, ε_z, circumferential strain, ε_θ, and shearing strain, γ, in the combined internal pressure and torsion creep test of 0·15% carbon steel under changing principal stress axes and variation of the equivalent stresses at 723 K in air. (Reprinted from ref. 14 by permission of Soc. Mater. Sci., Japan)

tension. In the following pages we will verify that this discrepancy results mainly from the effect of anisotropy of material crept with different strain histories, and also from mechanical recovery, particularly the Bauschinger effect.

Figure 4.41 [14] shows the experimental correlation between v and μ of Lode's parameters, during creep in two types of test under periodic variation of the equivalent stress levels. Here, tests A, B and D were made under combined axial tension σ_z and torsion τ, and tests C and E under combined circumferential tension σ_θ and torsion τ, where circumferential tension was imposed on the tubular test specimen by internal pressure. Test data showing the results for the material under a fixed equivalent stress level are also shown. The dashed lines show the flow-rule based on the Mises condition in which the relation $v = \mu$ is satisfied. Chained lines represent the flow-rule based on Prager type conditions, for which the yield condition is given by $f = J_2'^3 - \Omega J_3'^2$ where Ω is constant and J_3' is the third invariant of S_{ij} [26, 27]. We find that the variation of v during creep is observed not only immediately after the stress change but also during the respective path. In the present case of changing principal stress axes, we find that the data points are not

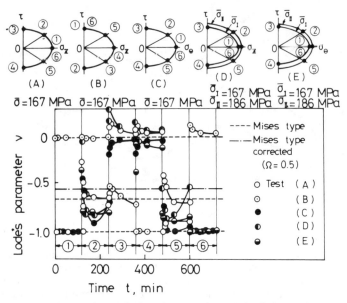

Fig. 4.41. Experimental correlation between μ and ν of the Lode parameters during creep in several types of test of 0·15% carbon steel under changing principal stress axes at 723 K in air. (Reprinted from ref. 14 by permission of Soc. Mater. Sci., Japan)

always close to the analytical ν and μ relation based on the Prager formula although it is satisfactory under the fixed principal stress axes [26, 27].

Figure 4.42 [14] shows an example of experimental results representing the effect of repeated torsional stress reversal on transient increases in shearing strain rate $\dot{\gamma}$ immediately after the stress reversal. The test was performed by using the same specimen as in Figs 4.39–4.41 under combined constant axial stress and cyclic torsional stress reversal. As we can see in the figure, it does not necessarily follow that the transient increase of the strain rate after the stress reversal becomes small with the progress of $|d\gamma|$. After reviewing the paper by Morrow *et al.* [28], we can conclude that the creep resistance of chemical lead is rather reduced after several cycles of torsional stress reversal and continues to decrease with additional stress reversals. This suggests that the Bauschinger effect is closely connected with the development of anisotropy of the material during creep as shown in the following analysis. Figure 4.43 [14] shows a graphical representation illustrating the kinematic hardening type of

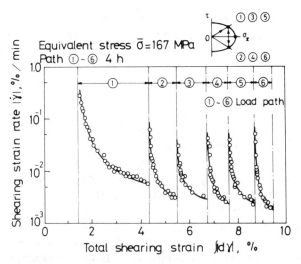

Fig. 4.42. Experimental shearing creep rate versus total shearing strain curves in the creep test of 0·15% carbon steel under combined constant axial tension and cyclic torsion reversals at 723 K in air. (Reprinted from ref. 14 by permission of Soc. Mater. Sci., Japan)

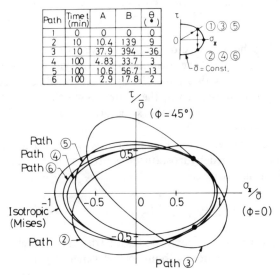

Fig. 4.43. Analytical illustration of kinematic hardening of the yield stress ellipse in the creep test of Fig. 4.42. Test conditions as for Fig. 4.42. (Reprinted from ref. 14 by permission of Soc. Mater. Sci., Japan)

yield curve with the repeat of torsional stress. From these data we can see that, at an early stage of creep after the stress reversal in paths 2 and 3, parameters A and B take larger values, and that marked rotation and the translation of the stress ellipse occurred, compared with those in low-cycle fatigue of the same material at room temperature shown in Fig. 4.17.

4.5.2 Influence of Hydrostatic Stress on Flow Stress and Creep at Room and Elevated Temperatures

In this section we will attempt to elucidate the influence of hydrostatic stress on plasticity laws, especially creep laws of polycrystalline metals at elevated temperatures. As we well know, approximation of the yield condition and the associated flow-rule of metals is independent of the first invariant of stress $J_1 = \sigma_{kk}$, and is very useful in advancing mathematical theories of plasticity. Concerning multiaxial creep problems, the theoretical analysis is also subject to fundamental assumptions in the theory of plasticity. However, it is apparent that the influence of hydrostatic stress is an unavoidable factor in the evaluation of stress dependence on metallic creep at room or elevated temperatures [29–38]. From this point of view it is possible to say that the study of metallic creep and fracture under combined hydrostatic pressure gives us important information for the formulation of creep laws for polycrystalline metals.

In discussing the influence of hydrostatic stress on the yield condition of polycrystalline metals, the formulation of the yield condition may be classified into two types. This division can be made according to physical properties illustrated by the degree to which hydrostatic stress is responsible for the change in the structure of metals and the plastic deformation mechanism.

In the present study, from the point of view of structural change of the metals under combined hydrostatic pressure, the following form of the yield condition is discussed [31]:

$$f(J_2') = F\left(J_1, \int d\varepsilon_{ij}\right) \text{ and } F = \sum_{i=1}^{\infty} a_i J_1^i \qquad (4.10)$$

where a_i is the parameter connected with $\int d\varepsilon_{ij}$. A concrete formulation of eqn (4.10) is represented without loss of generality as

$$J_2' = k(k^2 + CkJ_1 + DJ_1^2)^{1/2} \qquad (4.11)$$

This is not only a simple form but also can generally exhibit the plastic

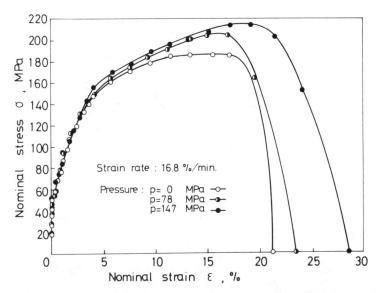

Fig. 4.44. Experimental tensile stress and strain curves of 0·002% C pure iron under combined hydrostatic pressures at 373 K in oil. (Reprinted from ref. 33a by permission of Soc. Mater. Sci., Japan)

behavior of the metals; k is the parameter that varies monotonically with work-hardening connected with the strain history $\int d\varepsilon_{ij}$. Considering that there will be an upper limit of tensile yield stress of the metals under hydrostatic or triaxial tension, the relation $C^2 \geqslant 4D$ must be satisfied in eqn (4.11). Therefore assuming $C^2 = 4D$ in order to simplify the formulation, eqn (4.11) can be written as

$$J'_2 = k(k + C'J_1) \tag{4.12}$$

where C' is taken as $C/2$.

Figure 4.44 [33a] shows the experimental tensile stress/strain curves of a cylindrical bar specimen (gauge length 25 mm, diameter 3·5 mm) of commercial pure iron under confining pressure at 373 K. Figure 4.45 [33a] shows a representation of octahedral shearing stress τ_{oct} versus octahedral normal stress σ_{oct} of Fig. 4.44. We can see from these figures that there is a tendency for flow stress to increase slightly with elevation of combined hydrostatic pressure and also that the numerical values of C and D gradually increase with the progress of plastic deformation. In these tests we paid special attention to ensure that the device would

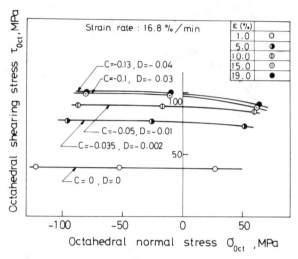

Fig. 4.45. Experimental and analytical representations of octahedral shearing stress, τ_{oct}, versus octahedral normal stress, σ_{oct}, relation of the flow stress of 0.002% C pure iron at 373 K in oil. (Reprinted from ref. 33a by permission of Soc. Mater. Sci., Japan)

maintain a high degree of precise, continuous and stable measurements of the load on the test specimen under high confining pressure at elevated temperatures. The apparatus shown in Figure 4.46 [32] was specially designed for the purpose of static tensile and tensile creep tests under hydrostatic pressure (<300 MPa) at elevated temperatures (<773 K). In the apparatus, for accurate detection of load which is applied to the test specimen, the displacement of the coil spring (17) which is connected longitudinally to the test specimen (20) is measured by DTF out of the pressure vessel. The precision in load measurement was to less than 0.2% under the conditions tested.

In 99.91% pure polycrystalline zinc there is a remarkable influence of hydrostatic stress on the torsional flow stress of the metal even at 288 K as shown in Fig. 4.47 [35], and the parameter takes larger values of $C = -1.83$ and $D = -0.05$ for shearing strain $\gamma = 150\%$. On the contrary. 99.75% polycrystalline pure aluminum was not influenced at all [33b]. These static torsional tests under confining pressure were performed by using a specially designed apparatus to exert combined loading of constant high hydrostatic oil pressure up to 10 GPa and the external torque T; it is shown in Fig. 4.48 [35]. Special care was also taken to

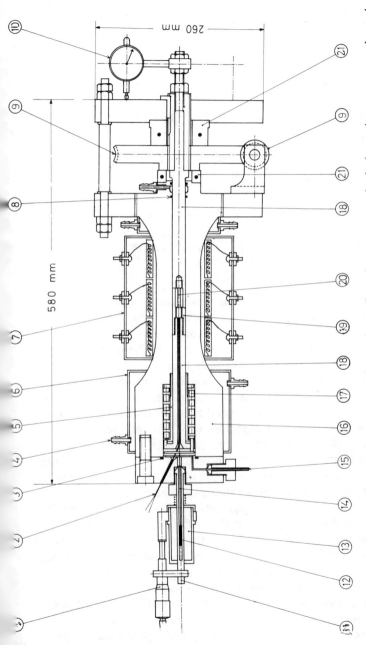

Fig. 4.46. Schematic diagram of the apparatus used for static tensile and creep tests under hydrostatic pressures at elevated temperatures. 1, Micrometer for detecting displacement of coil spring 17; 2, thermocouple which is attached on the end of specimen 20; 3, O-ring; 4, cooling water supply; 5, key; 6, cooling chamber; 7, electric furnace; 8, back-up ring; 9, worm and worm wheel; 10, dial gauge for measurement of combined displacement of coil spring and elongation of specimen; 11, pressure tube (titanium alloy); 12, DTF core; 13, DTF; 14, rod by which displacement of coil spring is transmitted to DTF core; 15, pressurized oil supply; 16, pressure vessel (Ni–Cr–Mo steel); 17, coil spring for loading; 18, plunger; 19, specimen holder; 20, test specimen; 21, bearings. (Reprinted from ref. 32 by permission of Soc. Mater. Sci., Japan)

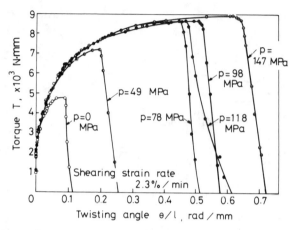

Fig. 4.47. Torque, T, versus twisting angle, θ, curves for 99·91% pure zinc at 228 K under hydrostatic pressures in oil. (Reprinted from ref. 35 by permission of Soc. Mater. Sci., Japan)

ensure that the device properly performed measurement of the torque load applied to the test specimen under high confining pressure, where only the DTF core is exposed to oil pressure while the DTF coil is set out of the vessel in the same manner as that in Fig. 4.46.

In discussing the influence of hydrostatic stress on plastic flow stress it is necessary to consider the relation between the influence of hydrostatic stress and that of strain history experienced in the materials. Figure 4.49 [38] is an example demonstrating nondimensional yield curves on the plane of $\sigma_{oct} = 0$ under the two conditions $C' = 0$ (Mises) and $C' = -0 \cdot 1$ in the sector between $\mu = 1$ of Lode's stress parameter and $\mu = -1$, where τ is the equivalent shearing stress introduced from τ_{oct} in eqn (4.12). In the figure, solid and dashed lines represent calculated results from eqn (4.12). It is necessary to note here that the shape and size of the estimated yield curve of Fig. 4.49 are influenced not only by the hydrostatic stress but also the strain history experienced in the materials. Therefore, in examining the influence of strain history on the plasticity law at elevated temperatures, it is recommended that the tests should be performed under a fixed condition of σ_{oct} for the purpose of separating the influence of strain history from that of hydrostatic stress [14].

Figure 4.50 [35] shows the partial torsional creep curves of a cylindrical bar specimen (gauge length 35 mm, diameter 8 mm) composed of polycrystalline 99·91% pure zinc at 288 K under a constant shearing

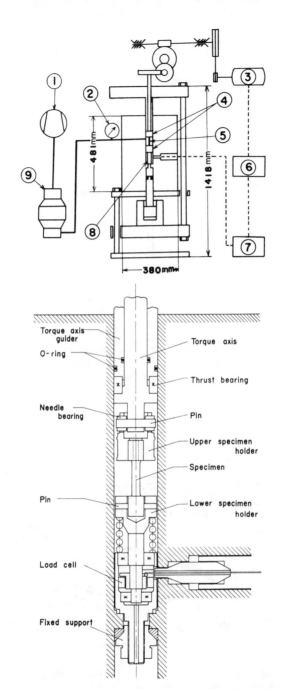

Fig. 4.48. Schematic diagram of the testing apparatus employed in the study on the creep behavior of metals under hydrostatic pressures. (a) 1, Compressor; 2, pressure gauge; 3, motor; 4, specimen holder; 5, test specimen; 6, controller; 7, millivolt meter; 8, load cell; 9, pneumatic driven hydraulic pump. (Reprinted from ref. 35 by permission of Soc. Mater. Sci., Japan)

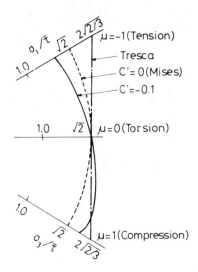

Fig. 4.49. Analytical representation illustrating the effect of hydrostatic stress on the yield curve of octahedral plane $\sigma_{oct} = 0$. (Reprinted from ref. 38 by permission of Soc. Mater. Sci., Japan)

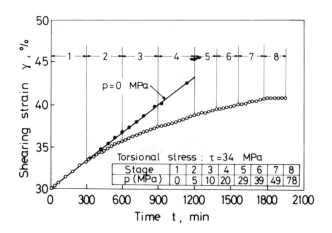

Fig. 4.50. Partial creep curves of 99·91% pure zinc at 288 K under a constant torsional stress of 34 MPa in oil, showing the effect of hydrostatic pressure on creep. (Reprinted from ref. 35 by permission of Soc. Mater. Sci., Japan)

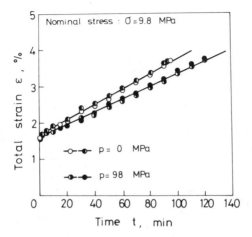

Fig. 4.51. Effect of combined hydrostatic pressure on tensile creep curves of coarse-grained pure aluminum crept at 473 K. (Reprinted from ref. 37 by permission of Soc. Mater. Sci., Japan)

stress of 34 MPa. The test was performed by using the apparatus in Fig. 4.48. A decrease in creep rate is observed with step-wise elevation of confining pressure. Figure 4.51 [37] shows the tensile creep curves of a sheet specimen (gauge length 30 mm, width 8 mm, thickness 1·5 mm) composed of coarse-grain 99·75% pure aluminum under combined hydrostatic pressure at 473 K, and the test was performed by using the apparatus shown in Fig. 4.46. As seen in these figures, the combined hydrostatic pressure brings about a decrease in creep rate. Furthermore, Fig. 4.52 [33b, 34] shows the effect of combined hydrostatic pressure on the tensile creep rupture of a cylindrical bar specimen (gauge length 30 mm, diameter 5·5 mm) composed of fine-grained 99·75% pure aluminum at 473 K. The above can be interpreted as follows.

Let us consider first the yield condition of eqn (4.12) for the sake of simplicity. In the state of uniaxial tensile stress σ_p under confining pressure p, the parameter k_p is determined from eqn (4.12) as

$$k_p = (1/2)[-C'\{1-3(p/\sigma_p)\} + \sqrt{C'^2\{1-3(p/\sigma_p)\}^2 + (4/3)}]\sigma_p \quad (4.13)$$

Uniaxial tensile stress $\bar{\sigma}_p$ under atmospheric pressure corresponding to k_p is also derived from eqn (4.12) as

$$\bar{\sigma}_p = (3/2)\{C' + \sqrt{C'^2 + (4/3)}\}k_p \quad (4.14)$$

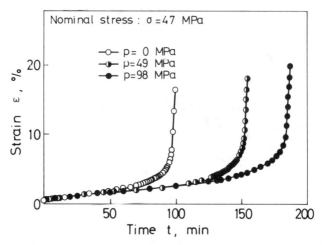

Fig. 4.52. Effect of combined hydrostatic pressure on tensile creep curves of 99·75% aluminum at 473 K. (Reprinted from ref. 33b by permission of Soc. Mater. Sci., Japan)

Figure 4.53 [34, 38] shows a schematic illustration of the apparent decrease in the creep stress of material subjected to combined confining pressure. We see that σ_p is smaller than σ_0 and that the tensile creep rate under pressure decreases compared with that in the atmosphere where σ_0 is a fixed axial stress.

In addition, by employing the yield condition of eqn (4.11) and assuming a relationship between shearing creep rate $\dot{\gamma}$ and shearing stress τ in an arbitrary stage of creep as $\dot{\gamma} = \alpha \tau^n$, the ratio of $\dot{\gamma}_p$ under hydrostatic pressure p to that in the atmosphere $\dot{\gamma}_0$ is given as [35]

$$(\dot{\gamma}_p/\dot{\gamma}_0)^{4/n} - 3C(p/\tau)(\dot{\gamma}_p/\dot{\gamma}_0)^{3/n} + 9D(p/\tau)^2(\dot{\gamma}_p/\dot{\gamma}_0)^{2/n} = 1 \qquad (4.15)$$

Figure 4.54 [35] is the representation of the relation between $\dot{\gamma}_p/\dot{\gamma}_0$ and p/τ for pure polycrystalline zinc at 288 K. The open circles indicate data points at shearing strain γ from 30% to 40% which were plotted from the data of Fig. 4.50, and the solid line was obtained from eqn (4.15) using values $C = 0.65$, $D = -0.021$ and $n = 5$. Analytical results of metallic creep at room temperature agree well with experimental results. The dashed line in Fig. 4.54 was calculated from an equation [31] similar to eqn (4.15) which is based on the yield condition of eqn (4.12) by using $C' = -0.13$.

Since the influence of combined hydrostatic pressure on metallic creep

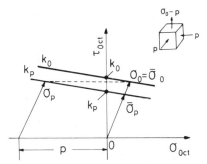

Fig. 4.53. Schematic diagram of the apparent decrease in the creep stress of material subjected to combined confining pressure. (Reprinted from ref. 38 by permission of Soc. Mater. Sci., Japan)

at room temperature is interpreted from the apparent decrease in creep stress, the ratio r of simple tensile creep rate $\dot{\varepsilon}_z$ to simple torsional creep rate $\dot{\gamma}$ at the same Mises equivalent strain is apparently influenced by hydrostatic stress. Figure 4.55 [35] shows the analytical variation of the intensity of the influence of hydrostatic stress affecting r with the numerical value of C'. Note that r nearly reaches unity in the range $C' < 10^{-2}$ but reaches a remarkably larger value than unity in the range $C' > 10^{-2}$.

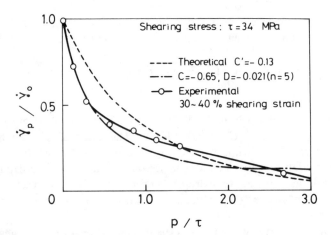

Fig. 4.54. Analytical and experimental representation of the effect of hydrostatic pressure on the simple torsional creep rate of 99·91% pure zinc under constant torsional stress at 288 K in oil. (Reprinted from ref. 35 by permission of Soc. Mater. Sci., Japan)

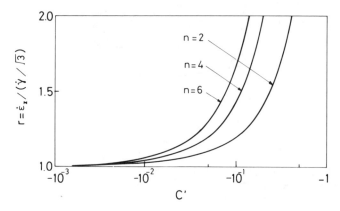

Fig. 4.55. Analytical variation of the ratio of simple tensile creep rate, $\dot{\varepsilon}_z$, to simple torsional creep rate, $\dot{\gamma}/\sqrt{3}$, at the same Mises equivalent stress with the numerical value C' for the intensity of hydrostatic pressure. (Reprinted from ref. 35 by permission of Soc. Mater. Sci., Japan)

In contrast, for metallic creep at elevated temperatures, the interpretation of the decrease in creep rate under confining pressure cannot always be made in the same way as above. Figure 4.56 [38] is a similar representation for fine-grained 99·75% pure aluminum at 473 K as that seen in Fig. 4.54. The open circles show the data points at the stage of steady-state tensile creep and the dashed line was calculated from the following equation based on eqn (4.12) [31] by using $C' = -0.013$:

$$\dot{\varepsilon}_p/\dot{\varepsilon}_0 = (\bar{\sigma}_p/\bar{\sigma}_0)^n = \lambda^n[-C'\{1-3(p/\sigma)\} + \sqrt{C'^2\{1-3(p/\sigma)\}^2 + (4/3)}]^n \tag{4.16}$$

where σ is tensile stress, n the stress exponent in the steady-state creep rate law and $\lambda = (3/4)\{C' + \sqrt{C'^2 + (4/3)}\}$. The value of C' used here is comparatively large while it is nearly zero for static flow stress at the same temperature. Therefore we can see that the decrease in creep rate of the metals at elevated temperatures results mainly from the physical pressure effect in the creep process rather than the apparent decrease in creep stress.

The solid line in Fig. 4.56 shows the calculated result determined from the relation $\dot{\varepsilon}_p/\dot{\varepsilon}_0 = \exp(-pV_c/kT)$ based on diffusional creep or Nabarro–Herring type creep at high temperatures [39], where V_c is the activation volume for creep and kT has the usual meaning. In this calculation, 6·3 cm^3/mol was used as the numerical value of V_c for pure

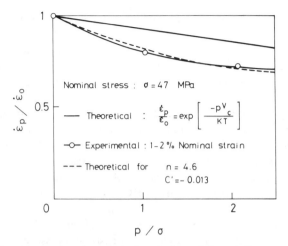

Fig. 4.56. Experimental and analytical variation of the tensile creep rate, $\dot{\varepsilon}_p$, of 99·75% pure aluminum at 473 K, with the elevation of combined hydrostatic pressure, p. (Reprinted from ref. 38 by permission of Soc. Mater. Sci., Japan)

polycrystalline aluminum [40]. The difference between the calculated result and the experimental result is attributed to the pressure effect on vacancy diffusivity in metallic creep. In general, the creep deformation of the metals at elevated temperatures proceeds primarily by dislocation motion, but the amount of dislocation motion is controlled by the flux of vacancies to the dislocations which allows them to move non-conservatively. This will be described in Chapter 5. The effect of hydrostatic stress on creep rupture and low-cycle fatigue fracture of metals at elevated temperatures will also be described in Chapter 7 from the viewpoint of fracture mechanics of solids with microstructure.

From the present analytical and experimental studies on creep laws of polycrystalline metals at room and elevated temperatures, we can make the following conclusions.

(1) No remarkable development of anisotropy was observed in polycrystalline metal crept under fixed principal stress axes at elevated temperatures.

(2) In general loadings, in which the principal axes of stress were changed and the level of the equivalent stress of Mises type was varied during creep, the creep curves under such conditions at elevated temperatures deviated somewhat from the analytical curves based on a strain-hardening hypothesis. In an analysis we saw that under such loadings the development of anisotropy of the crept material was more

remarkable than that under the fixed principal stress axes and it was also heavily dependent on strain history.

(3) In the creep tests under combined constant axial tension and reversed torsion reversals at elevated temperature, it did not necessarily follow that the transient increase in torsional creep rate immediately after the stress reversal gradually reduced with repetition of the reversals. Therefore the influence of the Bauschinger effect on metallic creep at elevated temperatures was remarkable when the stress reversal was imposed on material in which anisotropy was markedly developed.

(4) The effect of combined hydrostatic pressure on the static flow stress of pure polycrystalline metals was observable in the regions of large deformation but generally the intensity was small regardless of whether the tests were at room temperature or at elevated temperatures. We also found that the drastic decrease in creep rate of pure polycrystalline zinc at room temperature by combined hydrostatic pressure could be interpreted from the pressure intensity on the static flow stress.

(5) In contrast, at elevated temperatures, we found that the influence of hydrostatic stress on metallic creep could not be predicted from only the pressure intensity on the static flow stress at the same temperature. This discrepancy was attributed mainly to the difference in the presence of a pressure effect on vacancy diffusivity in metallic creep.

REFERENCES

1. T. Jimma, T. Murota and T. Ichbayashi, Stress–strain matrix of metal crystal and its application, *Journal of the Japan Society of Mechanical Engineers*, **75** (1972), 602 (in Japanese).
2. T. H. Lin and M. Ito, Theoretical plastic distribution of a polycrystalline aggregate under combined and reversed stresses, *Journal of Mechanics and Physics of Solids*, **13** (1965), 103, Pergamon Press.
3. E. Schmid and W. Boas, *Plasticity of Crystals*, Hughers, F. A. & Co., 1950.
4. S. V. Batdore and B. Budiansky, A mathematical theory of plasticity based on the concept of slip, *NACA Technical Note*, **187** (1949), 1871.
5. M. Ohnami and K. Shiozawa, The change in misorientation during the plastic deformation of polycrystalline metals, *Journal of the Society of Materials Science, Japan*, **21** (1972), 16 (in Japanese).
6. R. Aston, The tensile deformation of large aluminum crystals at crystal boundaries, *Transactions of the Cambridge Philosophical Society*, **23** (1927), 549.
7. M. Ohnami and M. Shikita, The relationship between crystal plasticity and plastic deformation of polycrystalline metallic materials under cyclic tensile stressing, *Proceedings of 12th Japan Congress on Materials Research* (1969), p. 45, Soc. Mater. Sci., Japan.

8. J. Saga, R. Koterazawa, Y. Miyoshi and S. Hashimoto, Inhomogeneous deformation in the interior of grain and the tensile property, *Transactions of the Japan Society of Mechanical Engineers*, **34** (1968), 842 (in Japanese).
9. K. Shiozawa, Thesis (Dr. of engineering), Ritsumeikan University, 1973, pp. 160–5 (in Japanese).
10. M. Ohnami and K. Shiozawa, Influence of strain history on plastic deformation of polycrystalline metals subjected to cyclic stressing, *Proceedings of 15th Japan Congress on Materials Research* (1972), p. 106.
11. M. Ohnami and K. Shiozawa, An X-ray study on the plastic deformation of polycrystalline metals subjected to cyclic stressing, *Journal of the Society of Materials Science, Japan*, **21** (1972), 295 (in Japanese).
12. M. Ohnami and K. Shiozawa, Plastic deformation of polycrystalline metallic materials subjected to cyclic stressing, *Proceedings of 13th Japan Congress on Materials Research* (1970), p. 115.
13. K. Yoshimura and Y. Takenaka, Yield condition and rate of working proper to a metal and their dependence on its strain history of extension and twist (1st report), *Transactions of the Japan Society of Mechanical Engineers*, **25** (1959), 133 (in Japanese).
14. M. Ohnami and N. Yoshida, Plasticity laws in creep of polycrystalline metallic materials at elevated temperatures, *Proceedings of 12th Japan Congress on Materials Research* (1969), p. 81.
15. M. W. Brown and K. J. Miller, Biaxial cyclic deformation behavior of steel, *Fatigue of Engineering Materials and Structures*, **1** (1979), 93, Pergamon Press.
16. M. Ohnami and N. Hamada, Biaxial low-cycle fatigue of a SUS 304 stainless steel at elevated temperatures, *Proceedings of 25th Japan Congress on Materials Research* (1982), p. 93.
17. M. Ohnami, M. Sakane and N. Hamada, Effect of changing principal stress axes on low-cycle fatigue life in various strain wave shapes at elevated temperatures, *ASTM Special Technical Publication*, **853** (1985), 622, Amer. Soc. Test. Mater.
18. E. M. Jordan, M. W. Brown and K. J. Miller, Fatigue under severe non-proportional loading, *ASTM Special Technical Publication*, **853** (1985), 569.
19. C. Sonsino and V. Grubisic, Fatigue behavior of cyclically softening and hardening steels under multiaxial elastic-plastic deformation, *ASTM Special Technical Publication*, **853** (1985), 586.
20. D. L. McDiarmid, Fatigue under out-of-phase biaxial stresses of different frequencies, *ASTM Special Technical Publication*, **853** (1985), 606.
21a. M. Sakane, M. Ohnami, S. Nishino and N. Hamada, Low-cycle fatigue and constitutive relation in biaxial stress state at elevated temperature, *Proceedings of International Conference on Creep* (1986, Tokyo), p. 477, Japan Soc. Mech. Engrs.
21b. M. Ohnami, M. Sakane and S. Nishino, Cyclic behavior of a type 304 stainless steel in biaxial stress states at elevated temperature, *International Journal of Plasticity*, University of Oklahoma, USA, in press.
22. M. Ohnami and N. Hamada, Biaxial low-cycle fatigue of SUS 304 stainless steel at elevated temperature, *Proceedings of 25th Japan Congress on Materials Research* (1982), p. 93.
23. S. Nishino, N. Hamada, M. Sakane, M. Ohnami, N. Matsumura and M.

Tokizane, Microstructural study of cyclic strain hardening behavior in biaxial stress state at elevated temperature, *Fatigue of Engineering Materials and Structures*, **9** (1986), 65.
24. For example, K. Kanazawa and K. J. Miller, Low-cycle fatigue under out-of-phase loading condition, *Journal of Engineering Materials and Technology*, **99** (1977), 229.
25. T. Mura, H. Shirai and J. R. Weertman, The elastic energy of dislocation structure in fatigued metals, *Proceedings of 2nd International Symposium and 7th Canadian Fracture Conference on Defects, Fracture and Fatigue* (1982), p. 67.
26. K. Motoie, C. Kang, M. Ohnami and S. Endo, Multiaxial creep deformation of metallic materials at elevated temperature, *Memoirs of the Research Institute of Science and Engineering, Ritsumeikan University*, **14** (1965), 97.
27. M. Ohnami, K. Motoie and N. Yoshida, Plasticity laws in high temperature creep of polycrystalline metallic materials (Influence of strain history in multiaxial creep under unsteady stressing), *Memoirs of the Research Institute of science and Engineering, Ritsumeikan University*, **15** (1966), 83.
28. JoDean Morrow and G. R. Halford, Creep under repeated stress reversals, *Proceedings of Joint International Conference on Creep* (1963, New York, London), 3–43, Inst. Mech. Engrs.
29. D. Hull and E. Rimmer, The growth of grain boundary voids under stress, *Philosophical Magazine*, **4** (1964), 673, Taylor and Francis, London.
30. B. M. Butcher and A. L. Ruoff, Effect of hydrostatic pressure on the high temperature steady state creep of lead, *Journal of Applied Physics*, **32** (1961), 2036, Amer. Institute of Physics.
31. M. Ohnami and K. Motoie, The effect of hydrostatic pressure on plastic deformation and creep of metals, *Proceedings of 12th Japan Congress on Materials Research* (1969), 78.
32. M. Ohnami, Influence of hydrostatic stress on plasticity laws of polycrystalline metallic materials, *Proceedings of 13th Japan Congress on Materials Research* (1970), p. 120.
33. M. Ohnami, K. Motoie and T. Yamakage: (a) The effect of hydrostatic pressure on plastic deformation and creep of polycrystalline metals at elevated temperatures, *Journal of the Society of Materials Science, Japan*, **20** (1971), 395 (in Japanese); (b) *Proceedings of 15th Japan Congress on Materials Research* (1972), p. 17.
34. M. Ohnami and T. Yamakage, The effect of hydrostatic stress on creep and creep rupture of polycrystalline metals at elevated temperatures, *Journal of the Society of Materials Science, Japan*, **21** (1972), 225 (in Japanese).
35. K. Motoie, T. Yamakage and M. Ohnami, Torsional creep of polycrystalline metallic materials under hydrostatic pressure at room temperature, *Journal of the Society of Materials Science, Japan*, **21** (1972), 782 (in Japanese).
36. M. Ohnami, K. Shiozawa and A. Kamitani, X-ray study on the influence of pressure soaking on the structural change of plastically deformed polycrystalline metal, *Proceedings of 15th Japan Congress on Materials Research* (1972), p. 50.
37. M. Ohnami, T. Yamakage, K. Shiozawa and M. Sakane, An X-ray study on the effect of hydrostatic stress on creep of polycrystalline metal at elevated

temperatures, *Journal of the Society of Materials Science, Japan*, **21** (1972), 1086 (in Japanese).
38. M. Ohnami, Study of plasticity laws of polycrystalline metals at elevated temperatures (especially the influence of hydrostatic stress on creep), *Proceedings of 1971 International Conference on Mechanical Behavior of Materials* (1st ICM), Vol. III (1972), p. 1, Soc. Mater. Sci., Japan.
39. F. R. N. Nabarro, Report of a Conference on the Strength of Solids (1954), p. 75.
40. B. M. Butcher, PhD Thesis, Cornell University, 1962.

Chapter 5

Plasticity Laws of Imperfect Crystals and Misorientation of Metals

The need for research on the microplasticity of metallic materials has been pointed out in connection with the study of macroplasticity. This chapter is one of a series of studies on the macroscopic measurement of plastic deformation in microscopically heterogeneous metallic material containing a great number of lattice imperfections; to that end, the misorientation of metal crystals is analyzed by means of the *geometrical aspect of the continuously dislocated continuum* and Laue X-ray diffraction. Applications of the continuously dislocated continuum to microstructural studies of metallic plasticity are described: they are the studies of work-hardening, cyclic constitutive relations, the influence of pressure soaking and the effects of hydrostatic stress on flow stress and high-temperature creep.

5.1 STUDY OF MISORIENTATION IN METALLIC PLASTICITY RELEVANT TO THE GEOMETRICAL ASPECTS OF CONTINUOUSLY DISLOCATED CONTINUUM (CDC)

5.1.1 Analysis of Material Space Based on Non-Riemannian Geometry

Under ideal conditions, the metal crystal exists as a solid having both a constant ordering relation and a constant lattice distance, with a regular arrangement of polyhedrons in which neighbors are attached to respective lattice points in three-dimensional Euclidean space. In actual metallic material, however, the so-called lattice defect exists, and nucleation, annihilation and translation of such so-called lattice defects occur when the material deforms plastically. Thus each respective lattice and its accompanying polyhedrons become mechanically and physically irregular and the lattice distance changes. For the imperfect crystal, therefore, it is necessary to consider non-Euclidean space such as Riemannian or non-Riemannian space. Geometrical studies of the continuously dislocated continuum, based on the so-called Cosserat continuum, have been made by many researchers. In particular, Bilby [1] and Kröner [2] recognized that the notion of the torsion of a space, as introduced by Eli Cartan, is isomorphic to crystal dislocation. Independent of this, Kondo, et al. [3] noted that a manifold of plastically deformed material is regarded as a *non-Riemannian space with Euclidean connection* from the viewpoint of the 'nonholonomic' property. The nonholonomic property is interpreted as follows.

We can do a 'mapping' in the local domain of an imperfect crystal of plastically deformed metal to the same local domain of a perfect crystal. When the deformed continuum is divided into small pieces without imperfections, we obtain a set of perfect crystal pieces. Each piece occupies Euclidean space, but the topological correspondences are not uniquely determined and the pieces are mechanically stable in length and form in the free state as shown in Fig. 5.1. This is Kondo's *'natural state'*. There is a linear relation between the coordinates $(dx)^a$ in the natural state of the piece and those in the actual state of the imperfect crystal dx^{κ}:

$$(dx)^a = B^a_{\kappa} dx^{\kappa} \quad \text{or} \quad dx^{\kappa} = B^{\kappa}_a (dx)^a \tag{5.1}$$

where B^a_{κ} and B^{κ}_a are functions of the actual coordinates x^{κ}. Since B^a_{κ} is different from respective pieces in the natural state, $(dx)^a$ is not, in general, a total differential. Coordinate system κ is called holonomic while system a is called nonholonomic. Provided that the components of the material vectors which represent the minute elements $\overline{OO'}$ shown in Fig. 5.2 are equal to each other in the natural state, the vector $V^{\kappa}(O)$ at

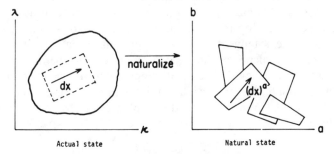

Fig. 5.1. Coordinate systems in actual and natural states of plastically deformed material.

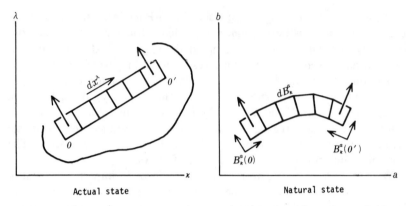

Fig. 5.2. Transformation of minute element of plastically deformed material from the actual state to the natural state.

point O parallels $V^\kappa(O')$ at point O' along dx^λ in the actual state, i.e.

$$V^a(x^\lambda) = V^a(x^\lambda + dx^\lambda) \tag{5.2a}$$

or

$$B_\kappa^a V^\kappa = (B_\kappa^a + dB_\kappa^a)(V^\kappa + dV^\kappa) \tag{5.2b}$$

Omitting the infinitesimal term of higher order in eqn. (5.2b), the following equation is obtained:

$$dV^\kappa = -B_a^\kappa (dB_\lambda^a) V^\lambda \tag{5.3}$$

where $\Gamma_{\mu\lambda}^\kappa dx^\lambda = -B_a^\kappa dB_\mu^a$ and $\Gamma_{\mu\lambda}^\kappa$ is called the *coefficient of connection* (it is not a tensor).

Consider an infinitesimal closed quadrangle developed to the natural state in non-Riemannian space with Euclidean connection as shown in Fig. 5.3. Here, 'develop to the natural state' refers to the mathematical operation where a tangent plane at point A is repeatedly attached to another tangent plane at the nearest point on the periphery of the quadrangle. It is believed that the discrepancy in the transformation of the element which accompanies the circuit along its periphery consists of two parts: (a) the translation or the displacement ΔA^κ of the position given by

$$\Delta A^\kappa = 2 S_{\mu\lambda}^{\cdot\cdot\kappa} \omega_2^\mu \omega_1^\lambda \tag{5.4}$$

and (b) the rotation of the orthogonal transformation of the frames given

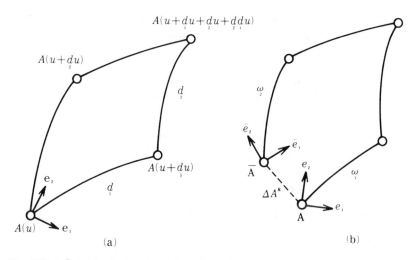

Fig. 5.3. Infinitesimal closed quadrangle (a) in non-Riemannian space and (b) in the natural state.

by

$$\Delta e_x = R_{\lambda\mu\nu}^{\cdots\kappa} e_\kappa \underset{2}{\omega}^\mu \underset{1}{\omega}^\nu \tag{5.5}$$

In these equations, $S_{\mu\lambda}^{\cdots\kappa}$ is the *torsion tensor*, $R_{\lambda\nu\mu}^{\cdots\kappa}$ is the *Riemann–Christoffel curvature tensor*, and $\underset{1}{\omega}$ and $\underset{2}{\omega}$ are the side lengths of the closed quadrangle in the natural state. The material vector $\mathbf{e}^\kappa$ at point A in the natural state is given as

$$\bar{\mathbf{e}}^\kappa = \mathbf{e}^\kappa + (R_{\lambda\mu\nu}^{\cdots\kappa} e^\lambda \underset{1}{\omega}^\mu \underset{2}{\omega}^\nu + 2S_{\nu\mu}^{\cdots\kappa} \underset{1}{\omega}^\nu \underset{2}{\omega}^\mu + O(\omega^3)) \tag{5.6}$$

where $\mathbf{e}^\kappa$ is the vector at point A in the actual state and $O(\omega^3)$ is Landau's notation. Rewriting this and discarding $O(\omega^3)$, the following equation is obtained:

$$\bar{\mathbf{e}}^\kappa - \mathbf{e}^\kappa = (R_{\lambda\nu\mu}^{\cdots\kappa} e^\lambda + 2S_{\nu\mu}^{\cdots\kappa}) \underset{1}{\omega}^\nu \underset{2}{\omega}^\mu \tag{5.7}$$

Thus, both the torsion tensor $S_{\mu\lambda}^{\cdots\kappa}$ and the curvature tensor $R_{\lambda\nu\mu}^{\cdots\kappa}$ correspond directly to the coefficients of connection $\Gamma_{\mu\lambda}^\kappa$ in the actual state and represent physical properties of a plastic manifold with infinitesimal area $\underset{1}{\omega}^\nu \underset{2}{\omega}^\mu (= f^{\nu\mu})$ [4].

Next, the physical significance of the torsion and curvature tensors will be discussed using a crystallographic model. Figure 5.4(a) [4] shows

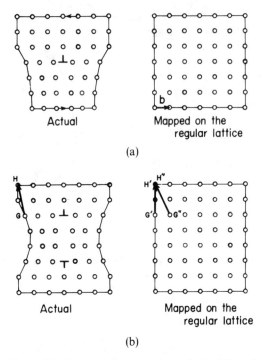

Fig. 5.4. Crystallographical model depicting (a) a single dislocation line and (b) multiple dislocation lines.

a conceptual model of the torsion tensor. When the irregular lattice is mapped to the regular one, the presence of only one dislocation line or of lines with the same sign will cause a discrepancy in the two ends of the Burgers circuit. This particular discrepancy is called Burgers vector **b**; edge and screw dislocations are assigned to different components of the torsion tensor. Denoting the Burgers vector as **b**, the infinitesimal area enclosed by the Burgers circuit as d**f**, and the *dislocation density tensor* as **S**, the following equation is obtained [1]:

$$\mathbf{b} = d\mathbf{f} \times \mathbf{S} \quad \text{or} \quad b^{\kappa} = S_{ji}^{\cdot \cdot \kappa} df_{ji} \tag{5.8}$$

In contrast, Fig. 5.4(b) [4] is a visualization of the concept of the curvature tensor. In the presence of dislocations with different signs, i.e. *dislocation dipole*, the Burgers circuit closes while the vector $\overrightarrow{GH}$ in the actual crystal changes to $\overrightarrow{G'H'}$ or $\overrightarrow{G''H''}$ in the natural state because of the stress or strain field around the dislocation. This demonstrates that

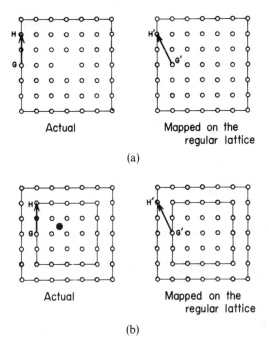

Fig. 5.5. Crystallographical model of (a) lattice vacancy and (b) interstitial atom.

there exists a lattice defect which is represented not only by the torsion tensor but also by the curvature tensor. In the same way, the existence of *lattice vacancy* and of *interstitial atom* shown in Fig. 5.5 (a) and (b) [5,6], results in the concept of a curvature tensor.

While crystal dislocation is a typical well known line defect, there is another lesser known line defect called *disclination*. Dislocation motion produces shearing strain and disclination denotes rotational strain. Let us consider the close-packed lattice of hexagons shown in Fig. 5.6(a). Taking off a triangular portion AOF in Fig. 5.6(a) and welding the plane OA to plane OF, the planes rotate as shown in Fig. 5.6(b). Denoting the lattice vectors at point A in Fig. 5.6(a) as $\mathbf{e}_1$ and $\mathbf{e}_2$ the transforming them along the circuit ABCDEFA, both the initial and terminal vectors agree with each other. In Fig. 5.6(b), however, in transforming the vectors $\mathbf{e}'_1$ and $\mathbf{e}'_2$ along the circuit $A_1B_1C_1D_1E_1A_1$, both vectors rotate. In the same way, rotation of both vectors also occurs as in Fig. 5.6(c). The rotation angle ω in the disclination is given as $\omega = 2p\pi/s$ [7], where s assumes the values 2, 3, 4, 6 and p represents both the strength and the

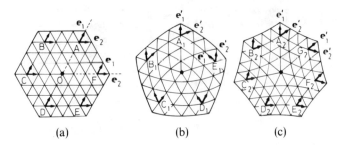

Fig. 5.6. Crystal model of disclinations, showing the line imperfection perpendicular to the plane (see Fig. 3.6 in Chapter 3). (a) Hexagonal (s = 6); (b) p = 1, s = 6; (c) p = −1, s = 6.

sign of disclination. When the disclination line parallels the rotation axis, as in the figures in Fig. 5.6, it is referred to as *screw* or *wedge disclination*. When the lines are at right-angles it is called *edge* or *twist disclination*. The *disclination density tensor*, $d_i^{\cdot jk}$, is defined as

$$\omega^{lk} = (1/2) d_i^{\cdot kq} n_q \, dS \tag{5.9}$$

where $\omega^{lk} = (1/2)\varepsilon^{lkn}\omega_n$, and is represented by the curvature tensor [8].

Now let us consider the climbing motion of dislocations. The scalar product of the Burgers vector, **b**, and the normal vector of the slip plane **n** is zero for slip dislocations on the slip plane. In the case of a product other than zero, non-conservative dislocation will occur. The Burgers vector which parallels **n** represents the climbing dislocation. Therefore it can be easily understood that the scalar product of the Burgers vector for climb dislocation, $\mathbf{b}_c$, and the vector for slip dislocation $\mathbf{b}_s$ is zero, i.e. the two vectors cross each other at right-angles. In order to consider the climbing motion of dislocations from the viewpoint of differential geometry, the large and small domains of an actual crystal are shown in Fig. 5.7(a) [6]. In the slip dislocation in the large domains, as shown in Fig. 5.7(b), a discrepancy of the two ends of the circuit encircling the large domain will result when both circuits are cut along the *y*-axis as shown in Fig. 5.7(c). Therefore the presence of climb dislocations in the dislocated crystal is represented by the concept of the curvature tensor. Figures 5.7(d) and (e) show the conceptual model of the torsion tensor only for those slip dislocations which exist in both domains.

5.1.2 Definition of Misorientation

The misorientation of metallic crystals will be analyzed by using Laue X-

IMPERFECT CRYSTALS AND MISORIENTATION OF METALS 191

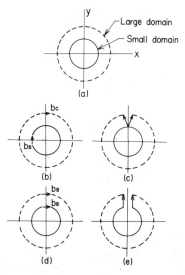

Fig. 5.7. Geometrical representation of dislocation climb motion. (Reprinted from ref. 6 by permission of Soc. Mater. Sci., Japan)

ray diffraction as follows. The X-ray condition is generally expressed by the equation

$$(\mathbf{S} - \mathbf{S}_0)/\lambda = \mathbf{H}_{hkl} = h\mathbf{b}^1 + k\mathbf{b}^2 + l\mathbf{b}^3 \tag{5.10}$$

where $\mathbf{S}_0$ is the unit vector of the incident X-ray beam, $\mathbf{S}$ that of the diffraction beam, λ the wavelength of the incident beam, $\mathbf{H}_{hkl}$ the normal vector to the (hkl) plane in the crystal and $\mathbf{b}^i$ ($i = 1,2,3$) the reciprocal vectors (see Fig. 5.8). Assuming no change in the incident beam, $\mathbf{S}_0$, the change in $\mathbf{S}$ is given as

$$\Delta \mathbf{S}/\lambda = \Delta \mathbf{H} + (\Delta \lambda/\lambda)\mathbf{H} \tag{5.11}$$

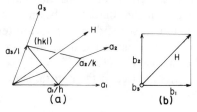

Fig. 5.8. (a) Geometrical relation and (b) reciprocal space, in diffraction plane.

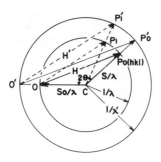

Fig. 5.9. Graphical representation of the change of the **H** vector using the Ewald reflection sphere.

Therefore $\Delta \mathbf{S}$ is represented by both the change in **H** itself, i.e. $\Delta \mathbf{H}$, and the variation of the wavelength, $\Delta \lambda$.

These relationships can be understood by using Ewald's reflection sphere shown in Fig. 5.9. Vector $\mathbf{S}_0/\lambda$ drawn $1/\lambda$ in length and parallel to the incident beam terminates at the starting point of the reciprocal vector. The diffraction condition on the (hkl) plane means that the reciprocal lattice point touches the sphere's surface (point P_0 in Fig. 5.9). If λ changes to λ' ($\lambda' = \lambda + \Delta\lambda$), the diffraction beam vector changes to $\mathbf{S}/\lambda'$ and is represented by $\overline{C\bar{P}'_0}$. Since $\overline{OP_0}$ and $\overline{O'P'_0}$ parallel each other, only the magnitude of **H**, represented by $O'P'_0$, changes. On the other hand, if the magnitude $\overline{CP_0}$ of diffraction vector $\mathbf{S}/\lambda$ changes to $\overline{CP_1}$ due to some infinitesimal change in Bragg angle θ, the vector **H** changes to $\overline{OP_1}$ and both direction and magnitude change. Thus, in eqn (5.11) the first term represents the change in the direction of vector **H** and the second term the change in magnitude. Therefore the variation of diffraction asterism, i.e. the misorientation, obtained by using the Laue X-ray method is a significant factor in the continuous translation of orientation on the diffraction plane. In other words, the misorientation is a reflection of the change in both the direction and the magnitude of vector **H**. Additionally, the change in the diffraction beam can be represented in terms of both the torsion tensor and the curvature tensor.

Next let us consider the geometrical change in crystal lattice in conjunction with the change in vector **H** [5, 6]. By setting the deformed crystal's reciprocal vectors $\mathbf{b}^i$ free in the natural state, the translation of the vector is given as

$$\Delta b^i = 2 S^{\cdot \cdot i}_{kl} du^k du^l \tag{5.12}$$

and the change in the vector itself is expressed by

$$\Delta \mathbf{b}^i = R_{jkl}^{\cdots i} \mathbf{b}^j du^k du^l \tag{5.13}$$

where $du^k du^l$ is the infinitesimal area of non-Riemannian space with Euclidean connection. In general, $S_{kl}^{\cdot\cdot i}$ has 27 components and $R_{jkl}^{\cdots i}$ has 81 components. Considering two-dimensional space for $k=1$ and $l=2$, eqns (5.12) and (5.13) are written

$$\Delta b^i = 2S_{12}^{\cdot\cdot i} du^1 du^2 \tag{5.12'}$$

$$\Delta \mathbf{b}^i = R_{j12}^{\cdot\cdot i} \mathbf{b}^j du^1 du^2 \tag{5.13'}$$

The change in $\mathbf{H}$ itself, drawn from eqn (5.10), is given as

$$\Delta \mathbf{H} = h\Delta \mathbf{b}^1 + k\Delta \mathbf{b}^2 + l\Delta \mathbf{b}^3 \tag{5.14}$$

By applying eqn (5.13') to eqn (5.14) the following equation is obtained:

$$\Delta \mathbf{H} = h(R_{112}^{\cdots 1}\mathbf{b}^1 + R_{212}^{\cdots 1}\mathbf{b}^2 + R_{312}^{\cdots 1}\mathbf{b}^3) du^1 du^2$$
$$+ k(R_{112}^{\cdots 2}\mathbf{b}^1 + R_{212}^{\cdots 2}\mathbf{b}^2 + R_{312}^{\cdots 2}\mathbf{b}^3) du^1 du^2$$
$$+ l(R_{112}^{\cdots 3}\mathbf{b}^1 + R_{212}^{\cdots 3}\mathbf{b}^2 + R_{312}^{\cdots 3}\mathbf{b}^3) du^1 du^2 \tag{5.15}$$

Multiplying both sides of eqn (4.15) by unit vectors a_i ($i=1, 2, 3$), the changes in vector $\mathbf{H}$ itself in various directions, i.e. $(\Delta H)_1$, $(\Delta H)_2$ and $(\Delta H)_3$, are written

$$(\Delta H)_1 = (hR_{112}^{\cdots 1} + kR_{112}^{\cdots 2} + lR_{112}^{\cdots 3}) dh^1 du^2 \tag{5.16a}$$

$$(\Delta H)_2 = (hR_{212}^{\cdots 1} + kR_{212}^{\cdots 2} + lR_{212}^{\cdots 3}) du^1 du^2 \tag{5.16b}$$

$$(\Delta H)_3 = (hR_{312}^{\cdots 1} + kR_{312}^{\cdots 2} + lR_{312}^{\cdots 3}) du^1 du^2 \tag{5.16c}$$

The change in the incident X-ray beam, brought about by the change in magnitude of $\mathbf{H}$ in the second term of eqn (5.11), can be written as the translation of the original point of the reciprocal vector:

$$(\Delta\lambda/\lambda)\mathbf{H} = \Delta b^i a_i \mathbf{H} \tag{5.17}$$

The change in the magnitude of $\mathbf{H}$ is given by $b^i a_i$. Therefore the change in the magnitude of $\mathbf{H}$, i.e. $(\Delta\lambda/\lambda)H$, is obtained from eqns (5.12') and (5.17) as

$$(\Delta\lambda/\lambda)H = 2(S_{12}^{\cdot\cdot i} a_i)(h\mathbf{b}^1 + k\mathbf{b}^2 + l\mathbf{b}^3) du^1 du^2 \tag{5.18}$$

Furthermore, employing the summation rule for i, $(\Delta\lambda/\lambda)H$ for an arbitrary direction is given as

$$(\Delta\lambda/\lambda)H = 2(hS^{\cdot\cdot1}_{12} + kS^{\cdot\cdot2}_{12} + lS^{\cdot\cdot3}_{12})\,du^1\,du^2 \qquad (5.19)$$

Thus, from eqns (5.16) and (5.19) the components of the change in the incident beam, $\Delta S/\lambda$, are written as

$$(\Delta S/\lambda)_1 = (hR^{\cdot\cdot\cdot1}_{1\,1\,2} + kR^{\cdot\cdot\cdot2}_{1\,1\,2} + lR^{\cdot\cdot\cdot3}_{1\,1\,2})\,du^1\,du^2$$
$$\quad + 2(hS^{\cdot\cdot1}_{12} + kS^{\cdot\cdot2}_{12} + lS^{\cdot\cdot3}_{12})\,du^1\,du^2 \qquad (5.20a)$$

$$(\Delta S/\lambda)_2 = (hR^{\cdot\cdot\cdot1}_{2\,1\,2} + kR^{\cdot\cdot\cdot2}_{2\,1\,2} + lR^{\cdot\cdot\cdot3}_{2\,1\,2})\,du^1\,du^2$$
$$\quad + 2(hS^{\cdot\cdot1}_{12} + kS^{\cdot\cdot2}_{12} + lS^{\cdot\cdot3}_{12})\,du^1\,du^2 \qquad (5.20b)$$

$$(\Delta S/\lambda)_3 = (hR^{\cdot\cdot\cdot1}_{3\,1\,2} + kR^{\cdot\cdot\cdot2}_{3\,1\,2} + lR^{\cdot\cdot\cdot3}_{3\,1\,2})\,du^1\,du^2$$
$$\quad + 2(hS^{\cdot\cdot1}_{12} + kS^{\cdot\cdot2}_{12} + lS^{\cdot\cdot3}_{12})\,du^1\,du^2 \qquad (5.20c)$$

Consequently the arbitrary changes in the diffraction angle of the incident beam are fully represented by both the torsion and curvature tensors. The physics of misorientation has been well researched by Frank [9] *et al.* Let us compare the above observations with his findings. Frank gave the following relation determining the resultant Burgers vectors for a set of dislocations:

$$\beta = \{(\mathbf{r} \times \mathbf{u})/|\mathbf{r} \times \mathbf{u}|\}2\sin(\chi/2) \qquad (5.21)$$

where $\beta = \mathbf{r}'' - \mathbf{r}'$, and $\mathbf{u}$ is the rotation axis by which β occurs as shown in Fig. 5.10. The two components β_{twist} and β_{tilt} were given as [9]

$$\beta_{\text{twist}} = \overrightarrow{AC} + \overrightarrow{DB} = -2\mathbf{N}\sin(\phi/2) \qquad (5.22)$$

$$\beta_{\text{tilt}} = \overrightarrow{CD} = 2\mathbf{n}\sin(\omega/2) \qquad (5.23)$$

where $\sin(\phi/2) = \sin(\chi/2)/\sqrt{1 + \tan^2\alpha\cos^2(\chi/2)}$ and $\sin(\omega/2) = \sin\alpha\sin(\chi/2)$. If $\alpha = 0$, then $\mathbf{U} = \mathbf{n}$, $\phi = \chi$, $\omega = 0$ and the grain boundary becomes a pure twist. On the other hand, if $\alpha = \pi/2$, then $\omega = \chi$, $\phi = 0$ and the boundary does a pure tilt. Amelinckz and Dekeyser [10] referred to a set of edge dislocations as a pure tilt boundary and to that of screw dislocations as a pure twist boundary. On the other hand, Read [11] concluded that the dislocation network of the grain boundary is composed of both types of dislocation.

Taking a_1, a_2 and $\mathbf{r}$ as the coordinate axes, it is believed that the misorientation $(\overrightarrow{AC} + \overrightarrow{DB})$ in eqn (5.22) signifies the change in $\mathbf{r}$ to the a_2 direction and $\overrightarrow{CD}$ in eqn (5.23) is the change for the a_1 direction. Since

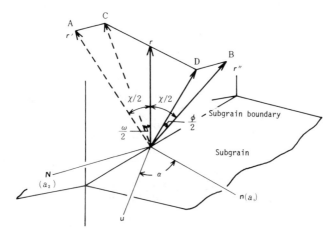

Fig. 5.10. Definition of misorientation by Frank, and the graphical representation of both tilt and twist.

the vector **r** is equal to **H**, eqns (5.22) and (5.23) are written as

$$\beta_{twist} = \overrightarrow{AC} + \overrightarrow{DB} = a_2(\Delta S/\lambda)_2 \tag{5.24}$$

$$\beta_{tilt} = \overrightarrow{CD} = a_1(\Delta S/\lambda)_1 \tag{5.25}$$

Therefore eqns (5.24) and (5.25) correspond to eqns (5.20b) and (5.20a) respectively, and β_{twist} and β_{tilt} contain both edge and screw dislocations. Furthermore, the misorientation represents not only both dislocations but also the quantities containing the crystalline defect in a global region, and we conclude that the physical picture drawn by Frank et al. represents but a part of the lattice defects in an actual crystal.

5.1.3 Analysis of Experimental Method [5,6]

Let us consider the coordinate system shown in Fig. 5.11 where the 1–0–2 plane is a test specimen surface and the 3-direction is normal to the surface. The change in the **H** vector, $\Delta H_{\phi,\psi}(\overline{OP})$, is written as

$$\Delta H_{\phi,\psi} = \gamma_1^2 \Delta H_1 + \gamma_2^2 \Delta H_2 + \gamma_3^2 \Delta H_3$$
$$= \sin^2 \psi(\Delta H_1 \cos^2 \phi + \Delta H_2 \sin^2 \phi) + \Delta H_3 \cos^2 \psi \tag{5.26}$$

where γ_1, γ_2 and γ_3 are the direction cosines, ψ is the angle between the 3-direction and the vector $H_{\phi,\psi}$ and ϕ is the angle between the 1-direction

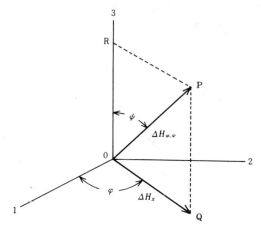

Fig. 5.11. Definition of coordinate system on the test specimen and direction of **H** vector change.

and the component $H_{\phi,\psi}$ on the surface of the specimen 1–0–2 plane, namely $\Delta H_x (=\overline{OQ})$. Applying $\Delta H_x = \Delta H_1 \cos^2 \phi + \Delta H_2 \sin^2 \phi$ to eqn (5.26), the following relation is obtained:

$$H_{\phi,\psi} = (\Delta H_x - \Delta H_3) \sin^2 \psi + \Delta H_3 \qquad (5.27)$$

If ϕ remains constant, the relation between $\Delta H_{\phi,\psi}$ and $\sin^2 \psi$ is represented by a straight line as shown in Fig. 5.12. The gradient of this straight line, M, is written as

$$M = \partial(\Delta H_{\phi,\psi})/\partial \sin 2\psi = \Delta H_x - \Delta H_3 \qquad (5.28)$$

and the resultant change in $\Delta \mathbf{H}$, i.e. ΔH_x, is

$$\Delta H_x = M + \Delta H_3 \qquad (5.29)$$

If $\sin^2 \psi = 0$ in eqn (5.27), then

$$\Delta H_{\phi,0} = \Delta H_3 \qquad (5.30)$$

and ΔH_3 is obtained from an intersecting point on the $\Delta H_{\phi,\psi}$ axis in Fig. 5.12. Moreover, if $\Delta H_{\phi,\psi} = 0$ in eqn (5.27), then

$$\sin^2 \psi = -\Delta H_3/(\Delta H_x - \Delta H_3) \qquad (5.31)$$

and we can find the direction where there is no change in the **H** vector. It is, however, believed that the translation in the **H** vector occurs even if

IMPERFECT CRYSTALS AND MISORIENTATION OF METALS 197

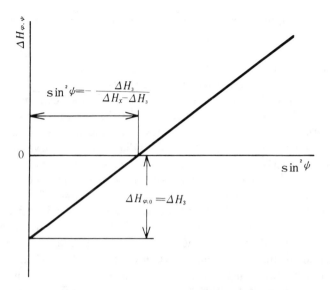

Fig. 5.12. Schematic representation of $H_{\phi,\psi}$ versus $\sin^2\psi$ diagram.

there is no change in **H** itself. Denoting the change in the **H** vector as H_{x0}, this is written as

$$\Delta H_{x0} = \Delta H_3 - (\Delta H_3/\sin^2\psi) \tag{5.32}$$

from eqn (5.28), and since $\Delta H_3/\sin^2\psi = M$,

$$\Delta H_{x0} = \Delta H_3 - M \tag{5.32'}$$

Therefore it is necessary to separate the change in the **H** vector itself from the translation, because ΔH_x contains both. So, from the difference between eqn (5.29) and eqn (5.32'), the following relation is obtained:

$$\Delta H_x = 2M \tag{5.33}$$

Thus we can distinguish the change in the **H** vector's position, shown in eqn (5.32'), from the change in the **H** vector itself, shown in eqn (5.33). For instance, from eqns (5.16), (5.19), (5.32') and (5.33), the change in the **H** vector in the 1-direction is written as

$$2M = (hR_{112}^{\cdots 1} + kR_{112}^{\cdots 2} + lR_{112}^{\cdots 3})du^1\,du^2 \tag{5.34}$$

$$\Delta H_3 - M = 2(hS_{12}^{\cdots 1} + kS_{12}^{\cdots 2} + lS_{12}^{\cdots 3})du^1\,du^2 \tag{5.35}$$

Therefore the components of the curvature tensor or the torsion tensor can be determined quantitatively from the simultaneous algebraic equations (5.34) and (5.35) with the three unknown quantities of these components relevant to the three diffraction planes of (hkl). In other words, it should be noted that the misorientation of plastically deformed metals obtained by using the Laue X-ray method can be perfectly represented by both the torsion tensor and the curvature tensor which characterize a non-Riemannian space with a Euclidean connection.

The following special cases are considered. If the curvature tensor $R_{112}^{\cdot\cdot\cdot i}$ in eqn (5.34) is zero, the gradient M becomes zero and the straight line in Fig. 5.12 parallels the $\sin^2\psi$ axis. This corresponds to the so-called *space with distant parallelism* proposed by Bilby et al. [1] where the orientation of the pieces in the natural state is fixed. This is convenient in the case of fixed principal axes of a crystal in crystallography. Also, if the torsion tensor $S_{12}^{\cdot\cdot i}$ is zero, then $\varDelta H_3 = M$ from eqn (5.35) and this corresponds to the Riemannian space in which only the curvature tensor exists. Furthermore, in the case where both the torsion tensor and the curvature tensor are simultaneously zero, $H_3 = M = 0$ from eqns (5.34) and (5.35), corresponding to the Euclidean space where the linear transformation from the actual state to the natural state always remains valid, i.e. an elastic body.

Based on the present analytical study of the misorientation of polycrystalline metal relevant to the geometrical aspect of the continuously dislocated continuum, the following conclusions were made.

(1) We assumed that the material space for plastically deformed metals is represented by a non-Riemannian space with a Euclidean connection, which has both the torsion tensor and the Riemann–Christoffel curvature tensor.

(2) Analysis based on differential geometry determined that the density of dislocations corresponds to the concept of the torsion tensor and that the existence of atomic vacancies, interstitial atoms and jogs in the crystal results in the concept of the curvature tensor.

(3) We found that the two components β_{twist} and β_{tilt} proposed by Frank et al. contain both edge and screw dislocations.

(4) We discovered that components of the curvature tensor or the torsion tensor can be determined quantitatively from the simultaneous algebraic equations (5.34) and (5.35) in which the three unknown quantities of these components are relevant to the three diffraction planes (hkl).

5.2 MICROSTRUCTURAL STUDY OF STATIC FLOW STRESS IN METALS BY MEANS OF CDC

Figure 5.13 [5, 6] shows the change in the misorientation, β, during the static tension test on a grain with the crystal orientation marked A in Fig. 5.14. A coarse-grained sheet specimen (gauge length 50 mm, width 15 mm, thickness 1 mm) of pure polycrystalline aluminum was used in this test. The average grain size in the specimen was about 8 mm. The misorientation, β, was measured using the Laue back-reflection X-ray method shown in Fig. 5.15 and β was calculated using the following equation:

$$\beta = [\cos^2\theta + \{(\sin^2 4\theta)/(16\tan^2\omega)\}]^{1/2}$$
$$\{\sin\omega/(R_0|\tan 2\theta|)\}S(\omega) \qquad (5.36)$$

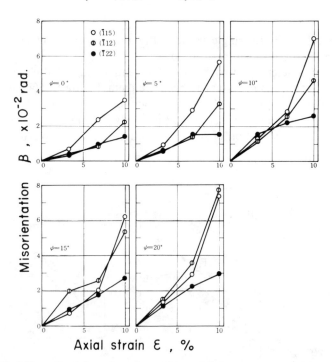

Fig. 5.13. Effect of axial strain, ε, in static tension test at room temperature, on the misorientation, β, of 99·45% pure aluminum with crystal orientation $\mu_G = 0\cdot493$. (Reprinted from ref. 5 by permission of Japan Soc. Mech. Engrs)

200 PLASTICITY AND HIGH TEMPERATURE STRENGTH OF MATERIALS

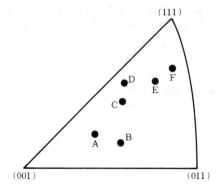

Fig. 5.14. Initial orientation of grains in coarse-grained 99·45% pure aluminum in the basic triangle.

In Fig. 5.15, ψ is the angle between the direction of an incident X-ray beam and the specimen surface normal.

Since β is determined by both the tangential and radial elongation of the Laue spot, it is necessary to separate them. Figure 5.16(a) shows the stretch of the spot on the film in the longest direction; $\overline{BA}$ and $\overline{SB}$ are the radial and tangential directions, respectively. Figure 5.16(b) shows the schematic representation of stereo-projection of the spot on Wulff's net;

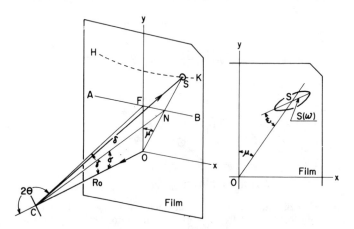

Fig. 5.15. Schematic representation in diffraction spots of the Laue back-reflection X-ray method, and diffraction of misorientation.

IMPERFECT CRYSTALS AND MISORIENTATION OF METALS 201

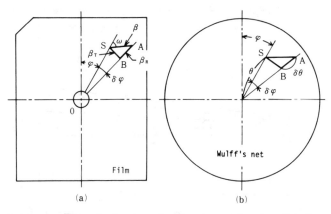

Fig. 5.16. Measurement of misorientation in tangential and radial directions on Wulff's net.

the tangential stretch $\overline{SB}$ is given as

$$\overline{SB} = \delta\phi \cos\theta = \sin\omega \cos\theta \cdot (\overline{SA}/\overline{OS}) \tag{5.37}$$

where $\delta\phi = \overline{SA}\sin\omega/\overline{OS}$. In the same way, the radial stretch $\overline{BA}$ is given as

$$\overline{BA} = \delta\theta = (\sin 4\theta)/(4\tan\omega)\sin\omega$$
$$\times (\overline{SA}/\overline{OS}) = (1/4)\sin 4\theta \cos\omega \cdot (\overline{SA}/\overline{OS}) \tag{5.38}$$

where $\delta\phi/\delta\theta = 4\tan\omega/\sin 4\theta$ is used. Denoting $\overline{SA} = S(\omega)$, $\overline{SB} = \beta_T$ and $\overline{BA} = \beta_R$, eqns (5.37) and (5.38) respectively are written

$$\beta_T = \sin\omega \cos\theta \cdot S(\omega)/\overline{OS} \tag{5.39}$$

$$\beta_R = (1/4)\cos\omega \sin 4\theta \cdot S(\omega)/\overline{OS} \tag{5.40}$$

Figure 5.17 [5, 6] shows the β versus $\sin^2\psi$ diagram obtained from the test data in Fig. 5.16 where the straight lines were drawn by using the least-squares method. In this figure, ψ is the angle between the direction of an incident X-ray beam and the specimen surface normal. It was deduced from the figure that the gradient angle of the straight line is acute in the early stage of deformation and gradually increases as deformation advances. Figures 5.18 and 5.19 [5, 6] show the results calculated using the components $S^{\cdot \cdot i}_{12}$ for the torsion tensor and $R^{\cdot \cdot \cdot i}_{112}$ for the curvature tensor, respectively, based on the data from Fig. 5.13. In these figures, the quantity $S^{\cdot \cdot k}_{\mu\lambda}$ is called the *dislocation density* and $R^{\cdot \cdot \cdot k}_{\nu\mu\lambda}$

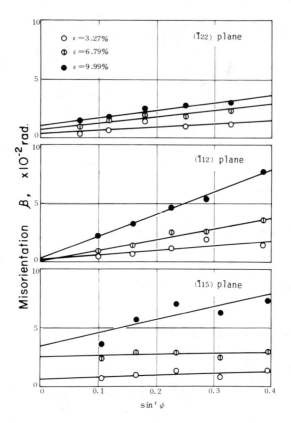

Fig. 5.17. Experimental relation between misorientation, β, and $\sin^2\psi$, for respective diffraction planes of 99·45% pure aluminum, in static tension test at room temperature. (Reprinted from ref. 5 by permission of Japan Soc. Mech. Engrs)

the *imperfection density*. $S^{\cdot\cdot 1}_{12}$ represents density of edge dislocation and $R^{\cdot\cdot\cdot 1}_{112}$ the component of the density of the lattice imperfection closely allied with the edge dislocation. The components $S^{\cdot\cdot i}_{12}$ and $R^{\cdot\cdot\cdot i}_{112}$, for $i = 2$ and 3, represent screw dislocation density and density of the lattice imperfection most closely allied with the screw dislocations. Figure 5.20 [6] shows the experimental relation of the dislocation density and the imperfection density to the Schmid factor μ_G of grains. From this figure we found that these densities are highly dependent on the ease of slip in the crystal grain.

Figures 5.21 and 5.22 [6] show the relationships, based on experimen-

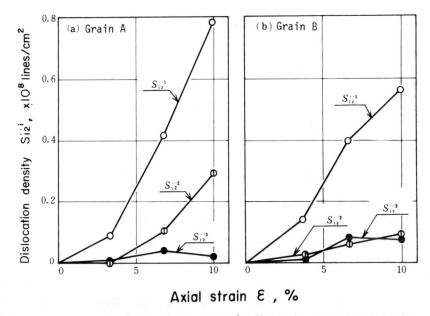

Fig. 5.18. Change in dislocation density, $S_{12}^{\cdot\cdot i}$, with axial strain, ε, in static tension test at room temperature, for 99.45% pure aluminum having crystal orientations A and B (see Fig. 5.14). (Reprinted from ref. 5 by permission of Japan Soc. Mech. Engrs)

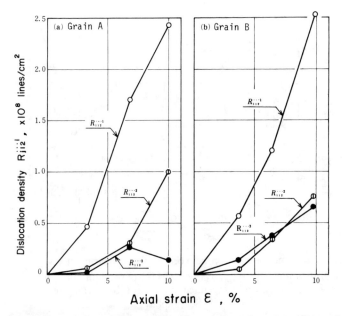

Fig. 5.19. Change in imperfection density with axial strain, ε, in static tension test at room temperature, for 99.45% pure aluminum having crystal orientations A and B (see Fig. 5.14). (Reprinted from ref. 5 by permission of Japan Soc. Mech. Engrs)

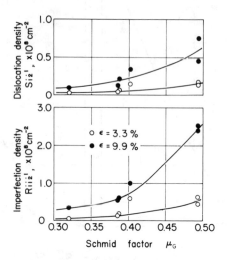

Fig. 5.20. Experimental relationship between dislocation density, $S_{12}^{\cdot\cdot 1}$, and imperfection density, $R_{112}^{\cdot\cdot 1}$, and crystal orientation, μ_G, of 99·45% pure aluminum, in static tension test at room temperature. (Reprinted from ref. 6 by permission of Soc. Mater. Sci., Japan)

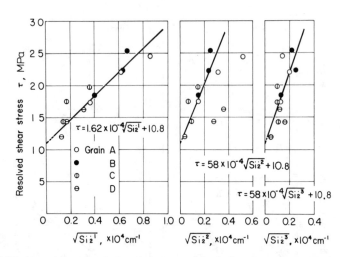

Fig. 5.21. Experimental relation between resolved shearing stress, τ, and square-root of dislocation density, $\sqrt{S_{12}^{\cdot\cdot i}}$, of 99·45% pure aluminum, in static tension test at room temperature. (Reprinted from ref. 6 by permission of Soc. Mater. Sci., Japan)

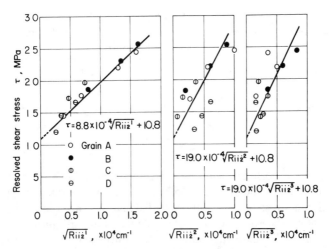

Fig. 5.22. Experimental relation between resolved shearing stress, τ, and square-root of imperfection density, $\sqrt{R_{112}^{\cdots i}}$, of 99·45% pure aluminum in static tension test at room temperature. (Reprinted from ref. 6 by permission of Soc. Mater. Sci., Japan)

tal results, between resolved shear stress and the square-roots of dislocation and imperfection densities, respectively. In this case the resolved shear stress, τ, in the grain is determined by the following equation:

$$\tau = F(\cos\chi_0/A_0)\{\sqrt{(1+\varepsilon)^2 - \sin^2\lambda_0}/(1+\varepsilon)\} \tag{5.41}$$

where A_0 is the cross-sectional area of the test specimen before deformation, χ_0 and λ_0 are the angles between the direction of tension and the slip plane normal and the slip direction, respectively, F is the axial load and ε is the tensile strain. These figures show the whole range of work-hardening. This relation is expressed by

$$\tau = \alpha G b \sqrt{S_{12}^{\cdots i}} + \tau_0 \tag{5.42}$$

$$\tau = \alpha' G b \sqrt{R_{112}^{\cdots i}} + \tau_0 \tag{5.43}$$

where G represents the shear modulus, b the Burgers vector, and α and α' the constants. Further, the flow stress and both the dislocation and the imperfection densities are thought to be mutually related, i.e. $\tau = \alpha G b \sqrt{S_{12}^{\cdots i}} + \alpha' G b \sqrt{R_{112}^{\cdots i}} + \tau_0$. Thus we can discuss the influence of dislocation and lattice imperfection densities on flow stress independently as shown in eqns (5.42) and (5.43). The resolved shear stress

Table 5.1
Value of α and α' for Pure Commercial Aluminum Determined from Eqns (5.42) and (5.43), respectively

Component	α	Component	α'
$S_{12}^{\cdot\cdot 1}$	0·189	$R_{112}^{\cdot\cdot\cdot 1}$	0·108
$S_{12}^{\cdot\cdot 2}$	0·677	$R_{112}^{\cdot\cdot\cdot 2}$	0·222
$S_{12}^{\cdot\cdot 3}$	0·677	$R_{112}^{\cdot\cdot\cdot 3}$	0·222

τ_0 at zero for both dislocation and imperfection densities is 1.10 kgf/mm² (11 MPa), as shown in Figs 5.21 and 5.22. This value is regarded as the frictional stress for the motion of dislocations in the material. Table 5.1 [6] shows the values of α and α' calculated according to eqns (5.42) and (5.43). In this case the values of the Burgers vector and the shear modulus were 2.858×10^{-7} mm and 26 GPa (2.7×10^3 kgf/mm²), respectively. We found from this table that there are various values of α and α' depending on the types of torsion and curvature tensor components. It is therefore clear that the flow stress may be governed not only by the dislocation density but also by the imperfection density, and the influence of each of these densities on the flow stress of the metal is different.

A great deal of theoretical research concerning the above value of α has been done, choosing various mechanisms as the cause of the work-hardening of the metals. For instance, Li [12] proposed $\alpha = 0.375$ from the viewpoint that subgrains break and the dislocations disperse, while Seeger [13] suggested $\alpha = 0.2$ for fcc metals based on the assumption of randomly distributed dislocations. In addition, actual measurement of α has been made, with the typical results for polycrystalline aluminum summarized in Table 5.2 [14].

Assume that the relation between shear strain γ and both the dislocation and imperfection densities is defined by the following equation:

$$\gamma = (AR + C_1) + (BS + C_2) \tag{5.44}$$

In this equation, R represents the imperfection density, S the dislocation density and A, B, C_1, C_2 are material constants. Differentiating eqns (5.42), (5.43) and (5.44), the work-hardening rate is obtained as follows:

$$\theta = d\tau/d\gamma = Gb/[2\{(A/\alpha')\sqrt{R} + (B/\alpha)\sqrt{S}\}] \tag{5.45}$$

Table 5.2
Measurement of α in Polycrystalline Aluminum (Reprinted from ref. 14 by permission of BAIFUKAN).

Metal	Dislocation density (cm^{-2})	α	Measurement method	Ref.
99.9% Al	$7.3 \times 10^8 - 9.7 \times 10^{10}$	0.51	X-ray diffraction	15
99.99% Al	$2.6 \times 10^6 - 25 \times 10^8$	0.36	Etching and X-ray diffraction	16
99.99% Al	$(4-25) \times 10^4$	0.54	Etching	17
99.99% Al	$2.7 \times 10^7 - 4.2 \times 10^8$	0.1–0.3	Etching and X-ray diffraction	18

To simplify the discussion, assuming that only the dislocation density is connected with work-hardening of the metal, the following equation may be introduced from eqn (5.45):

$$\theta_1 = (d\tau/d\gamma)_1 = (\alpha G b/2B)S^{-1/2} \qquad (5.46)$$

In the same way, if the work-hardening rate of the metal is represented by only the imperfection density, the following equation is obtained:

$$\theta_2 = (d\tau/d\gamma)_2 = (\alpha' G b/2A)R^{-1/2} \qquad (5.47)$$

Figures 5.23 and 5.24 [6] show the experimental relationships between the resolved shear strain γ and the dislocation and imperfection densities, respectively. In these figures, experimental plots were adjusted to the same orientation factor, $\mu_G = 0.493$, marked A and B in Fig. 5.14. Within the limits of the experiments we found that throughout the entire work-hardening process the densities of the dislocations and imperfections increased proportionally with resolved shear strain. Figure 5.25 [6] shows the calculated variations of the change in work-hardening rate with the advance of resolved shear strain. (These calculations were made using eqns (5.46) and (5.47) on the basis of the slope of the straight lines A and B in Figs 5.22 and 5.23.) In this figure, the experimental plots indicate the results in static tension as shown in Fig. 5.26 [6] for a crystal with a Schmid factor $\mu_G = 0.493$. We found from this figure that the work-hardening rate obtained using the component of the torsion tensor is in agreement with the experimental result of the second stage of work-hardening. In the third stage of deformation, the work-hardening rate is determined by the component of the curvature tensor. Therefore

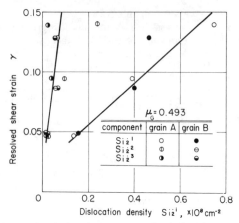

Fig. 5.23. Experimental relation between resolved shearing strain, γ, and dislocation density, $S_{12}^{\cdot\cdot i}$, of 99·45% pure aluminum, in static tension test at room temperature. (Reprinted from ref. 6 by permission of Soc. Mater. Sci., Japan)

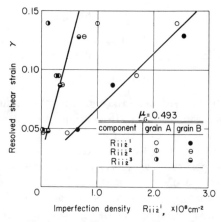

Fig. 5.24. Experimental relation between resolved shearing strain, γ, and imperfection density, $R_{112}^{\cdot\cdot\cdot i}$, of 99·45% pure aluminum, in static tension test at room temperature. (Reprinted from ref. 6 by permission of Soc. Mater. Sci., Japan)

we concluded that the work-hardening mechanism in the third stage of deformation of metal is explained by the Riemann–Christoffel curvature tensor which includes the nonconservative motion of dislocations and the distribution of lattice defects in crystals. However, it may also be practical to discuss the expression with the torsion tensor of the work-hardening mechanism in the early stage of deformation.

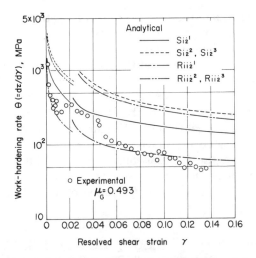

Fig. 5.25. Experimental and analytical relations between work-hardening rate, θ, and resolved shearing strain, γ, of 99·45% pure aluminum, in static tension test at room temperature. Fine lines indicate the variation of work-hardening rate, when constants C_1 and C_2 are zero in eqn (5.44). (Reprinted from ref. 6 by permission of Soc. Mater. Sci., Japan)

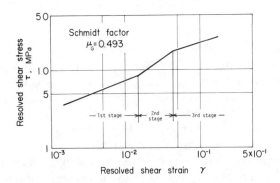

Fig. 5.26. Experimental result of resolved shearing stress, τ, versus resolved shearing strain, γ, of 99·45% pure aluminum, in static tension test at room temperature. (Reprinted from ref. 6 by permission of Soc. Mater. Sci., Japan)

From the X-ray study of flow stress in coarse-grained pure polycrystalline aluminum under static tension, the following conclusions were obtained.

(1) We discovered that the resolved shear stress in the metal is a linear

function of the square-root of either the dislocation density, which is represented by the components of the torsion tensor, or the imperfection density represented by the components of a Riemann–Christoffel tensor.

(2) Furthermore, we found that the proportional constants α and α' in these equations have various values depending on the kind of lattice defects. It was also found from an examination of the work-hardening rate of metal in connection with the torsion and curvature tensors that the work-hardening mechanism in the latter stages of deformation must be explained by the curvature tensor.

(3) We pointed out that misorientation is a useful tool for the microscopic study of the mechanism of work-hardening in metals and that it is also important to consider the contribution of the curvature tensor.

5.3 MICROSTRUCTURAL STUDY OF THE CYCLIC CONSTITUTIVE RELATION IN METALS BY MEANS OF CDC

5.3.1 Cyclic Constitutive Relation and Misorientation

In this section we will elucidate the mechanism of plastic deformation under conditions of cyclic stress from the microstructural change in coarse-grained pure polycrystalline aluminum measured by the Laue X-ray method.

Figure 5.27 [19] shows the hysteresis loop in one cycle and the change with the advancing cycle in a coarse-grained pure aluminum sheet specimen (length 120 mm, width 20 mm, thickness 1·5 mm) subjected to deflection controlled bending cycling. As seen from the figure, a drastic decrease of yield stress in reversed loading, i.e. the so-called Bauschinger effect, was observed. Figure 5.28 [19] shows the change in the intensity of the Bauschinger effect $n = |M_R/M_F|$, where M_F is the maximum bending moment at the tension side and M_R is the bending moment of departure from the elastic line in reversed loading. These figures show that there is a remarkable amount of cyclic hardening in an early stage of the cycle, and furthermore that the intensity of the Bauschinger effect decreases as the cycle advances. Therefore we can note that there is a close relation between cyclic work-hardening and the Bauschinger effect, and also that the contribution of the Bauschinger effect to cyclic work-hardening decreases with the advance of the cycle. This correlates well with the strain history for the Bauschinger effect [20]. Furthermore, the change in the Bauschinger effect concurrent with the advancing cycle is similar to the change for a single crystal. Buckley and Entwistle pro-

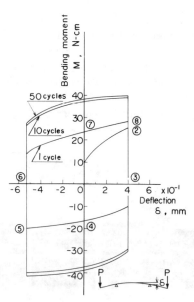

Fig. 5.27. Change in bending moment/deflection curves of commercial pure aluminum at room temperature, as repeated bending advances. (Reprinted from ref. 19 by permission of Japan Soc. Mech. Engrs)

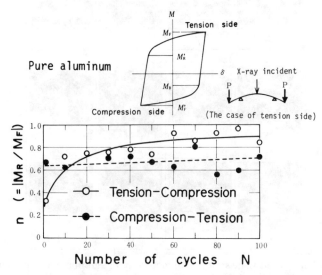

Fig. 5.28. Change in Bauschinger effect with repeated bending. (Reprinted from ref. 19 by permission of Japan Soc. Mech. Engrs)

posed that the Bauschinger effect consists of two stages, one influential and the other inconsequential [21]. In Section 4.2 of Chapter 4, crystal plasticity in an early stage of stress cycling was analyzed using the 'constant stress model', when it behaved like the 'constant strain model' in an advanced cycle. These studies show that early in the cycle the ease with which slip occurs and the intensity of the Bauschinger effect are great, while later, in the advanced stages of the cycle, due to complexities

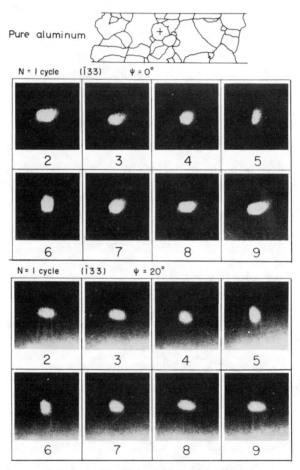

Fig. 5.29. Changes in Laue spot of commercial pure aluminum at room temperature, during one cycle of plane bending. +, X-ray. (Reprinted from ref. 19 by permission of Japan Soc. Mech. Engrs)

like cross-slip, this correspondence is not in evidence. A discussion of this microscopic mechanism follows.

Figure 5.29 [19] shows an example of the change in asterism on film obtained using the Laue X-ray method in one cycle shown in Fig. 5.27. In the figure, ψ is the angle between the direction of an incident X-ray beam and the specimen surface normal; the stages 1 to 9 correspond to the marks in Fig. 5.27. We observed that the Laue spot under tension

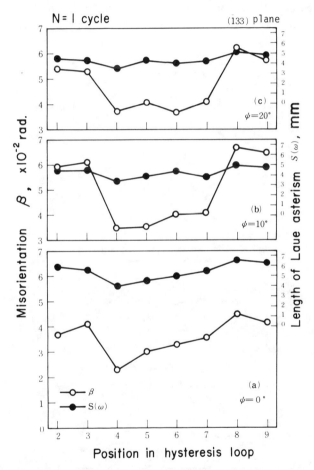

Fig. 5.30. Change in misorientation, β, of commercial pure aluminum at room temperature, during one cycle of plane bending ($N = 1$ cycle). (Reprinted from ref. 19 by permission of Japan Soc. Mech. Engrs)

stretches in a direction different from that when under compression, and the change in shape reverses when loading goes over zero strain (4 or 7) where asterism is nearly circular. Figure 5.30 [19] shows an example of misorientation β determined using eqn (5.36). We found that β changes remarkably from stage to stage in one cycle and that the direction of the change is significant. Figure 5.31 [19] shows separately the tangential misorientation β_T and the radial misorientation β_R calculated using eqns (5.39) and (5.40). The figure shows that under tension β_R is considerable, but decreases with compression after unloading, while β_T behaves in the opposite manner. Figures 5.32 and 5.33 [19] show an example of the

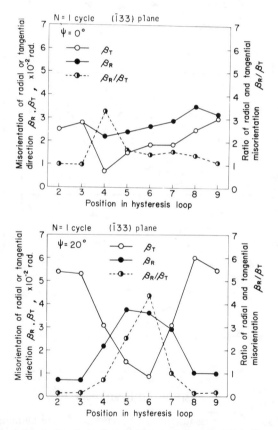

Fig. 5.31. Change of misorientation in tangential and radial directions, β_T and β_R, of commercial pure aluminum at room temperature, during one cycle of plane bending. (Reprinted from ref. 19 by permission of Japan Soc. Mech. Engrs)

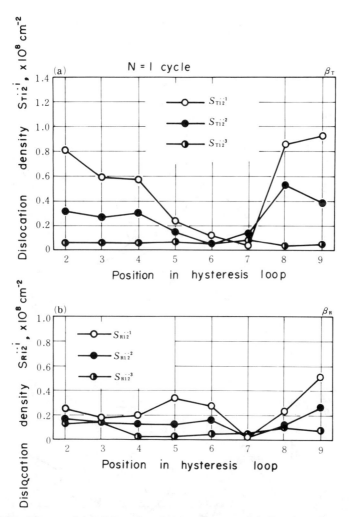

Fig. 5.32. Change of (a) dislocation density in tangential direction, S_T, and (b) dislocation density in radial direction, S_R, of commercial pure aluminum at room temperature, during one cycle of plane bending. (Reprinted from ref. 19 by permission of Japan Soc. Mech. Engrs)

dislocation and imperfection densities for both tangential and radial directions. Figure 5.34 [19] summarizes the structural change in one cycle, where S and R represent the dislocation and imperfection densities, respectively. From these results we can proceed to consider the mechanism of hysteresis.

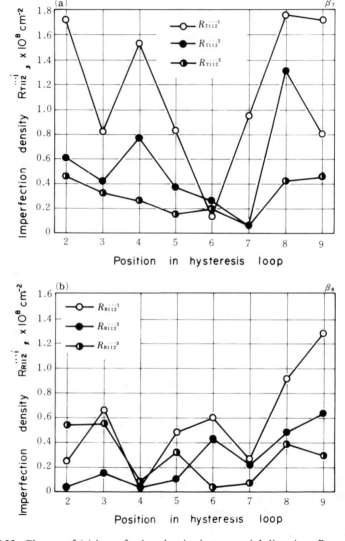

Fig. 5.33. Change of (a) imperfection density in tangential direction, R_T, and (b) imperfection density in radial direction, R_R, of commercial pure aluminum, during one cycle of plane bending. (Reprinted from ref. 19 by permission of Japan Soc. Mech. Engrs)

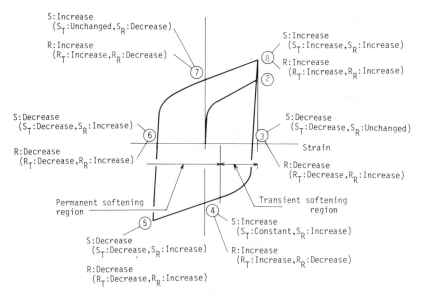

Fig. 5.34. Change of microstructure of commercial pure aluminum at room temperature, during one cycle in low-cycle fatigue. (Reprinted from ref. 19 by permission of Japan Soc. Mech. Engrs)

5.3.2 Cyclic Constitutive Relation and Microstructure

As was previously stated, the material vector in plastically deformed material is represented by eqn (5.6). In this equation the torsion tensor is the tensor of the third order, the curvature tensor that of the fourth order, and the first term on the right-hand side of the equation contains the vector $\mathbf{e}_x$ which indicates the direction of the defects. Amari [22] pointed out that, while the torsion tensor expresses only dislocation density, the curvature tensor denotes slips which are influential upon the plastic flow. Furthermore, the decrease in the Riemann–Christoffel curvature tensor $R_{jkl}^{\cdots i}$ means that the material reaches a state of distant parallelism geometrically. Especially, when $R_{jkl}^{\cdots i} = 0$, it indicates Levi-Civita's parallelism and implies a localized Euclidean space and also uniform distribution of dislocations.

Generally, an early stage of reversed loading after unloading is referred to as a transient softening region [19], which is controlled mainly by reversible dislocations and corresponds to the structural change in the interior of a cell. In the present experiment, a decrease in dislocation density, particularly the tangent dislocation density S_T, was observed in the unloading stage. This resulted from the fact that dislocation pile-up occurs in one direction during the loading stage and then reverses direction and eventually vanishes when loading is released. Furthermore, in stage 4 of reversed loading, S_T remains almost unchanged while the radial dislocation density S_R increases, thus resulting in a slight increase of the overall dislocation density. This region is called the *permanent softening region* [19] and corresponds to the change in the cell structure itself. The phenomenon that S_T remains constant and S_R increases implies the nucleation of dislocations different from those which contributed to the previous plastic deformation. At the same time, the increase in the tangential imperfection density R_T suggests the redistribution of the dislocations which existed previously. In stage 5, where deformation advances in the permanent softening region, S_T decreases and both S_R and the radial imperfection density R_R increase. This results from the fact that both the process of nucleation and that of dislocation pile-up advance while the dislocation shown by S_T vanishes.

That the Bauschinger effect first occurs in the unloading path may be understood from the fact that pile-up dislocations disappear and internal stress in the cell relaxes. It is, however, believed that not all pile-up dislocations vanish; some move in the reverse direction and then, due to a resistant stress field, immobilize in the course of slip. When the load is reversed, dislocation generates anew and moves easily in the direction where resistance to both generation and mobility of dislocations is comparatively small. Since the direction of dislocation mobility changes, it is necessary to consider the deformation mechanism as an individualized phenomenon dependent upon a particularly resistant stress field. Furthermore, the resistance increases with advanced deformation under reversed loading and results in cyclic work-hardening.

Figure 5.35 [19] shows an example of the changes in both dislocation and imperfection densities in one cycle at $N = 100$ cycles. The figure illustrates that the change is small compared with that at $N = 1$ cycle, because the structural change with advanced stress cycling is hindered by the decrease in the intensity of the Bauschinger effect shown in Fig. 5.27.

From the present experiment where we subjected pure polycrystalline aluminum to cyclic bending, the following conclusions were drawn.

(1) The cyclic Bauschinger effect was remarkable in an early stage of strain cycling, but decreased with advanced work-hardening.

(2) A change in the stretching direction of the Laue spot was observed in one cycle. We surmised that this illustrates the direction of structural

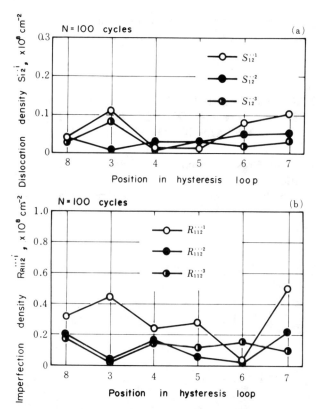

Fig. 5.35. Change in (a) dislocation density, $S_{12}^{\cdots i}$, and (b) imperfection density, $R_{112}^{\cdots i}$, of commercial pure aluminum at room temperature, during one cycle of plane bending ($N=100$ cycles). (Reprinted from ref. 19 by permission of Japan Soc. Mech. Engrs)

change in the metal which correlates with the direction of dislocation mobility in the metal.

(3) The Bauschinger effect was first observed in the unloading path, and it was considered that, with advanced loading in the opposite direction, both translation of pile-up dislocations and annihilation of some other dislocations occur. Furthermore, dislocations regenerate and pile up in the direction where the stress field resistance is comparatively weak.

(4) We also surmised that in an advanced cycle the stress field resistance to dislocation mobility becomes uniformly distributed and

the dislocations immobilize. In consequence, both the intensity of the Bauschinger effect and the number of dislocations which correlate with it decreased with strain cycling.

5.4 MICROSTRUCTURAL STUDY OF THE INFLUENCE OF PRESSURE SOAKING ON PLASTICALLY DEFORMED METALS BY MEANS OF CDC

5.4.1 Effects of Pressure Soaking on Dislocation and Imperfection Densities

The effect of hydrostatic pressure soaking on metal plasticity was studied from the viewpoint of inhomogeneity in both local and global lattice imperfections.

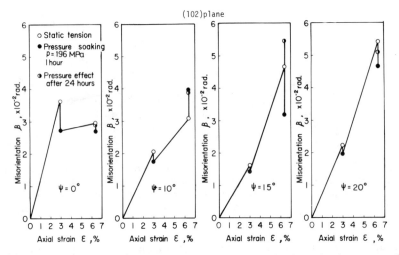

Fig. 5.36. Change in misorientation, β, of 99·5% pure aluminum having a crystal orientation of $\mu_G = 0.4$ with pressure soaking of 196 MPa after prestraining, with increase of axial strain, ε, in the atmosphere. (Reprinted from ref. 23 by permission of Soc. Mater. Sci., Japan)

Figure 5.36 [23] shows the change in misorientation of a coarse-grained 99·5% pure aluminum sheet specimen (gauge length 30 mm, width 8 mm, thickness 1·5 mm) under static tension and pressure soaking of the plastically deformed metal which was performed in pressurized oil

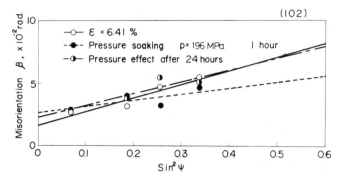

Fig. 5.37. Experimental relation between misorientation, β, and $\sin^2 \psi$, of 99·5% pure aluminum in static tension test in atmosphere, immediately after pressure soaking of 196 MPa and 24 h later, respectively. (Reprinted from ref. 23 by permission of Soc. Mater. Sci., Japan)

at 196 MPa (2000 kgf/cm^2) for 1 h. Measurement of the misorientation was made by using the Laue back-reflection X-ray method and ψ denotes the angle between the direction of the incident X-ray beam and the specimen surface normal. In order to examine the structural recovery for the pressure effect, X-ray measurement was also performed 24 h later after the pressure soaking. From this figure we found that, while misorientation in the plastically deformed aluminum decreased after the pressure soaking, the major part of the decrease recovered within 24 h after the pressure soaking. Figure 5.37 [23] shows the β versus $\sin^2 \psi$ diagram obtained from the data of Fig. 5.36, where the straight lines were drawn by using the least-squares method. In this figure, ψ is the angle between the direction of an incident X-ray beam and the specimen surface normal.

Figure 5.38 [23] shows results calculated using the torsion tensor $S_{1\overset{..}{2}}^{..i}$ and the curvature tensor $R_{11\overset{..}{2}}^{...i}$, according to the information in Fig. 5.37. In the figure, $S_{1\overset{..}{2}}^{..1}$ represents edge dislocation density and $R_{11\overset{..}{2}}^{...1}$ is lattice imperfection density related to edge dislocation. $S_{1\overset{..}{2}}^{..i}$ and $R_{11\overset{..}{2}}^{...i}$ for $i = 2$ and 3 represent screw dislocation density and lattice imperfection density related to screw dislocation density, respectively. We can see that pure commercial aluminum subjected to pressure soaking exhibits no discernible change in dislocation density during the minor deformation stage under static tension, and only a slight decrease in advanced stages of deformation. On the contrary, imperfection density shows a remarkable decrease after pressure soaking. This decrease is largely dependent on the

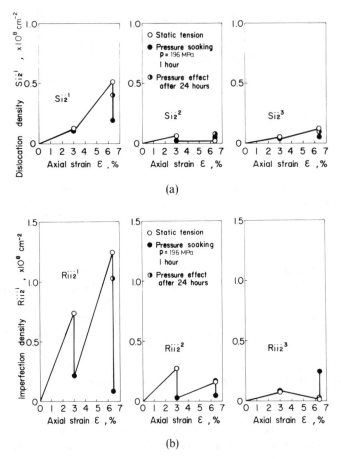

Fig. 5.38. Effect of pressure soaking, and 24 h lapse of time after pressure soaking, on (a) dislocation density, $S_{12}^{\cdot\cdot i}$, and (b) imperfection density, $R_{112}^{\cdot\cdot\cdot i}$, of 99·5% pure aluminum, when $\mu_G = 0\cdot4$. (Reprinted from ref. 23 by permission of Soc. Mater. Sci., Japan)

ease of slip in the crystal grain as shown in Fig. 5.39 [23]. It has also been found that within 24 h after pressure soaking the bulk of the decrease in dislocation and imperfection densities reverted to normal. Based on these findings, it may be said that the remarkable decrease of imperfection density in metal under pressure soaking is a result of dispersed dislocation redistribution, the annihilation of atom vacancies and the collapse of voids and cavities.

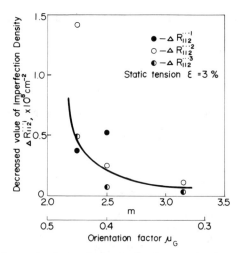

Fig. 5.39. Experimental relation between the decrease in the value of imperfection density, $R_{112}^{\cdots i}$, of 99·5% pure aluminum, immediately after pressure soaking, and the orientation factor of the grain, μ_G. (Reprinted from ref. 23 by permission of Soc. Mater. Sci., Japan)

5.4.2 Change in Conceptual Material Space of Plastically Deformed Metals with Pressure Soaking Effect

It is known that the occurrence of microplasticity in metals subjected to hydrostatic pressure soaking is due to the inhomogeneity of the metals. Therefore it is useful to consider the effects of pressure soaking on plastically deformed metals from the viewpoint of a conceptual material space.

Let us consider a material space and denote the displacement at an arbitrary point as u^i. This displacement can be divided into the mean displacement $\bar{u}^i$ and the deviatoric one $u^{i\prime}$, where i represents the rectangular coordinates. If a material point x_0^i translates to $x_0^i + \bar{u}^i + u^{i\prime}$ due to deformation, the line element after inhomogeneous deformation is divided into two parts; one is the mean element

$$d\bar{x}^i = dx_0^i + d\bar{u}^i \tag{5.48}$$

and another the deviatoric element

$$dx^{i\prime} = du^{i\prime} \tag{5.49}$$

Therefore the displacement ds is given as

$$ds^2 = (d\bar{x}^i + dx^{i\prime})(d\bar{x}_i + dx_i^{\prime}) = d\bar{x}^i d\bar{x}_i + dx^{i\prime} d\bar{x}_i + d\bar{x}^i dx_i^{\prime} + dx^{i\prime} dx_i^{\prime} \tag{5.50}$$

Assuming random change in dx^i and dx_i the second and third terms of the right-hand side of eqn (5.50) vanish but the fourth term does not, provided that there is some correlation between dx^i and dx_i. Therefore eqn (5.50) is written as

$$ds^2 = dx^i dx_i + dx^i dx_i \tag{5.51}$$

Since $d\bar{s}^2 = dx^i dx_i$ and $d\hat{s}^2 = dx^i dx_i$, eqn (5.51) is rewritten as

$$ds^2 = d\bar{s}^2 + d\hat{s}^2 \tag{5.52}$$

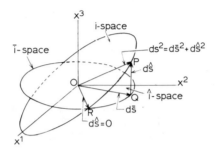

Fig. 5.40. Schematic representation of change in material space of deformed material subjected to hydrostatic pressure. (Reprinted from ref. 24 by permission of Soc. Mater. Sci., Japan)

Thus, inhomogeneously deformed material space may be represented as shown in Fig. 5.40 [24]. The i-space of actual material is at an angle to $\bar{i}$-space which is represented by the mean displacement. Furthermore, $\hat{i}$-space of the deviatoric displacement is at right-angles to the $\bar{i}$-space. Tensile deformed material space is regarded as i-space in which displacement consists of the mean $\overline{OP}$ and the deviator $\overline{QP}$. In the case of hydrostatic pressure soaking, it seems that the deviatoric displacement $\overline{QP}$ decreases and reaches $\overline{OR}$ where $d\hat{s} = 0$. For instance, mobile dislocations accumulate at the grain boundary but disperse under hydrostatic pressure with the redistribution resulting in the homogeneous material space. As seen from Figs 5.38 and 5.39, the lack of change in the torsion tensor is largely the result of pressure soaking, while the remarkable change in the curvature tensor indicates that the material space reaches the so-called distant parallelism space. This idea is applicable to the plastic behavior of metals under combined hydrostatic pressure which will be described in the next section.

5.4.3 Effects of Pressure Soaking on Both Friction Stress and Locking Force of Dislocations in Metals

Figure 5.41 [25] shows an experimental relationship between the decrease in yield stress, $\Delta\sigma_y$, and prestrain ε in a pure commercial aluminum sheet specimen (gauge length 60 mm, width 15 mm, thickness 1·5 mm) at reloading after pressure soaking of 196 MPa (2000 kgf/cm^2) for 1 h in oil. This figure shows that $\Delta\sigma_y$ increases with the advance of prestrain. It was also found that $\Delta\sigma_y$ is larger for the fine-grained test specimen than for the coarse-grained one.

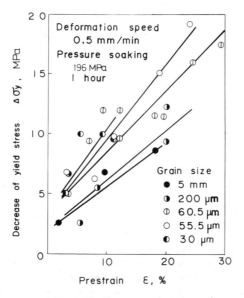

Fig. 5.41. Decrease in yield stress, $\Delta\sigma_y$, as a function of prestrain, ε, of commercial pure aluminum at reloading in the atmosphere, immediately after pressure soaking of 196 MPa for 1 h. (Reprinted from ref. 25 by permission of Soc. Mater. Sci., Japan)

Figure 5.42 [25] shows an experimental relation between $\Delta\sigma_y$ and $d^{-1/2}$, where d is the average grain diameter of the specimen. This figure illustrates that the following proportional relationship is satisfied, which is derived by differentiating both sides of Petch's relation, $\sigma_y = \sigma_0 + k_y d^{-1/2}$:

$$\Delta\sigma_y = \Delta\sigma_0 + \Delta k_y d^{-1/2} \tag{5.53}$$

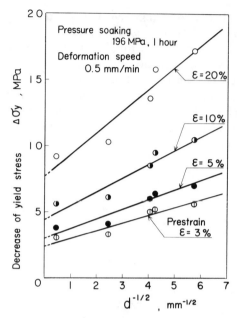

Fig. 5.42. Experimental relation between $\Delta\sigma_y$ and $d^{-1/2}$ of pure aluminum. (Reprinted from ref. 25 by permission of Soc. Mater. Sci., Japan)

where σ_0 represents friction stress in the crystal and k_y is the parameter representing the locking force of dislocations. Figure 5.43 [25] shows the change, with the advance of prestrain, in both k_y and σ_0 in a specimen subjected to pressure soaking. In this figure it is obvious that both the friction stress and the value of the locking force parameter decrease due to the pressure soaking (this at a remarkable rate with the advance of prestrain in the specimen pulled in the atmosphere). The decrease in the locking force of dislocations suggests that pressure soaking converts some locked dislocations into free ones. As mentioned in the previous paragraph, there was a remarkable change in the imperfection density of plastically deformed polycrystalline aluminum subjected to pressure soaking. Therefore we concluded that, in the prestrained specimen pressure soaking contributes to a partial change from pile-up dislocations to movable dislocations rather than to the nucleation of movable dislocations from the dislocation sources. It is notable that the decrease in yield stress of metal at reloading after pressure soaking results from the rapid movement of these dislocations.

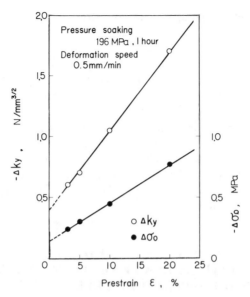

Fig. 5.43. Change of both k_y and σ_0, with advanced prestrain in the atmosphere, of commercial pure aluminum subjected to pressure soaking at 196 MPa for 1 h. (Reprinted from ref. 25 by permission of Soc. Mater. Sci., Japan)

From the experimental results regarding the influence of hydrostatic pressure soaking on a pure commercial aluminum sheet specimen, the following conclusions can be made.

(1) No remarkable change in the dislocation density of plastically deformed coarse-grained metal subjected to pressure soaking was observable from the determination of the quantities of the torsion tensor in non-Riemannian geometry.

(2) The imperfection density evaluated by calculating the quantities of the Riemann–Christoffel curvature tensor decreased remarkably immediately after pressure soaking and this was largely dependent on the ease of slip in the interior of the grain. This is probably due to the dispersed redistribution of dislocations, annihilation of vacancies, and the collapse of voids and cavities.

(3) The bulk of dislocation and imperfection densities decreases in metals subjected to pressure soaking recovered within 24 h after pressure soaking.

(4) In the unidirectional tension test under atmospheric pressure, at tensile reloading immediately after pressure soaking, yield stress de-

creased considerably compared to flow stress, but then returned asymptomatically to the original flow curve as deformation progressed. We also found that pressure soaking does not influence the fracture ductility of the metal.

(5) We observed that the yield stress at reloading immediately after pressure soaking decreased remarkably with the prestrain when pulled in the atmosphere. This was also the case for the fine-grained test specimen.

(6) Concerning the influence of pressure soaking on the structural changes in metal, we surmised that the temporary increase in the number of movable dislocations results from the decrease of both frictional stress and locking force. It may be that the decrease in yield stress at reloading immediately after the pressure soaking is caused by the accelerative effect of movable dislocations.

5.5 MICROSTRUCTURAL STUDY OF EFFECTS OF HYDROSTATIC PRESSURE ON FLOW STRESS AND HIGH-TEMPERATURE CREEP IN METALS BY MEANS OF CDC

Figure 5.44 [26] contrasts tensile stress and strain curves of a coarse-grained 99·45% pure aluminum sheet specimen (gauge length 30 mm,

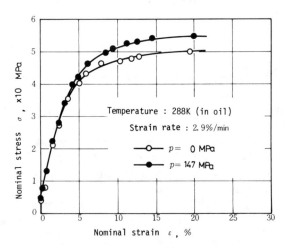

Fig.5.44. Increase of tensile flow stress of 99·45% pure aluminum under combined hydrostatic pressure of 147 MPa in oil. (Reprinted from ref. 26)

width 8mm, thickness 1·5 mm) under atmospheric pressure and again under hydrostatic pressure of 147 MPa (1500 kgf/cm^2) at room temperature. We can see from this figure that in the advanced stages of plastic deformation the flow stress of the metal is greater under hydrostatic pressure than it is under atmospheric pressure. The discrepancy between the two curves is not large but the difference between the two microstructural changes is remarkable as demonstrated by the following.

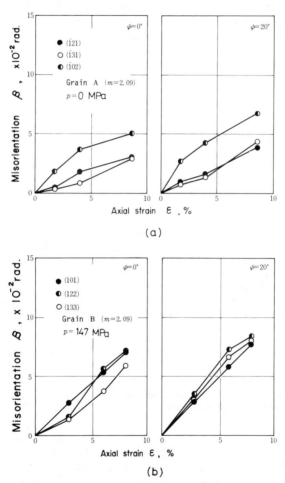

Fig. 5.45. Change of misorientation, β, of 99·45% pure aluminum with advancing axial strain, ε, in static tension test at room temperature, under (a) atmospheric pressure and (b) 147 MPa, when $\mu_G = 0.48$ ($m = 2.09$). (Reprinted from ref. 26)

Figure 5.45 [26] shows an example of change in misorientation β during static tension tests, under atmospheric and hydrostatic pressures respectively, for grains with identical crystal orientation, $\mu_G = 0.48$ ($m = 2.09$). In this figure, ψ is the angle between the direction of the incident X-ray beam and the specimen surface normal. Figure 5.46 [26] shows a typical example of the β versus $\sin^2 \psi$ diagrams obtained from the test data of Figs 5.45, where the straight lines were drawn by using the least-squares method. In the early stages of deformation there is little difference between the gradient of the straight line in the atmosphere and that under hydrostatic pressure; however, with advanced deformation the gradient under hydrostatic pressure is smaller than that under atmospheric pressure.

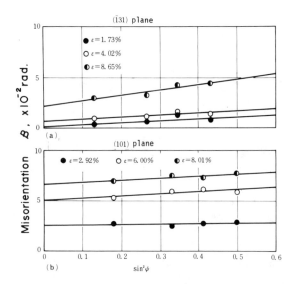

Fig. 5.46. Experimental relation between the misorientation, β, and $\sin^2 \psi$ of 99·45% pure aluminum, in static tension test at room temperature, under (a) atmospheric pressure and (b) 147 MPa. (Reprinted from ref. 26)

Figure 5.47 [26] shows the results of calculations for the torsion and curvature tensors, $S_{12}^{\cdot\cdot i}$ and $R_{112}^{\cdot\cdot\cdot i}$, based on data from Fig. 5.46. It was found that the imperfection density $R_{112}^{\cdot\cdot\cdot i}$ of metal in the atmosphere increases with the advance of deformation while it exhibits a remarkable decrease in the advanced stage of deformation under hydrostatic pres-

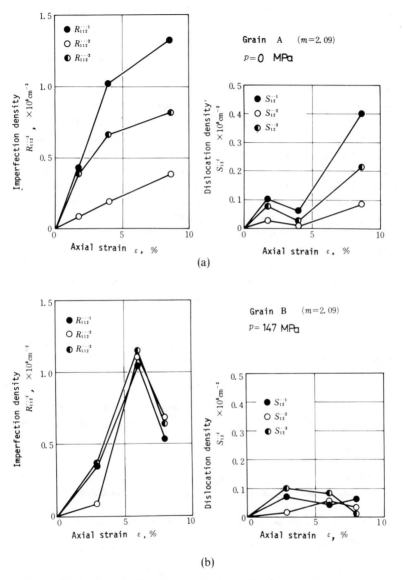

Fig. 5.47. Change in both imperfection density, $R_{112}^{\cdots i}$, and dislocation density, $S_{12}^{\cdot \cdot i}$, of 99·45% pure aluminum, under (a) atmospheric pressure and (b) 147 MPa. (Reprinted from ref. 26)

sure. This was also observed for other grains with different crystal orientations.

We can note here that the decrease in the imperfection density of metal under hydrostatic pressure shows that the material space approaches the distant parallelism concept proposed by Bilby et al. [1]. Considering that the distant parallelism space is uniform space, the deformation of metal under confining pressure is uniform. Therefore the increase in the flow stress of metal under combined hydrostatic pressure results from a decrease in the mobility of dislocations. Additionally, the decrease in the imperfection density not only implies a decrease in vacancy density but also results in decreased mobility of screw dislocations with jog. Hanafee et al. [27] also examined this experimentally. Based on these considerations, it would appear that a redistribution of vacancies will occur with advanced plastic deformation in metal under combined hydrostatic pressure.

In order to clarify the relation between the redistribution of dislocations and flow stress under combined hydrostatic pressure, the following observations were made using a model of slip plane dislocation redistribution (Fig. 5.48 [26]). In this model, for the purpose of simplification, the following assumptions were made: (i) dislocations move on a given slip plane; (ii) there is no interaction between dislocations on a given slip plane; (iii) the distance, r, of individual dislocation from the dislocation source is represented as

$$r = L(K/N)^n \tag{5.54}$$

where L is the distance between the source and the obstacle, K the numbering symbol of the number of N dislocations, and n a constant.

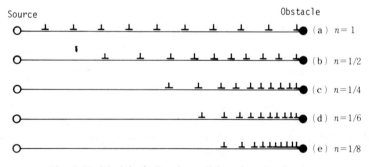

Fig. 5.48. Model of slip plane dislocation distribution.

Next, the stress field τ' around a dislocation is generally written as

$$\tau' = Gb/\alpha r \tag{5.55}$$

where G is the shear modulus, b the Burgers vector and α a constant which nearly assumes the numerical value of 5. Therefore the back-stress through pile-up dislocations to the dislocation source is written as

$$\tau = (GbN^n/\alpha L) \sum_{K=1}^{N} (1/K^n) \tag{5.56}$$

Using the approximation

$$\sum (1/K^n) \cong \int (dK/K^n) = [1/(1-n)](N^{1-n} - 1) \tag{5.57}$$

in eqn (5.56), τ is given by

$$\tau = (Gb/\alpha L)[1/(1-n)](N - N^n) \tag{5.58}$$

In eqn (5.56), τ for $n=1$, i.e. τ_0, is written as $\tau_0 = (GbN/\alpha L)\sum_{K=1}^{N}(1/K)$ and is approximated as $\tau_0 = (GbN/\alpha L)\ln N$. Therefore eqn (5.58) is rewritten as

$$(\tau/\tau_0) = (1 - N^{n-1})/[(1-n)\ln N] \tag{5.59}$$

Figure 5.49 [26] shows a typical example of calculation results using eqn (5.59). These calculations demonstrate that the numerical value of (τ/τ_0) increases corresponding to the increase of n to 1, e.g. the advance of the uniform dislocations redistribution. Therefore it is significant that the flow stress increase in polycrystalline metal under combined hydrostatic

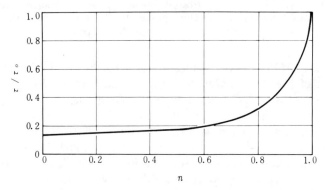

Fig. 5.49. Analytical representation of increase in back-stress caused by uniform distribution of dislocations. (Reprinted from ref. 26)

pressure results both from the increase of back-stress through pile-up dislocations to the dislocation source and from the decrease in dislocation mobility caused by uniform redistribution.

Next we consider the effects of hydrostatic stress on high-temperature creep of metals.

In Section 4.5 of the previous chapter, we pointed out that the effect of hydrostatic pressure on tensile creep of polycrystalline metals at elevated temperatures could not be explained simply by the extension of so-called continuum mechanics of the materials under confining pressure.

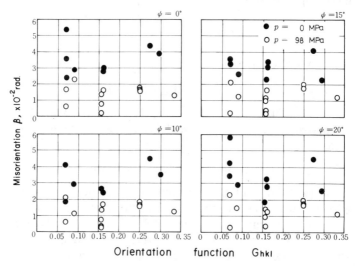

Fig. 5.50. Effect of combined hydrostatic pressure, p, on the misorientation, β, of coarse-grained commercial pure aluminum at 473 K. (Reprinted from ref. 28 by permission of Soc. Mater. Sci., Japan)

Figure 5.50 [28] shows the measurement results of misorientation for the data of Fig. 4.51 in Chapter 4. We found that the decrease in the creep rate of a coarse-grained pure aluminum sheet specimen under combined hydrostatic pressure at 478 K (Fig. 4.51) closely correlated with the decrease in the misorientation (Fig. 5.50). Figure 5.51 [28] shows the changes in the dislocation density, $S_{12}^{\cdot\cdot i}$, and the imperfection density, $R_{112}^{\cdot\cdot\cdot i}$, calculated from the data in Fig. 5.50. It is clear that the density of lattice defects decreases drastically by the combination of uniaxial loading and hydrostatic pressure. This may be interpreted as follows.

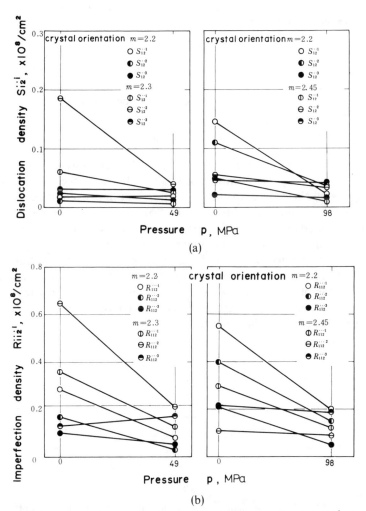

Fig. 5.51. Effect of confining pressure, p, on (a) dislocation density, $S_{12}^{\cdots i}$, and (b) imperfection density, $R_{112}^{\cdots i}$, of commercial pure aluminum crept at 473 K. (Reprinted from ref. 28 by permission of Soc. Mater. Sci., Japan)

The steady-state creep of metal at elevated temperature results from the equilibrium of strain-hardening and recovery. Both cross-slip of screw dislocations and climb of edge dislocations are considered as the mechanism of the recovery, with the latter predominating. Therefore the climbing rate of the dislocation element which constitutes the dislocation network is given as [28]

$$dl/dt \simeq Gb^3 C_j D/lRT \qquad (5.60)$$

where G is the shear modulus, b the Burgers vector, l the average distance between dislocations which make up the dislocation network, C_j the jog density, and D the self-diffusion coefficient. From eqn (5.60) the creep rate is written as

$$\dot{\varepsilon} \propto (\tau^3 C_j/RT)\exp[-(H_d + pV_d)/RT] \qquad (5.61)$$

where τ is the flow stress of the crystal. The pressure effect on $\dot{\varepsilon}$ is explicitly given by the term in eqn (5.61) and also implicitly proven by the pressure effect on both the dislocation and imperfection densities of the crystal as follows. As shown in Fig. 5.51, the drastic decrease in $R_{112}^{\cdot\cdot\cdot i}$ which contains atom vacancies suggests a decrease in jog density C_j contributing to the dislocation climbing. Therefore we concluded from eqn (5.61) that $\dot{\varepsilon}$ decreases under confining pressure. It also appears that the decrease in the dislocation density under confining pressure results in the increase of the average distance l; as a result, the recovery rate dl/dt of eqn (5.60) and the dislocation mobility decrease. Thus the creep rate decreases under combined hydrostatic pressure.

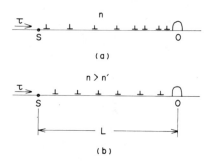

Fig. 5.52. Model of the dislocation density decrease in metal crept under hydrostatic pressure. (a) $p=0$; (b) $p=p$.

Finally, let us consider the model shown in Fig. 5.52 [28] in order to see why the dislocation density of crept metals under confining pressure decreases compared with that in the atmosphere. The model illustrates distribution of dislocations on a slip plane, where S and O show dislocation source and obstacle, respectively. Figure (a) shows the equilibrium state where back-stress due to n pile-up dislocations balances the stress in the source. In Fig. (b) a smaller number of dislocations n' will balance the stress in the source, since the dislocations locate more uniformly under hydrostatic pressure than they do in the atmosphere. Therefore it is surmised that the dislocation-density decrease under pressure compared with that in the atmosphere.

From the X-ray study of the effects of hydrostatic pressure on flow stress and high-temperature creep in coarse-grained pure aluminum, we arrived at the following conclusions.

(1) A remarkable decrease in the density of lattice imperfections in an advanced stage of tensile plastic deformation results from uniform redistribution of the dislocations and also from the decreased density of atom vacancies.

(2) In particular, the former results in the increase of back-stress to the dislocation source from pile-up dislocations, and also in decreased dislocation mobility. The latter results in a decrease of screw dislocations with jog.

(3) A decrease in the creep rate under combined hydrostatic pressure as compared with that in the atmosphere was clearly observed. The misorientation of crept metal measured by the Laue back-reflection X-ray method decreased with confining pressure, and it was found that there was a clear distinction between microstructural change under atmospheric and hydrostatic pressures.

(4) Both the dislocation density and the global lattice imperfection density, which contains atom vacancies, drastically decreased under pressure. This would result not only from the pressure effect in self-diffusion in high-temperature creep, but also from decreases in jog density which is closely related to atom vacancy density and flow stress of the crystal due to a widening of the average distance between dislocations.

(5) The dislocation density of crept metal decreased under confining pressure. We surmised that with confining pressure the dislocations distributed uniformly and the increase in back-stress to the dislocation source reduced the nucleation of dislocation.

REFERENCES

1. B. A. Bilby, R. Bullough and E. Smith, Continuous distributions of dislocations: a new application of the methods of non-Riemannian geometry, *Proceedings of the Royal Society*, **A231** (1955), 263, Royal Society, London.

2. E. Kröner, Die Versetzung als Elementare Eigenspannungsquells, *Zeitschrift für Naturforschung*, **11a** (1956), 969, Verlag der Zeitschrift für Naturforschung, Tübingen.
3. K. Kondo, *RAAG Memoirs*, **I** (1955), **II** (1958), **III** (1962), Gakujitsubunken-Fukyukai, Tokyo.
4. K. Kondo and M. Yuki, On the current viewpoint of non-Riemannian plasticity theory, *RAAG Memoirs*, **2**, **D-VII** (1958), 124.
5. M. Ohnami and K. Shiozawa, Definition of misorientation and the application to strength of materials, *Transactions of the Japan Society of Mechanical Engineers*, **39** (1971), 769 (in Japanese).
6. K. Shiozawa and M. Ohnami, Study of flow stress of metal by means of the geometrical aspect of the continuously dislocated continuum, *Proceedings of 1973 Symposium on Mechanical Behavior of Materials* (1974), p. 93, Soc. Mater. Sci., Japan.
7. F. R. N. Nabarro, *Theory of Crystal Dislocation*, Clarendon Press, Oxford, 1967, p. 120.
8. R. W. Lardner, *Mathematical Theory of Dislocations and Fracture*, University of Toronto Press, 1974, pp. 264–70.
9. F. C. Frank, *Report of Symposium on Plastic Deformation of Crystal Solid* (1950), p. 150, Carnegie Institute of Technology.
10. S. Amelinckx and W. Dekeyser, The structure and properties of grain boundaries, *Solid State Physics*, **8** (1959), 325, Pergamon Press.
11. W. T. Read, *Dislocation in Crystals*, McGraw-Hill, New York, 1953, p. 178.
12. J. C. M. Li, Petch relation and grain boundary sources, *Transactions of Metallurgical Society of AIME*, **277** (1963), 239, Amer. Soc. Metals.
13. A. Seeger, *Dislocation and Mechanical Properties of Crystals*, John Wiley and Sons, New York, 1957.
14. M. Ohnami and K. Shiozawa, *Strength and Fracture of Polycrystalline Body*, BAIFUKAN, 1976, p. 126 (in Japanese).
15. F. Hultgren, Grain boundary and substructure hardening in aluminum, *Transactions of Metallurgical Society of AIME*, **230** (1964), 898.
16. D. Ye. Ovsiyenko and I. Ye. Sonnina, *Fiz. Metal. i Metalloved*, **14** (1962), 252 (English trans., *Physics Metals Metalog. (USSR)*).
17. M. Lauriente and R. B. Pond, Effect of growth imperfections on the strength of aluminum single crystals, *Journal of Applied Physics*, **27** (1956), 950, Amer. Institute of Physics.
18. J. T. McGrath and G. B. Craig, The effect of striation-type substructure on the deformation of aluminum single crystals, *Transactions of Metallurgical Society of AIME*, **215** (1959), 1022.
19. M. Ohnami and K. Shiozawa, Change in misorientation of polycrystalline metal in the cyclic stress/strain constitutive, *Transactions of the Japan Society of Mechanical Engineers*, **40** (1974), 2135 (in Japanese).
20. Y. Gokyu and T. Kishi, Effect of metallurgical factors on Bauschinger effect (Review), *Journal of the Japan Society for Technology of Plasticity*, **10** (1969), 863 (in Japanese), CORONA-Sha, Tokyo.
21. S. N. Buckley and K. M. Entwistle, The Bauschinger effect in super-pure aluminum single crystals and polycrystals, *Acta Metallurgica*, **4** (1956), 352, Pergamon Press.

22. S. Amari, On some primary structures of non-Riemannian plasticity theory, *RAAG Memoirs*, **III**, D–I (1962), 99.
23. M. Ohnami, K. Shiozawa and A. Kamitani, X-ray study on the influence of pressure soaking on the structural change of plastically deformed polycrystalline metal, *Proceedings of 15th Japan Congress of Materials Research* (1972), p. 50, Soc. Mater. Sci., Japan.
24. M. Ohnami, K. Shiozawa and A. Kamitani, X-ray study on the influence of hydrostatic pressure soaking on the structural change of deformed polycrystalline metals, *Journal of the Society of Materials Science, Japan*, **21** (1972), 109 (in Japanese).
25 M. Ohnami, M. Ohmura, K. Shiozawa and A. Kamitani, Effect of hydrostatic pressure soaking on the plastic deformation of polycrystalline aluminum, *Proceedings of 16th Japan Congress of Materials Research* (1973), p. 115, Soc. Mater. Sci., Japan.
26. M. Ohnami and K. Shiozawa, An X-ray study on the structural change of plastically deformed polycrystalline metal under combined hydrostatic pressure, *Proceedings of 4th International Conference on High Pressure* (Kyoto, 1974), p. 228.
27. J. E. Hanafee and S. V. Radcliffe, Effect of hydrostatic pressure on dislocation mobility in lithium fluoride, *Journal of Applied Physics*, **18** (1967), 4248, Amer. Institute of Physics.
28. M. Ohnami, T. Yamakage, K. Shiozawa and M. Sakane, An X-ray study on the effect of hydrostatic stress on creep of polycrystalline metal at elevated temperatures, *Journal of the Society for Materials Science, Japan*, **21** (1972), 1086 (in Japanese).

Chapter 6

Fracture Laws in a Continuum Mechanical Approach

In this chapter we describe the drastic difference between the crack behavior and failure life of metallic creep and that of metallic low-cycle fatigue at elevated temperatures in the following continuum mechanical aspects: the effect of hydrostatic pressure, the effect of stress biaxiality, the effect of load waveform and sequence, and the notch effect. We will also examine the creep–fatigue interaction of polycrystalline metals and heat-resisting alloys from the viewpoint of the differences mentioned above. Thermal fatigue of a circular disk specimen will also be described. The requirements for the life prediction of structural heat-resisting metals and alloys under creep–fatigue interacted conditions are becoming more sophisticated as technology rapidly advances. Creep–fatigue interaction is one of the most important technological problems in the fracture of structural materials.

In this chapter special attention is paid to the effects of biaxial stresses on the high-temperature low-cycle fatigue criterion in the creep range. Historically there have been many studies on low-cycle fatigue in multiaxial stress states, and many strain and stress parameters have been proposed to arrange the low-cycle fatigue data in these stress states [1, 2]. However, there is no parameter which can arrange the data in a variety of stress states at elevated temperatures. We must consider the following aspects in order to examine the validity of such parameters. First, we must examine the case where there is a wide range in the stress (or strain) ratio of the minimum principal stress (or strain) to the maximum stress (or strain), which covers the range -1 to 1. Second, we must examine the effect of crack mode and varying principal stress axes on the failure life data correlation with strain and stress parameters, and lastly, creep–fatigue conditions. This chapter aims at clarifying these problems.

6.1 EFFECT OF HYDROSTATIC STRESS ON CRACK BEHAVIOR AND FRACTURE LIFE UNDER CREEP, LOW-CYCLE FATIGUE AND THE INTERACTION OF THESE CONDITIONS AT ELEVATED TEMPERATURES

In this section, as a means of illustrating the drastic difference between the effect of hydrostatic pressure on the metallic creep rupture life and that on the metallic low-cycle fatigue failure life at elevated temperatures, we describe the creep–fatigue interaction of commercial pure copper in three types of test under both hydrostatic and atmospheric pressures at 543 K. The test apparatus, shown in Fig. 4.46 in Chapter 4, was used for both creep rupture and low-cycle fatigue tests, and a centrally notched plate specimen, having a 30/10 mm gauge length, 8 mm width and 1 mm thickness, was used for the tests under both a hydrostatic oil pressure of 98 MPa and atmospheric pressure.

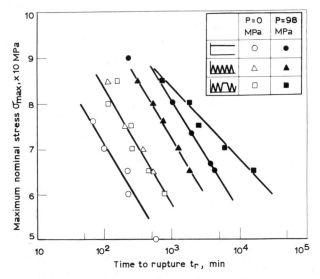

Fig. 6.1. Time to rupture of commercial pure copper in three stress waves under both atmospheric pressure and hydrostatic pressure of 98 MPa at 543 K. (Reprinted from ref. 3 by permission of Japan Soc. Mech. Engrs)

Time to rupture t_r of the metal tested in the three types of test, under both hydrostatic and atmospheric pressures at 543 K, it shown in Fig. 6.1 [3], where the types of test are the static tensile creep rupture test and the zero-to-tension low-cycle fatigue tests with and without a hold-time of 30 min at peak tensile stress. The figure shows a remarkable increase in the time to rupture of the metal under combined hydrostatic pressure for the three types of test. As is seen from the figure, the rupture life of the metal is in an order relative to each type of pressure. Under atmospheric pressure, the creep test shows the shortest time to rupture, and both the pure fatigue test and the fatigue test with hold-time show almost the same rupture time which is longer than that in the creep test. On the other hand, under hydrostatic pressure, the shortest rupture time is observed in the pure fatigue test, the intermediate time is observed in the creep test, and the longest time is observed in the fatigue test with hold-time. It seems that this rupture life phenomenon of the metal results from the difference in the effect of pressure on metallic creep rupture and the low-cycle fatigue at elevated temperatures.

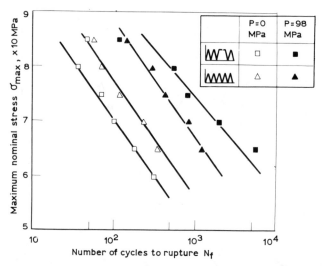

Fig. 6.2. Experimental relation between the maximum nominal stress, σ_{max}, and the number of cycles to rupture, N_f, of commercial pure copper under both types of pressure. Test conditions as for Fig. 6.1. (Reprinted from ref. 3 by permission of Japan Soc. Mech. Engrs)

Figure 6.2 [3] also shows the experimental relation based on the number of cycles to rupture N_f in both the pure fatigue and the fatigue test with hold-time. Under atmospheric pressure, the rupture time under pure fatigue is almost the same as that under fatigue with hold-time, but the number of cycles to rupture under fatigue with hold-time is far smaller than that under pure fatigue. This reduction in the number of cycles to rupture due to the hold-time is recognized as a common phenomenon at elevated temperatures. This is a very important design problem if we are to evaluate the failure life of fatigue with hold-time on the basis of either the time to rupture or the number of cycles to rupture.

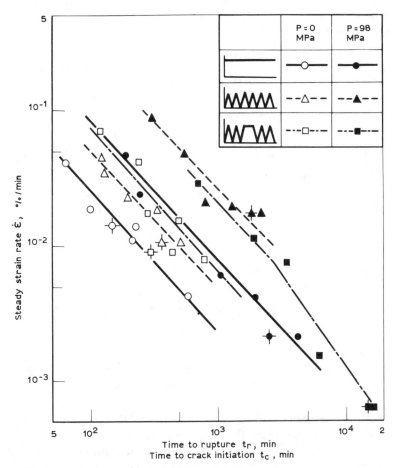

Fig. 6.3. Experimental relation between the steady-state rate, $\dot{\varepsilon}$, and the time to rupture, t_r, or the time to crack initiation, t_c, of the material under both types of pressure at 543 K. The absence of a dashed mark indicates time to rupture, t_r, and a dashed mark represents time to crack initiation, t_c. (Reprinted from ref. 3 by permission of Japan Soc. Mech. Engrs)

Figure 6.3 [3] shows the experimental relation between the steady-state strain rate, $\dot{\varepsilon}$, and the rupture time, t_r, or the crack initiation time, t_c, of the metal in the three types of test under both types of pressure. The figure clearly shows that the superposition of hydrostatic pressure results in an increase in the rupture ductility B of the metal, where B is the apparent ductility in the relation $\dot{\varepsilon} t_r = B$ [4], and where the crack initiation time is defined as the time when a crack reaches a length of 0·5 mm. Since the slope in the figure is nearly -1, we should note that the relation holds without the distinction of the stress waveforms. On the other hand, the growth of a crack in the metal with time under $\sigma_{max} = 64$ MPa and both types of pressure is shown in Fig. 6.4 [3]. Under hydrostatic pressure, both the retardation of crack initiation and the deceleration of crack propagation rate are observed when compared to results observed in the atmosphere.

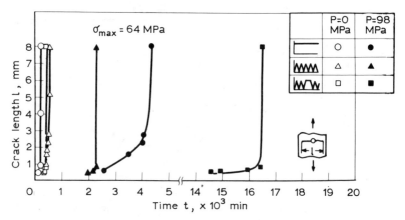

Fig. 6.4. Crack growth curves in the three types of test under $\sigma_{max} = 64$ MPa, atmospheric and hydrostatic pressures. Test conditions as for Fig. 6.1. (Reprinted from ref. 3 by permission of Japan Soc. Mech. Engrs)

Both the retardation of crack initiation and the deceleration of the crack propagation rate by the superposed pressure will be discussed in connection with the pressure dependence of the strain rate and the rupture strain near the crack tip. Mechanical factors which control the crack propagation rate are not necessarily understood under the condition of applied stress exceeding the yield stress of the metal, but it seems that the crack propagation rate is closely related to the strain rate and the strain gradient near the crack tip. Assume that the local strain rate, $\dot{\varepsilon}_{loc}$, and the local rupture strain, ε_{rloc}, at the crack tip are proportional to the global strain rate and the rupture strain, ε_r, of the specimen, respectively, i.e.

$$\dot{\varepsilon}_{loc} \propto \dot{\varepsilon}$$
$$\varepsilon_{rloc} \propto \varepsilon_r \tag{6.1}$$

We can assume from eqns (6.1) that the decrease in the crack propagation rate of the metal under pressure can be attributed to both the decrease of the local strain rate and the increase of local rupture strain at the crack tip.

The crack propagation rate under both pressures will be discussed with reference to the J-integral proposed by Rice [5]. Now, assume that there are two bodies A and B identical except for a distinction in the crack length. Denoting the crack lengths of the bodies A and B as l and $l+dl$, respectively, the stress versus displacement relation of the two bodies is shown schematically in Fig. 6.5(a). The hatched part in the figure represents the decrease of potential energy U of the body, and the relation between ΔU and the J-integral is given by

$$J = -\Delta U / \Delta l \tag{6.2}$$

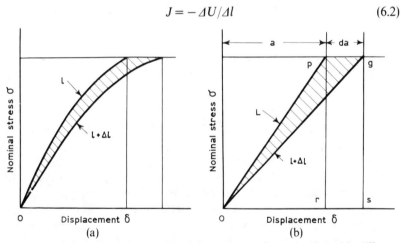

Fig. 6.5. Nominal stress versus elongation curves of two bodies with different crack lengths.

Now, consider the application of the J-integral to the crack problem in creep. In the first place, we approximate the hatched part of Fig. 6.5(a) by utilizing a triangular shape, as shown in Fig. 6.5(b). Assume that a body with a crack length l, in a creep state, is in the state of point p of

Fig. 6.5(b), and also assume that the crack propagates instantaneously over the length dl, then a virtual stress drop $\Delta\sigma$ will occur. But such a virtual stress drop will be compensated at once by the creep stress and the body will elongate by da during the compensation. Therefore the potential energy, ΔU, is released during the increase as crack length dl is represented by the area of the triangle opq in Fig. 6.5(b), and is given by

$$\Delta U = (1/2)\sigma\, \mathrm{d}a \tag{6.3}$$

Employing the conventional sign of energy and substituting eqn (6.3) into eqn (6.2), the following modified J-integral, J', by the unit thickness of the specimen under creep conditions, is derived as

$$J' = (1/2)\sigma(\mathrm{d}a/\mathrm{d}l) \tag{6.4}$$

In the case of pure fatigue and fatigue with hold-time, the modified J-integrals J'_f and J'_{cf} are calculated in the same manner. As is seen from eqn (6.4), the modified J-integral, J', is calculated from both the strain–time and the crack length–time curves. The physical meaning of J' is quite different from the usual crack extension force G. We think that J' represents the sum of the crack extension force, G', and the force, E^p, to extend the plastic strain region of the whole specimen where, in general, G' is the sum of the energy release rate G in linear fracture mechanics and the force which extends the plastic strain region near the crack tip which is directly related to crack extension. Therefore J' is written as

$$J' = G' + E^p \tag{6.5}$$

The micromechanical meaning of eqn (6.5) will be described in the following chapter.

Figures 6.6 (a), (b) and (c) [3] show the experimental relation between the modified J-integral, J', calculated in the three types of test and the rupture time ratio, t/t_r. We found, from the figure, that the effect of hydrostatic pressure on J' of the metal is quite different with the distinction of stress waveforms. In the case of creep, J'_c under atmospheric pressure is greater than that under hydrostatic pressure, but as regards J'_{fc} in the fatigue test with hold-time this relationship is quite the contrary. Therefore it is worthy of note that the effect of hydrostatic pressure on crack behavior is quite different with the distinction of stress waveforms.

Next, by using the modified J-integral, J', we discuss the change in the crack extension force G' of the metal by applying superposed pressure. In pure fatigue, it is considered that E^p in eqn (6.5) will take almost the

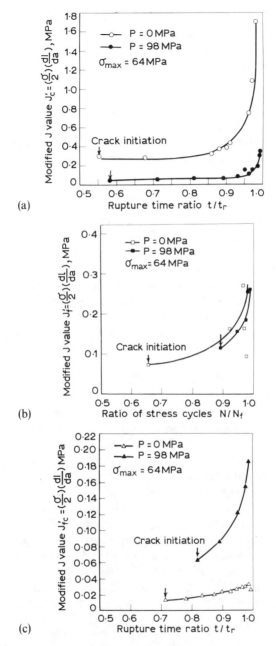

Fig. 6.6. Comparison of the modified J-integral, J', under atmospheric pressure with that under hydrostatic pressure, under the conditions of $\sigma_{max} = 64$ MPa at 543 K. (a) Creep test; (b) pure fatigue test; (c) fatigue test with hold-time. (Reprinted from ref. 3 by permission of Japan Soc. Mech. Engrs)

same value under both pressures because the strain rate does not decrease by applying superposed pressure. Because we know that the magnitude of J' of the metal tested under hydrostatic pressure is greater than that under atmospheric pressure, it is noted that G' under hydrostatic pressure will be greater than that under atmospheric pressure. This agrees well with the tensile tear test results [6] conducted on commercial pure copper. The crack extension force G' of the metal under hydrostatic pressure is larger than that under atmospheric pressure (see Chapter 7). Therefore it is noted that a remarkable decrease in the fatigue crack propagation rate under pressure corresponds to an increase in G' under pressure from the viewpoint of fracture mechanics.

In the creep test, on the other hand, it seems that E^p of the metal under hydrostatic pressure will be smaller than that under atmospheric pressure due to the fact that the strain rate of the metal is decreased remarkably by the superposed pressure as shown in Fig. 4.51 in Chapter 4. Additionally, both imperfection and dislocation densities of commercial pure aluminum (99·75%) also remarkably decrease under superposed hydrostatic pressure, as shown in Fig. 5.51 in Chapter 5. It seems that a decrease in the J' value under hydrostatic pressure results from a decrease in the strain rate or in the imperfection and dislocation densities of the metal. Besides, as the creep crack propagation rate of the metal under hydrostatic pressure is smaller than that under atmospheric pressure, it is estimated that G' of the metal tested under hydrostatic pressure will be greater than that under atmospheric pressure.

Figure 6.7 [3] shows the experimental relation between the creep damage, ϕ_c ($=\Sigma_i t/t_r$), and the fatigue damage, ϕ_f ($=\Sigma_i N/N_f$), of the metal in the fatigue test with hold-time under both types of pressure. The sum of the creep and fatigue damages of the metal is apparently greater than unity under both pressures. We found from the figure that the creep damage is predominant under atmospheric pressure, but the fatigue damage is dominant under hydrostatic pressure. The shift in the amount of damage under the combined loading of hydrostatic pressure can be attributed to the fact that the effect of superposed hydrostatic pressure on the rupture life is greatly affected by the stress waveform. In fact, the rupture life is increased 22 times under superposed hydrostatic pressure, but the ratio increases by only a multiple of 4 in the pure fatigue test. Thus the amount of damage of the material accumulated in a unit of time is decreased to 1/22 in the creep test but only to 1/4 in the pure fatigue test. The remarkable decrease in the accumulated rate of creep damage to the material in hold-time under pressure can be considered to

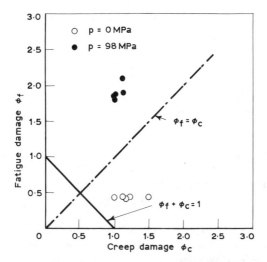

Fig. 6.7. Experimental relation between the fatigue damage, ϕ_f, and the creep damage, ϕ_c, of the material in the fatigue test with hold-time, calculated by the linear cumulative damage rule under both types of pressure. Test conditions as for Fig. 6.1. (Reprinted from ref. 3 by permission of Japan Soc. Mech. Engrs)

result in a shift of the amount of damage, as shown in Fig. 6.7.

In this section, three types of test, i.e. static tensile creep test and zero-to-tension low-cycle fatigue test with and without hold-time, were performed on a commercial pure copper plate specimen under both atmospheric pressure and hydrostatic pressure of 98 MPa at 543 K. From the test results we concluded the following.

(1) Both the time to rupture and the number of cycles to rupture of the metal increased in the three types of test conducted under hydrostatic pressure, in contrast to those under atmospheric pressure. The increase in the ratio of the rupture life under hydrostatic pressure to that under atmospheric pressure is greatest in the creep test. In the fatigue test this ratio is smallest, and in the fatigue test with hold-time the ratio takes an intermediate value. Consequently it is suggested that this increased ratio of the rupture life becomes smaller as the stress waveform frequency changes.

(2) The cause of the increase in the rupture life under hydrostatic pressure mainly lies in both the retardation of crack initiation and the deceleration of the crack propagation rate by superposed pressure. These crack behaviors under hydrostatic pressure mainly result from both the

decrease of the local strain rate and the increase of the local rupture strain at the crack tip.

(3) Total damage, $\phi_c + \phi_f$, calculated by the linear cumulative damage rule for fatigue with hold-time under both types of pressure, is apparently greater than unity. Under hydrostatic pressure the fatigue damage, ϕ_f, of the metal is predominant and under atmospheric pressure the creep damage, ϕ_c, is dominant.

6.2 EFFECT OF STRESS BIAXIALITY ON CRACK BEHAVIOR AND FAILURE LIFE IN HIGH-TEMPERATURE LOW-CYCLE FATIGUE OF HEAT-RESISTING METALS AND ALLOYS IN THE CREEP RANGE

6.2.1 Comparison Between Crack Behavior and Failure Life in Push-Pull Tests and Those in Reversing Torsional Tests: Preliminary Research

As preliminary research concerning the effect of biaxial stresses, a comparison between the strain-controlled push-pull and the reversed torsional data at 873 K in air is described for solution heat-treated type 304 austenitic stainless steel. The tests were performed by using a hollow cylindrical specimen with a circular notch 1 mm in diameter and a gear-mechanism type combined push-pull and reversed torsional testing machine.

Figure 6.8 [7] shows a great decrease in the failure life, N_f, of the metal due to the presence of hold-time at the peak tensile strain in both the push-pull and the torsional tests on the basis of the von Mises equivalent total strain range, $\Delta\bar{\varepsilon}$. The figure shows that $\Delta\bar{\varepsilon}$ is an adequate comparative parameter of the failure life of the material between the push-pull and the torsional tests, both with and without hold-time, if we admit the scatter band which is shown in the figure. To be even more precise, N_f in the push-pull pure fatigue test is slightly greater than that in the torsional test, when compared in the same strain range. On the other hand, in the fatigue test with hold-time, the phenomenon mentioned above is quite the reverse. The failure life in the push-pull test is shorter than that in the torsional test. The hold-time reduces N_f in both the push-pull and the torsional tests, but the ratio of decrease due to the presence of hold-time is larger in the push-pull test. The difference in the ratio of the life reduction between the two loading histories is discussed from the viewpoint of crack initiation and the propagation behavior in the following.

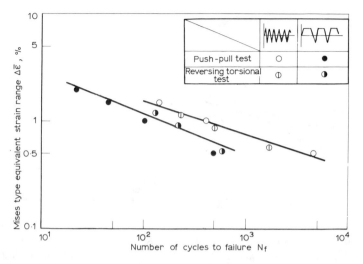

Fig. 6.8. Esperimental relation between Mises strain range, $\Delta\bar{\varepsilon}$, and the number of cycles to failure, N_f, for a tubular specimen of type 316 stainless steel at 873 K in air. (Reprinted from ref. 7 by permission of Japan Soc. Mech. Engrs)

Figure 6.9 [7] shows the experimental relation between $\Delta\bar{\varepsilon}$ and the crack propagation rate when the crack length is 1 mm. We found that the acceleration of the crack propagation rate occurs due to the hold-time, which is introduced in the strain waveform. In the pure fatigue test, the crack growth rate in the push-pull test is lower than that in the reversed torsion test. However, in the test with hold-time, the data are quite the contrary. The ratio of the acceleration of the propagation rate in the test without hold-time to that with hold-time is greater in the push-pull test, and the ratio increases as $\Delta\bar{\varepsilon}$ decreases. Prior to this discussion, the stress state near the crack tip is considered in both the push-pull and the torsion tests.

In the push-pull test the specimen with the central notch was fractured in mode I (crack opening type, Fig. 6.10(a)); on the other hand, in the torsion test, in which the crack propagates at nearly a 45° angle to the specimen axes (Fig. 6.10(b)), the fracture mode is also mode I, which was also observed in the push-pull. Ignoring the initial circular notch and also one of the two types of crack as shown in Fig. 6.10(c), the stress state of Fig. 6.10(c) becomes equivalent to that of Fig. 6.10(d). σ_1 denotes the applied stress near the crack tip in the push-pull test and that in the torsional test is σ'_1 ($=\sigma'_2=\tau$) as shown in the figure. Comparing the

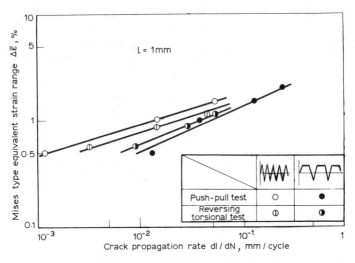

Fig. 6.9. Crack propagation rate in both the push-pull and the reversed torsion tests with and without hold-time, when the crack length is 1 mm. Test conditions as for Fig. 6.8. (Reprinted from ref. 7 by permission of Japan Soc. Mech. Engrs)

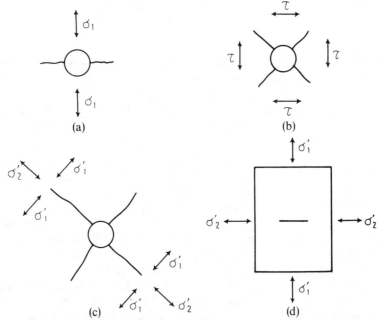

Fig. 6.10. Stress amplitude and biaxiality of stress near the crack tip: (a) push-pull test; (b) applied shear stress in the reversed torsion test; (c) applied principal stresses near the crack tip in case (b); (d) biaxiality of stress in the crack plate subjected to reversed torsion.

magnitude of stress σ_1 with that of σ'_1 under the same equivalent total strain range, $\Delta\bar{\varepsilon}$, it is noted that σ_1 is greater than σ'_1 ($=\sigma_1/\sqrt{3}$), since this material has the same cyclic stress–strain relation in both the push-pull and the torsion tests on the basis of the von Mises criterion as shown in Fig. 4.21 in Chapter 4 under fixed principal stress axes. However, the test data show a quite contrary tendency: in the test without hold-time the crack propagation rate in the torsional test is larger than that in the push-pull tests.

The cause of the contradiction in the crack propagation rate seems to be related to the existence of stress σ'_2 parallel to the crack in the torsional test. In linear fracture mechanics, additional stress parallel to the crack in mode I has no effect on the stress intensity factor, but when the applied stress exceeds the yield stress the additional stress has a great effect on the stress intensity factor. In fact it has been reported by Hilton [8] that the additional stress σ'_2, which is parallel to the crack and also applied in an opposite direction to the crack propagation direction, results in a great increase in the plastic strain intensity factor. Therefore it is worthy of note that in the case of fatigue without hold-time, though the principal stress in the torsional test is smaller than that in the push-pull test ($\sigma'_1 = \sigma_1/\sqrt{3}$), the crack propagation rate of the material is accelerated by additional stress parallel to the crack. Consequently, since additional stress accelerates the crack propagation rate in the torsional test, the failure lives of the metal in both the push-pull and the torsional tests without hold-time do not differ greatly. Nevertheless, the applied stress near the crack tip in the torsional test is smaller than that in the push-pull test. These will be described again in the following section from the alternative criterion in which the effect of additional stress parallel to the crack on the crack behavior is taken into consideration.

As to the fatigue test with hold-time, we can consider that the hold-time in the push-pull test results in acceleration of the crack propagation rate more so than in the torsional test due to the following reasons. The hydrostatic component of stress does not exist in the torsional test but the positive component of stress does exist in the push-pull test, and the negative component of hydrostatic stress remarkably decreases the creep damage of the material during the hold-time, as mentioned in the previous section. Consequently the convention occurs in the crack propagation rate between the tests without and with hold-time. In this section, the push-pull and the reversed torsional low-cycle fatigue tests were performed by using a solution heat-treated type 316 stainless steel with and without hold-time at 873 K in air. We concluded the following.

(1) The number of cycles to failure for the fatigue test with hold-time was much smaller than that for the pure fatigue test in both the push-pull and the torsional tests. The ratio of the reduction of the failure life of the metal by the hold-time was slightly larger in the push-pull test than in the torsional test.

(2) Comparing the failure life in the push-pull test with that in the torsional test, the von Mises equivalent strain range was found to be a good parameter for making a comparison in both the pure fatigue test and the fatigue test with hold-time. However, to be more precise, in the pure fatigue test the failure life in the push-pull test was slightly larger than that in the torsional test in the von Mises criterion, but in the fatigue test with hold-time the life with the torsional fatigue was larger.

(3) The equivalent strain range of the von Mises type was an adequate parameter in comparing the number of cycles to crack initiation and the crack propagation rate in the push-pull test with those of the torsional test. But a slight difference existed between them in the von Mises criterion. In the pure fatigue test, the crack propagation rate in the torsional test is larger than that in the push-pull test, but in the fatigue test with hold-time the rate in the push-pull test is larger.

(4) The crack behavior mentioned above was discussed in relation to the fracture mode and hydrostatic stress near the crack tip. It was clarified that, in the pure fatigue test, an additional negative principal stress parallel to the crack accelerated the crack propagation rate. On the other hand, in the fatigue test with hold-time it was estimated that the positive component of hydrostatic stress in the push-pull test accelerated the rate of creep damage in the material at the crack tip.

(5) The conclusions (1)–(3) indicate the necessity of examining the alternative criterion in which the effect of the principal stress parallel to the crack on the crack propagation rate in high-temperature low-cycle fatigue is considered. It must be examined from the critical aspects of the effect of principal stress parallel to the fatigue crack and that of the crack direction, i.e. crack mode, on the crack behavior and failure life, and it must cover a wide range of minimum principal strain to maximum principal strain ratio, $\varepsilon_3/\varepsilon_1$, -1 to 1, for the strain ratio, as shown in the following.

6.2.2 High-temperature Low-cycle Fatigue Criterion in Biaxial Stress States

In this section we will propose a new equivalent stress and strain parameter including the principal stress parallel to the crack based on the crack opening displacement (COD) concept. We will examine the applicability of the parameter in a low-cycle fatigue test without hold-time by using the same shape and dimensions of the tubular specimen with and without a circular notch for a solution heat-treated type 304 austenitic stainless steel at temperatures ranging from 923 K to 293 K in air. An originally developed biaxial electro-hydraulic servo-type fatigue testing machine, as shown in Fig. 4.24 in Chapter 4, which can apply an axial load in combination with a torsional load, was employed for performing a maximum principal stress-controlled and the Mises equivalent strain-controlled tests at 0·1 Hz under a triangular stress/strain waveform. Stress ratio, λ, and strain ratio, ϕ, in the study are defined as $\lambda = \sigma_3/\sigma_1$ and $\phi = \varepsilon_3/\varepsilon_1$ respectively, where σ_1 and ε_1 are the maximum principal stress and strain while σ_3 and ε_3 are the minimum principal stress and strain, respectively, in a hollow cylindrical specimen. Note that in the notched specimen all macrocracks initiated on the circular hole propagated in only mode I, the direction normal to the maximum principal stress, but in the unnotched specimen the transition to mode II, the direction of the maximum shear stress, occurred at a specific range of strain ratio, which will be described in Section 6.2.4.

Figures 6.11 (a) and (b) [9] show the arrangement of biaxial $\Delta\sigma_1$-controlled low-cycle fatigue test data with (a) the maximum principal stress range $\Delta\sigma_1$ and (b) the von Mises equivalent stress range $\Delta\bar{\sigma}$. The fatigue data for the reversed torsion test demonstrate a smaller fatigue strength as determined by the arrangement of $\Delta\sigma_1$ while $\Delta\bar{\sigma}$ reverses the arrangement where the push-pull data exhibit smaller fatigue strength. Both parameters cannot arrange the biaxial fatigue data within a scatter band with a factor of 2. However, the ratio of the fatigue life in the torsional test to that in the uniaxial test under $\Delta\sigma_1$-controlled conditions takes on a nearly constant value ranging from 0·1 to 0·3 and independent of temperature levels and maximum principal stress range tested.

Figure 6.12 [9] is an arrangement of the crack propagation rate with crack opening displacement (COD) defined as the displacement at the notch root. The figure shows that the correlation of the data is so good that, if we can formulate COD by a simple equation using the applied stresses, the equation will provide an adequate comparative parameter for arranging the biaxial fatigue data. Below we formulate this equation by inelastic finite element analysis.

Figure 6.13 [9] shows the results of the finite element analysis of COD for a centrally cracked plate subjected to monotonic biaxial loading, and

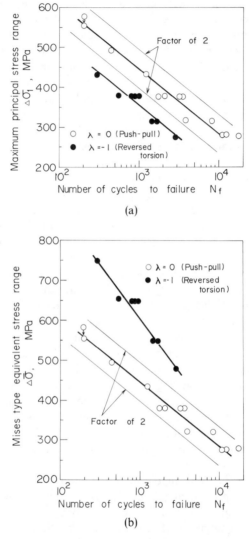

Fig. 6.11. Arrangement of the biaxial low-cycle fatigue life of type 304 stainless steel at 823K, by (a) the maximum principal stress range, $\Delta\sigma_1$, and (b) the Mises equivalent stress range, $\Delta\bar{\sigma}$. (Reprinted from ref. 9 by permission of Japan Soc. Mech. Engrs)

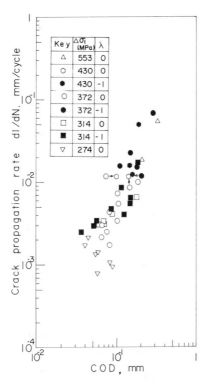

Fig. 6.12. Crack propagation rate, dl/dN, versus crack opening displacement, COD, plot. Test conditions as for Fig. 6.11. (Reprinted from ref. 9 by permission of Japan Soc. Mech. Engrs)

it shows analytical relations between (COD/σ_1) and $(2-\lambda)$. In this analysis, the constitutive equation $\varepsilon = (\sigma/E) + C\sigma^{1/n}$ was used, where E and n are Young's modulus and the strain-hardening exponent of the material tested at a tensile strain rate of $10^{-3}\,\text{s}^{-1}$ and a temperature of 823 K. If the data are approximated by a straight line at each COD, we can get

$$(COD/\sigma_1) = A(2-\lambda)^m \tag{6.6}$$

where m is 0·5 for $0 \geqslant \lambda \geqslant -1$ and A is a function of COD, i.e. a function of applied stress. The material constant m depends on the strain-hardening exponent n but takes the value 0·5 for a type 304 stainless steel at temperatures ranging from 923 K to 293 K [10].

In order to apply the equation to the low-cycle fatigue test, the

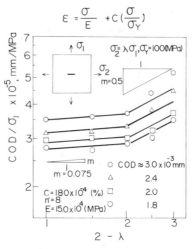

Fig. 6.13. Analytical relation between COD/σ_1 and $(2-\lambda)$ by inelastic finite element analysis. (Reprinted from ref. 9 by permission of Japan Soc. Mech. Engrs)

principal stress σ_1 is converted into the principal stress range $\Delta\sigma_1$ and the coefficient A is modified as $\alpha A'$. Equation (6.6) is thus reduced to

$$(\text{COD}/A') = \alpha \Delta\sigma_1 (2-\lambda)^m \qquad (6.7)$$

Equation (6.7) shows the crack opening displacement in biaxial stress states. Considering that the crack propagation rate in the biaxial stress state can be correlated with COD, as shown in Fig. 6.12, and the major part of the fatigue failure life is concerned with the crack propagation period, eqn (6.7) is an appropriate mechanical parameter for describing low-cycle fatigue lives in a biaxial stress state.

The right-hand side of eqn (6.7) is the principal stress range modified by the principal stress ratio, ϕ. So we define an equivalent stress range as

$$\Delta\sigma^* = \alpha \Delta\sigma_1 (2-\lambda)^m \quad (\alpha = 1/2^m) \qquad (6.8)$$

The nondimensional coefficient α is determined by satisfying the relation $\Delta\sigma^* = \Delta\sigma_1$, for the uniaxial case, hence $\alpha = 1/\sqrt{2}$ for $m=0.5$. Thus $\Delta\sigma^*$ is the equivalent stress taking account of the effect of the stress parallel to the crack, and it can be calculated only by knowing the principal stresses applied to the specimen.

Figure 6.14 [9] is the rearrangement of Fig. 6.11 using the new equivalent stress range $\Delta\sigma^*$. The arrangement of the data is fairly

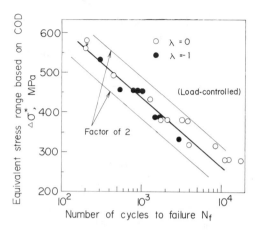

Fig. 6.14. Arrangement of the low-cycle biaxial fatigue failure data by COD stress range, $\Delta\sigma^*$. Test conditions as for Fig. 6.11. (Reprinted from ref. 9 by permission of Japan Soc. Mech. Engrs)

improved and all the data fall into a scatter band with a factor of 2. Thus it is concluded that the equivalent stress is applicable to the biaxial low-cycle fatigue data in the maximum principal stress-controlled tests at temperatures ranging from 923 K to 293 K [10].

Most of the biaxial low-cycle fatigue tests are carried out in a strain-controlled mode so that the parameter for arranging the fatigue lives are developed on a strain basis. Below we will derive the equivalent strain corresponding to $\Delta\sigma^*$.

Equivalent stress developed, and equivalent strain to be derived, are the parameters relating the stress and strain intensities at the crack tip, so we can assume the following equation on the analogy of the relation $K_\varepsilon = K_\sigma^{1/n}$ in nonlinear fracture mechanics [11]:

$$\varepsilon^* = \sigma^{*1/n} \tag{6.9}$$

where K_σ is the inelastic stress intensity factor, K_ε the inelastic strain intensity factor and n the strain-hardening exponent in the constitutive equation on the von Mises basis. Thus we can obtain the following equivalent strain range $\Delta\varepsilon^*$ which corresponds to eqn (6.8):

$$\Delta\varepsilon^* = C'\Delta\varepsilon_1(\phi^2 + \phi + 1)^{1/2}[(\lambda^2 - \lambda + 1)^{-1/2}(2-\lambda)^m]^{1/n}$$

or
$$\Delta\varepsilon^* = C\Delta\varepsilon_1(\phi^2 + \phi + 1)^{1/2}\{(\phi+2)[3(\phi^2+\phi+1)]^{-1/2} [3/(\phi+2)]^m\}^{1/n} \tag{6.10}$$

where $C = 2^{(n-m)n}/\sqrt{3}$. The constant C is determined such that $\Delta\varepsilon^*$ reduces to $\Delta\varepsilon_1$ in the push-pull test. For example, the value of C is 0·62 in the case of $m = 0.5$ and $n = 0.56$ for a type 304 stainless steel at 923 K.

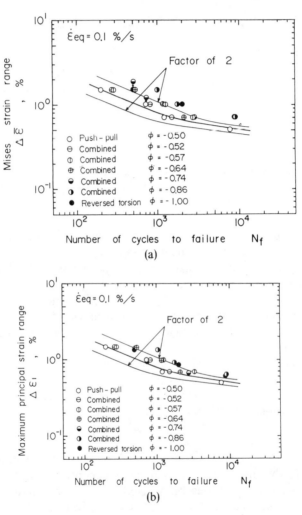

Fig. 6.15. Arrangement of the low-cycle biaxial fatigue failure data of type 304 stainless steel at 923 K, with (a) Mises strain range, $\Delta\bar{\varepsilon}$, (b) maximum principal strain range, $\Delta\varepsilon_1$, (c) maximum shear strain range, $\Delta\gamma_{max}$, (d) BMK strain range, $\Delta\bar{\gamma}$, (e) LE strain range, $\Delta[(\gamma_{max}/2) + 0.2\,\varepsilon_n]$, and (f) COD strain range, $\Delta\varepsilon^*$. (Reprinted from ref. 12a by permission of Japan Soc. Mech. Engrs)

Figures 6.15 (a)–(f) [12a, b] are the arrangements of the Mises equivalent strain-range controlled low-cycle fatigue data for the unnotched tubular specimen at 923 K in air with the following strain parameters: (a) the Mises equivalent strain range $\Delta \bar{\varepsilon}$; (b) maximum principal strain range $\Delta \varepsilon_1$; (c) maximum shear strain range $\Delta \gamma_{max}$; (d) equivalent shear strain range, $\Delta \bar{\gamma} = [A(\Delta \varepsilon_n)^j + (\Delta \gamma_{max}/2)^j]^{1/j}$ [13]; (e) Γ^*-plane parameter, $\Gamma^* = (\gamma^*_{max}/2) + K\varepsilon^*_n$ [14]; (f) the equivalent strain range based on COD, $\Delta \varepsilon^*_I$ in eqn (6.10) and $\Delta \varepsilon^*_{II} = 0.5|\Delta \gamma_{max}|$. In these parameters, γ_{max} is the maximum shear strain, ε_n is the normal strain on that shear plane, γ^*_{max}

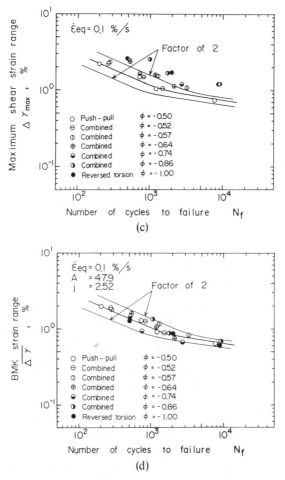

Fig. 6.15. — contd.

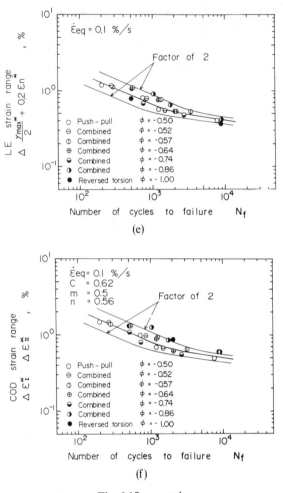

Fig. 6.15.—*contd.*

and ε_n^* are those for the shear plane which intersects the free specimen surface at an angle of 45°, and A, j and K are material constants; ε_I^* and ε_{II}^* are for mode I and mode II fracture respectively. Both $\Delta\bar{\gamma}$ and Γ^* parameters were modified from the Γ-plane parameter originally proposed by Brown and Miller [15]; the material constants in $\Delta\bar{\gamma}$ were determined as $A=47.9$ and $j=2.57$ from the present data, and K has the value 0.2. Four strain bases, maximum principal strain range, Γ-plane parameter, Γ^*-plane parameter and the equivalent strain range based on

COD, are effective to correlate the biaxial low-cycle fatigue data for the unnotched tubular specimen. However, the Mises equivalent strain range and maximum shear strain range yield a poor correlation. As to the stress parameter, as shown in Fig. 6.16 (a)–(d) [12a, b], only the equiva-

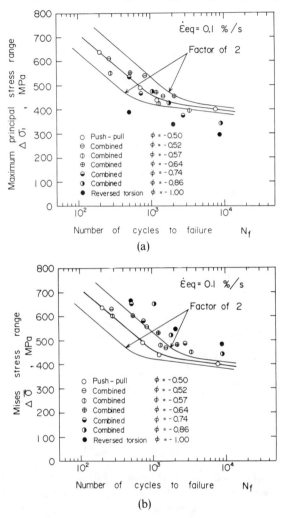

Fig. 6.16. Arrangement of the low-cycle biaxial fatigue failure data, with (a) Mises stress range, $\Delta\bar{\sigma}$, (b) maximum principal stress range, $\Delta\sigma_1$, (c) maximum shear stress range, $\Delta\tau_{max}$, and (d) COD stress range, $\Delta\sigma^*$. Test conditions as for Fig. 6.15. (Reprinted from ref. 12a by permission of Japan Soc. Mech. Engrs)

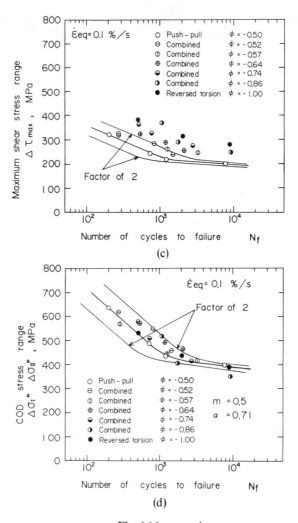

Fig. 6.16.—contd.

lent stress range based on COD (eqn (6.8)) gives a satisfactory result while the Mises equivalent stress range $\Delta\bar{\sigma}$, maximum principal stress range $\Delta\sigma_1$ and maximum shear stress range $\Delta\tau_{max}$ cannot correlate the biaxial low-cycle fatigue data within a factor of 2 scatter band as well as the strain parameters, where $\Delta\sigma_{II}^* = 1\cdot4|\Delta\tau_{max}|$ is for mode II fracture.

Table 6.1 [12a, b] summarizes the test data together with the values of the parameters shown in Figs 6.15 and 6.16. The values at the half-life

Table 6.1

Results of Low-cycle Biaxial Fatigue of Type 304 Stainless Steel at 923 K in Air (Reprinted from ref. 12a by permission of Japan Soc. Mech. Engrs).

Specimen no.	ϕ	Strain (%)								Stress (MPa)							N_f
		$\Delta\varepsilon$	$\Delta\gamma$	$\Delta\varepsilon_1$	$\Delta\gamma_{max}$	$\Delta\varepsilon_{eq}$	$\Delta\varepsilon_1^*$	$\Delta\varepsilon_{11}^*$		$\Delta\sigma$	$\Delta\tau$	$\Delta\sigma_1$	$\Delta\tau_{max}$	$\Delta\sigma_{eq}$	$\Delta\sigma^*$	$\Delta\sigma_{11}^*$	
419	−0·50	1·50	0	1·50	2·25	1·50	1·50			638	0	638	319	638	638		200
420	−0·50	1·00	0	1·00	1·50	1·00	1·00			492	0	492	246	492	492		720
418	−0·50	0·70	0	0·70	1·00	0·70	0·70			439	0	439	219	439	439		1200
432	−0·50	0·50	0	0·50	0·75	0·50	0·50			403	0	403	201	403	403		7600
423	−0·52	1·45	0·67	1·50	2·28	1·50	1·48			572	153	611	325	631	620		270
424	−0·52	0·98	0·45	1·00	1·52	1·00	0·98			513	124	542	285	557	549		820
416	−0·52	0·68	0·31	0·70	1·06	0·70	0·69			427	113	455	241	470	462		1480
421	−0·57	1·30	1·30	1·50	2·34	1·50	1·43			469	213	552	317	597	572		280
415	−0·57	0·87	0·87	1·00	1·56	1·00	0·95			322	206	423	261	481	447		1240
413	−0·57	0·61	0·61	0·70	1·10	0·70	0·67			304	192	397	245	450	419		3280
433	−0·64	1·06	1·84	1·48	2·43	1·50	1·34			459	225	551	321	602	573		530
439	−0·64	0·71	1·23	0·99	1·62	1·00	0·89			364	224	470	289	532	496		1180
438	−0·64	0·50	0·86	0·69	1·14	0·70	0·62			397	161	454	256	485	468		2100
428	−0·74	0·75	2·25	1·45	2·52	1·50	1·20			345	320	536	363	652	581		510
406	−0·74	0·50	1·50	0·96	1·68	1·00	0·80			280	295	466	326	582	511		730
414	−0·74	0·35	1·05	0·67	1·17	0·70	0·56			202	256	377	276	488	418		2720
435	−0·86	0·39	2·51	1·39	2·58	1·50		1·29		207	356	474	371	651		519	1050
436	−0·86	0·26	1·67	0·92	1·72	1·00		0·86		270	258	426	291	521		407	1750
437	−0·86	0·18	1·17	0·65	1·20	0·70		0·60		175	237	340	253	446		354	8900
404	−1·00	0	2·60	1·30	2·60	1·50		1·30		18	383	391	383	663		536	500
418	−1·00	0	1·73	0·87	1·73	1·00		0·87		49	314	339	315	546		441	2000
431	−1·00	0	1·21	0·61	1·21	0·70		0·61		32	279	295	279	484		391	8600

cycle are employed in calculating the stress and strain parameters.

It should be emphasized that the Mises equivalent stress and strain, which are excellent parameters to express the cyclic yield condition and constitutive equation in the biaxial stress state (see Section 4.4 in Chapter 4), and have been most commonly used to express the equivalent base in the multiaxial stress state, are revealed to be not a good parameter for correlating the biaxial low-cycle fatigue data from the viewpoint of a sophisticated life assessment of structural materials at high temperatures. Thus, the poor correlation of the Mises strain is reported by the author and other researchers but some reports support the applicability of the Mises equivalent strain parameter. The lack of sufficient data in the most serious point in finding the proper stress and strain parameters. From this point of view we will describe some critical data in the following sections.

In this section, the following conclusions were obtained from the biaxial low-cycle fatigue tests under combined axial and torsional stress using a centrally notched and unnotched hollow cyclindrical specimen of solution heat-treated type 304 stainless steel at temperatures ranging from 923 K to 293 K in air.

(1) Maximum principal stress, maximum shear stress and the von Mises equivalent stress cannot properly arrange the biaxial low-cycle fatigue failure life data at elevated temperatures for the notched and unnotched tubular specimens. The poor arrangement is caused by the maximum principal stress criterion. It does not include a parallel stress effect on the crack behavior which increases the crack propagation rate while the von Mises equivalent stress includes too much of a parallel stress effect.

(2) As to the stress parameter, only the equivalent stress based on COD can arrange the biaxial low-cycle fatigue failure data properly for both specimens.

(3) The equivalent strain based on COD is also derived from the equivalent stress. The equivalent strain parameter is effective to correlate the biaxial low-cycle fatigue failure data as well as the maximum principal strain parameters, Γ-plane parameter and Γ^*-plane parameter. However, the Mises equivalent strain range and the maximum shear strain range yield a poor correlation.

6.2.3 High-temperature Biaxial Low-cycle Fatigue Tests for Cruciform Specimens

In the previous combined push-pull and reversed torsional low-cycle fatigue tests conducted at elevated temperatures, the test condition of the strain ratio of the minimum principal strain to the maximum strain, expressed as $\phi = \varepsilon_3/\varepsilon_1$, is limited to a range of $-1 \leqslant \phi \leqslant -v$, where v is Poisson's ratio in the plastic range and is assumed to be 0·5. Therefore it is necessary to perform the tests under the additional condition $-0.5 \leqslant \phi \leqslant 1$. Performing a biaxial push-pull test using a cruciform specimen is the most adequate testing method for satisfying the condition $-1 \leqslant \phi \leqslant 1$ of the strain ratio for the purpose of examining crack behavior in the biaxial high-temperature low-cycle fatigue test.

Before designing the biaxial low-cycle fatigue testing machine, we performed an inelastic analysis of the cruciform specimen by which the uniform stress/strain state is realized within the central portion of the cruciform specimen by using the cyclic constitutive equation of the material tested at elevated temperatures. The analysis was performed by using the isoparametric two-dimensional finite elements with eight nodes, as shown in Fig. 6.17 [16]. The material tested was 1 Cr–1 Mo–0·25 V steel with a Young's modulus of 205 GPa and a yield stress of 410 MPa in the cyclic constitutive equation at 823 K in the atmosphere. The material was extracted from the steam turbine rotor. The multilinear approximation method was used for constructing the cyclic stress/strain relation. Figure 6.18 [16] shows the stress/strain distributions in the x-

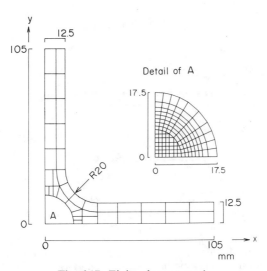

Fig. 6.17. Finite element mesh.

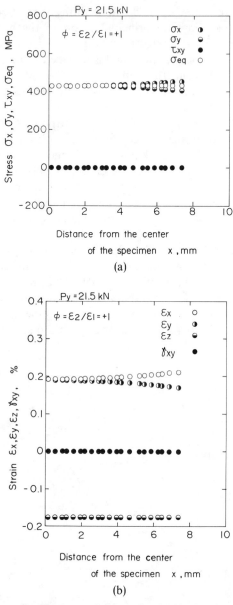

Fig. 6.18. (a) Stress distribution and (b) strain distribution in central portion of cruciform specimen, by inelastic finite element analysis. (Reprinted from ref. 16 by permission of Japan Soc. Mech. Engrs).

and y-directions of the central portion of the specimen, and we found that the distributions are uniform in the domain of a 6 mm diameter within the central portion with a 15 mm diameter and 1 mm thickness. Therefore we concluded that the cruciform specimen, as shown in Fig. 6.19 [16, 17], is adequate for the purpose of the present test.

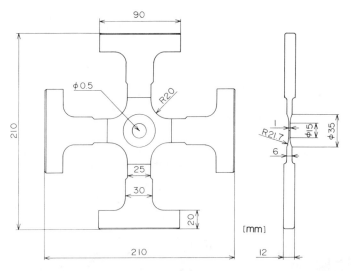

Fig. 6.19. Shape and dimensions of the cruciform specimen.

Figure 6.20 [16, 17] shows the block diagram of the testing machine and Fig. 6.21 [16, 17] is the overview. The testing machine is composed of four electro-hydraulic actuators and servo-valves and has a load capacity of 5 tons. The most important point was to construct software and control circuits which could constantly maintain the central point of the specimen. Basically, the analogue control for stroke in the actuators was adopted and additionally a digital feedback control circuit was used for performing the strain-controlled test, by which the strain in the central portion of the specimen is maintained in each cycle by changing the function wave in the servo-amplifiers through the microcomputer. The measurement of strain in the central portion was made by using a pair of X–Y differential transducer-type devices, as shown in Fig. 6.22 [16]. Figure 6.23 [16] shows an example of hysteresis loops for $\Delta \varepsilon_x = \Delta \varepsilon_y = 0.78\%$ under the principal strain ratio of 1 at 823 K. We can see that the strain in the central portion of the specimen remains

270 PLASTICITY AND HIGH TEMPERATURE STRENGTH OF MATERIALS

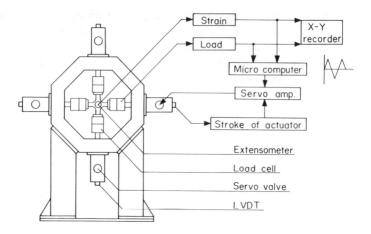

Fig. 6.20. Block diagram of the high-temperature biaxial low-cycle fatigue testing machine for cruciform test specimen.

Fig. 6.21. Overview of the testing machine.

FRACTURE LAWS IN A CONTINUUM MECHANICAL APPROACH 271

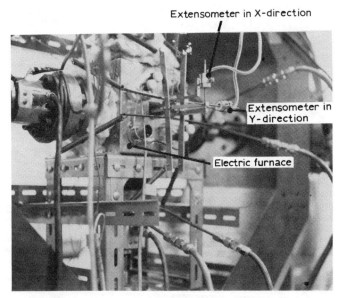

Fig. 6.22. X-Y differential transducer-type extensometer and electric furnace for heating the specimen.

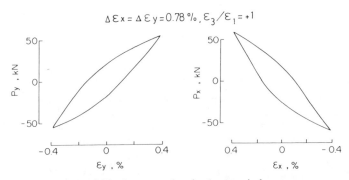

Fig. 6.23. An example of a hysteresis loop.

constant even if the material behavior is that of cyclic hardening or softening during the test. A specially designed electric furnace was used for heating the central portion of the cruciform specimen.

Figure 6.24 (a) [16, 17] shows the low-cycle fatigue failure life N_f and N_c of a centrally circular notched cruciform specimen under triangular strain waveform with strain rate $10^{-3}\,\mathrm{s}^{-1}$ at 823 K in air. The test was

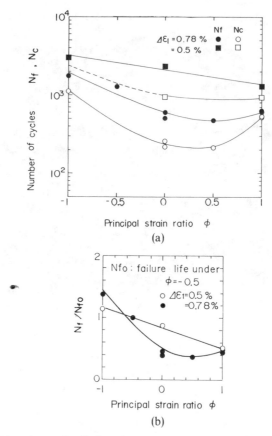

Fig. 6.24. Experimental relation between the number of cycles to failure, N_f/crack initiation, N_c, and principal strain ratio, ϕ, of 1Cr–1 Mo–0·25V steel at 823 K in air: (a) N_f and N_c versus ϕ plot; (b) normalized failure life, N_f/N_{f0}, versus ϕ plot. (Reprinted from ref. 16 by permission of Japan Soc. Mech. Engrs)

performed under the strain ratio $-1 \leqslant \phi \leqslant 1$ at a fixed maximum principal total strain range $\Delta\varepsilon_1$ of 0·5% and 0·78%. Here N_f is defined as the number of cycles at which the tensile load amplitude in the x- and y-directions of the specimen decreases to 3/4 of the maximum value, and N_c is the number of cycles when the cracks reach about 100 μm. We can see that the low-cycle fatigue crack initiation life and the failure life of the metal depend on the biaxiality of the stress. Figure 6.24(b) [16] is the normalized representation of Fig. 6.24(a) by N_f/N_{f0}, where N_{f0} is the failure life under uniaxial push-pull, i.e. $\phi = -0·5$.

Figures 6.25 (a), (b) and (c) [16] show the comparison between the

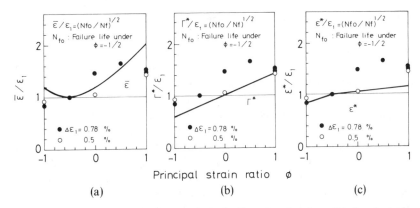

Fig. 6.25. Comparison between analytical effect on the three kinds of strain parameters and that of the experimental results: (a) Mises strain, $\bar{\varepsilon}$; (b) LE strain, Γ^*; (c) COD strain, ε^*. Test conditions as for Fig. 6.24. (Reprinted from ref. 16 by permission of Japan Soc. Mech. Engrs)

analytical effects of the strain ratio ϕ on the three kinds of strain parameter and the experimental data. The three parameters are (a) the Mises equivalent strain $\bar{\varepsilon}$, (b) the Γ^*-plane parameter and (c) the equivalent strain ε^* based on COD. Here ε^* was derived directly from the analytical relation of (COD/ε_1) versus $(2-\phi)$ using inelastic finite element analysis as

$$\Delta\varepsilon^* = \beta\Delta\varepsilon_1(2-\phi)^{m'} \quad (\beta=1\cdot5^{-m'}) \tag{6.11}$$

where $m'=-1$ and $\beta=2\cdot5$ in $-1\leqslant\phi<-0\cdot5$, and $m'=-0\cdot16$ and $\beta=1\cdot16$ in $-0\cdot5<\phi\leqslant1$ for 1 Cr–1 Mo–0·25 V steel at 823 K in air. On the other hand, the experimental data points were plotted from the relation $(\Delta\varepsilon/\Delta\varepsilon_1)=(N_f/N_{f0})^{-1/2}$, where $\Delta\varepsilon$ shows the strain parameter of $\bar{\varepsilon}$, Γ^* and ε^*, respectively, and N_f/N_{f0} is given as in Fig. 6.24(b). We found from these comparisons that there is a difference between them and the difference is large in the range $-0\cdot5<\phi\leqslant1$ compared to that for $-1\leqslant\phi<-0\cdot5$ which corresponds to combined axial and torsional stresses in the previous sections. This suggests the necessity for research in the ϕ value range from $-0\cdot5$ to 1. We also found that the von Mises equivalent strain $\bar{\varepsilon}$ is inadequate for the arrangement of the actual fatigue failure life for a wide range of the strain ratio, particularly for the equi-biaxial loading $\phi=1$. Besides, it is seen that the Γ^*-plane parameter and the equivalent strain based on the COD concept are more adequate for this type of arrangement.

It was also observed that all macroscopic through-cracks, in the centrally notched cruciform specimen, propagate in mode I irrespective of the strain ratio tested. In the cruciform specimen subjected to biaxial push-pull, the maximum and minimum principal strains are produced on the plane parallel to the specimen surface for ϕ values of -1 to -0.5, assuming that Poisson's ratio in the plastic region is 0.5. However, in the ϕ value range -0.5 to 1, the minimum principal strains are produced on the plane normal to the specimen surface. Therefore it is necessary to observe the cracks not only on the specimen surface but also throughout the thickness, for the determination of the crack mode in the ϕ value range -0.5 to 1. It was observed that the surface cracks originated from the circular notch root in mode I and also propagated throughout the thickness in mode I. Thus, we concluded for the notched cruciform specimen that the main cracks propagate in mode I irrespective of the ϕ values tested. In the following section, the direction of macrocrack propagation in biaxial low-cycle fatigue will be described in connection with microcrack behavior.

We conclude that the failure life data in biaxial low-cycle fatigue in the range of strain ratio ϕ from -1 to 1 for a cruciform specimen of 1Cr–1Mo–0.25V steel, used for steam turbine rotors, are arranged well by the Γ^*-plane criterion and the COD criterion, using a newly designed biaxial fatigue testing machine for cruciform specimens. Also, the cracks propagated in mode I (growth on the plane normal to the direction of the maximum principal stress) irrespective of the ϕ values tested for the centrally notched cruciform specimen.

6.2.4 Crack Direction in High-temperature Biaxial Low-cycle Fatigue

Brown and Miller [18] surveyed the theories concerning the cause of the so-called stage I or mode II (growth on the plane of the maximum shearing stress) to stage II or mode I (growth on the plane normal to the direction of the maximum principal stress) transition in the low-cycle fatigue and high-cycle fatigue of metals. In biaxial low-cycle fatigue under combined torsion and tension (or bending), they noted the following. Cracks occurred close to the maximum shear planes except during uniaxial loading; on the other hand, mode I cracking in low-carbon steel at 723 K was observed. Mode II cracking for a hollow cylindrical specimen of heat-treated 1 Cr–Mo–V steel at room temperature was observed under out-of-phase loading, i.e. cracks initiated on those planes experiencing the greatest range of shear strain, irrespective of the varying principal stress axes. Brown and Miller [18] observed the following. The mode I–mode II transition occurred at a strain ratio, $\psi = \gamma/\varepsilon$, of about 1·5 (in other words, principal strain ratio ϕ of the minimum principal strain to the maximum principal strain is -0.62 assuming Poisson's ratio of 0·5) in 1 Cr–Mo–V steel at 293 K and AISI 316 stainless steel at 673 K in air. However, in 1 Cr–Mo–V steel at 838 K in air, cracks were initiated by oxidation on mode I planes for all states of strain, which enabled a transition from mode I to mode II to occur for a ψ value of 4 (ϕ value of -0.79). They concluded that, at high temperatures in an oxidizing environment, fatigue cracks originate from pits and grow on mode I planes and a transition to mode II propagation occurs for the ψ value above 4. They also discussed the fact that the transition depends not only on the strain state but also on the strain amplitude, higher ranges making transition more likely. Thus there is a body of conflicting data concerning the stage I to stage II transition.

In this section we will describe this type of transition for micro- and macro-crack propagation behavior, using a hollow cylindrical specimen, with and without a small circular notch, composed of solution heat-treated type 304 stainless steel under combined push-pull and reversed torsion at 923 K in air.

Figure 6.26 [12a, b] shows the crack direction of the three types of crack, which is classified regarding the crack length, in biaxial low-cycle fatigue for the unnotched hollow cylindrical specimen. The largest is the main crack which leads to fracture of the specimen and the length is larger than 1 mm. The second is the subcrack whose length is between 0·1 mm and 1 mm, and the smallest is the microcrack whose length is less than 0·1 mm. The testing method is the same as that mentioned in previous chapters. The crack modes are as follows. Propagation of the main cracks for the unnotched specimen changes from mode I (growth on the plane normal to the direction of the maximum principal stress, i.e. crack opening mode) to mode II (growth on the plane of the maximum shearing stress, i.e. in-plane shear mode) at the ϕ value of -0.64 to -0.86. Under combined push-pull and reversed torsion the intermediate principal strain is in a radial direction relative to the tubular specimen, and both the maximum and minimum principal strains are produced in the plane parallel to the tubular specimen surface. Therefore it is noted that the crack growth planes in modes I and II under reversed torsion and combined push-pull/reversed torsion are normal to the specimen surface and that the crack mode can be determined from the optical microscope observation of surface cracks.

After failure, each specimen, upon examining photographic evidence,

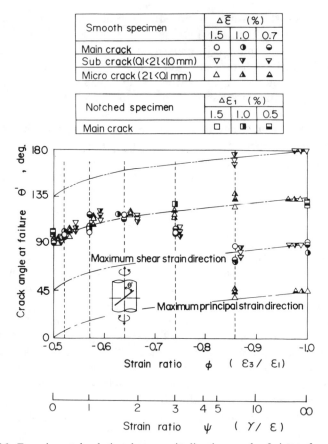

Fig. 6.26. Experimental relation between inclination angle, θ, in surface cracks and principal strain ratio, ϕ, for a tubular specimen of type 304 stainless steel at 923 K in air. (Reprinted from ref. 12a by permission of Japan Soc. Mech. Engrs)

showed multiple cracking all over the surface in a gauge length (20 mm). The angle of the cracks in relation to the specimen axis was measured for a large number of cracks on each specimen to determine typical crack directions. A similar transition for the unnotched specimen is also observed in the subcracks but not for the microcracks. We observed that microcracks smaller than 0·1 mm propagate in the mode I direction irrespective of the strain ratio tested. However, we observed that, for centrally notched and precracked hollow cylindrical specimens, the main cracks propagate in the mode I direction irrespective of the states of strain. This was observed in the other data for the centrally notched tubular specimen of type 304 stainless steel at 823 K in air [19], and for the centrally notched cruciform specimen of 1 Cr–1 Mo–0·25 V steel at 823 K in the atmosphere [16, 17].

Figure 6.27(a) [12a, b] shows an example of growth curves for surface microcracks around the main crack in the unnotched specimen under the von Mises equivalent strain range $\Delta\bar{\varepsilon}=1\cdot0\%$ in reversed torsion ($\phi=-1$). The main crack in reversed torsion propagated in mode II, as shown in Fig. 6.26. It was found that the majority of surface cracks coalesce with the main crack when the surface cracks reach a length of about 200–500 μm. Figure 6.27(b) [12a, b] shows the variation of the number of surface cracks within a 2 mm × 2 mm area on the unnotched specimen surface. We found that coalescence of cracks occurs when the surface crack density reaches about 30 mm^{-2}. Figure 6.27(c) [12a, b] shows the experimental relation between the number of surface cracks larger then 100 μm and the inclination angle relative to the specimen axis during the test under reversed torsion.

Figure 6.28(a) [12a, b] shows the optical microscope observations of the data of Fig. 6.26. Figure 6.28(b) [12a, b] shows high magnification photographs of the subcrack that propagates on the specimen surface and into the specimen in the reversed torsion test ($\phi=1$) of Fig. 6.28(a). The figure clearly shows that the subcrack propagation is in zigzag manner and the subcrack appears to be composed of mode II microcracks. Whether the subcrack is made by linking of microcracks or the zigzag shape is made in the propagation process of the subcrack is not yet apparent since a definite conclusion requires detailed observation of the propagation process of subcracks. The photographs only show the result after the linking or the propagation. In elevated temperature experiments, surface oxidation affects the crack morphology. In order to examine the effect of oxidation on the three types of crack, the reversed torsion test was carried out at room temperature. The macrocrack and the subcrack direction at room temperature are the same as those at 923 K in air and are mode II, but the shape of the subcrack is different. Figure 6.28(c) [12a, b] shows the shape of the subcrack at room temperature; it is rather straight and no zigzag pattern is found. Mode I microcracks were not found at room temperature. In conclusion, the surface oxide film is brittle and is the cause of the opening type mode I microcrack. The oxidation does not affect the crack direction of the subcrack essentially but affects the shape of the subcrack.

In this section, from the data for strain-controlled combined axial and

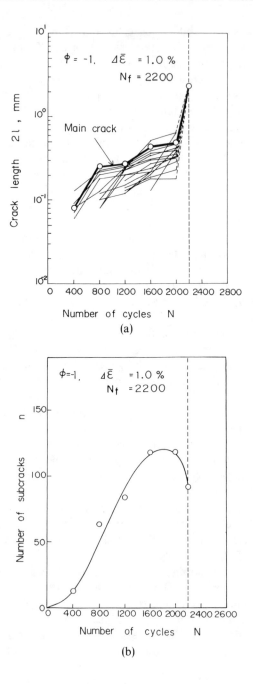

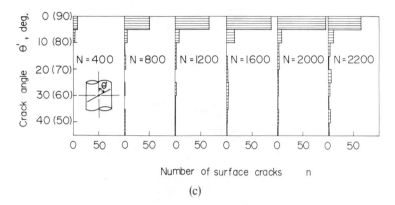

Fig. 6.27. (a) Surface crack growth curves, (b) change in the number of surface cracks with number of strain cycles, and (c) distribution of the angle of surface crack, having a length larger than 100 μm, in relation to the specimen axis, under $\Delta\bar{\varepsilon} = 1.0\%$ in reversed torsion ($\phi = -1$). Test conditions as for Fig. 6.26. (Reprinted from ref. 12a by permission of Japan Soc. Mech. Engrs).

torsional low-cycle fatigue for a centrally notched and unnotched tubular specimen of solution heat-treated type 304 stainless steel at 923 K and 1 Cr–1 Mo–0·25 V steel at 823 K in the atmosphere, we have made the following conclusions.

(1) In the region of principal strain ratio, $\phi = \varepsilon_3/\varepsilon_1$, ranging from -0.5 to -0.64, the main crack of the unnotched specimen propagates in the mode I direction but the mode II direction is observed for the ϕ value range -0.86 to -1. On the other hand, in the notched and pre-through cracked specimens the main crack propagates in the mode I direction irrespective of the ϕ value tested. Notably, in the reversed torsion test the unnotched specimen cracks in the mode II direction which is quite different behavior from that observed in the case of the notched and precracked specimens.

(2) Subcracks of the unnotched specimen, whose length is between 0·1 mm and 1 mm, exhibit the same propagation direction as the macrocrack. Microcracks, of which the length is less than 0·1 mm, always are mode I. The fracture feature of the unnotched specimen is the linking of many mode II type subcracks in the reversed torsion test. The shape of the subcracks is zigzag at 923 K in air but is rather straight in mode II type at room temperature. The zigzag shape of the subcrack is attributed to the microcrack in mode I type being formed by oxidation.

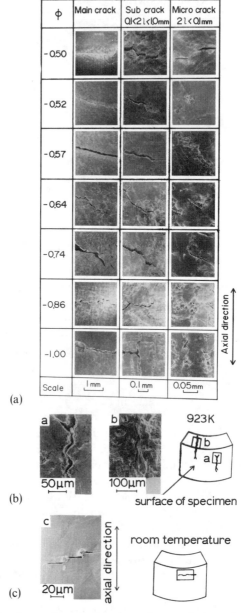

Fig. 6.28. (a) Summary of crack direction of unnotched specimen in the strain ratio, ϕ, ranging from -1 to -0.5 at 923 K in the atmosphere, and (b) high-magnification photographs of subcrack in the reversed torsion ($\phi = -1$) at 923 K in air. (c) Shows the shape of the subcrack in the reversed torsion at room temperature. (Reprinted from ref. 12a by permission of Japan Soc. Mech. Engrs)

6.3 EFFECT OF LOAD WAVEFORM AND SEQUENCE ON HIGH-TEMPERATURE LOW-CYCLE FATIGUE OF HEAT-RESISTING METALS AND ALLOYS FROM THE VIEWPOINT OF CRACK BEHAVIOR

6.3.1 Effect of Strain Wave Shapes on Creep–Fatigue Interaction

In the present section, the effect of strain wave shapes on the creep–fatigue interaction of a cobalt-based superalloy X-40 (C 0·51, Si 0·20, Ni 9·81, Cr 25·91, W 7·45, Fe 0·219, Co Bal, wt%), commonly used for gas turbine rotors, is described by using four types of total strain-range controlled fatigue tests at 1073 K in the atmosphere. The four types of wave shape are (a) push-pull continuous low-cycle fatigue test, (b) the low-cycle fatigue test with tensile hold-time in each cycle, (c) the low-cycle fatigue test with compressive hold-time in each cycle and (d) the low-cycle fatigue test with tensile hold-time in every four cycles, as shown in Fig. 6.29 [20]. Strain waveforms (b)–(d) have equivalent creep and fatigue components in the strain wave and only the combinations of the components are different from one another. In the strain waveform of (d), a cycle which contains the hold-time is denoted by 'cycle 1', that following cycle 1 is denoted by 'cycle 2', in a block of four cycles, and so on. The test was performed by using a hollow cylindrical specimen with a circular notch of diameter 1 mm and an electro-hydraulic servo-type push-pull testing machine.

Figure 6.30 [20] shows the test results, where the number of cycles to failure N_f is defined as the number of cycles at which the stress amplitude is decreased to 3/4 of the maximum amplitude. It is clear from the figure that a smaller number of cycles to failure was observed for strain wave (b) than for (a), but in strain wave (c) a decrease in the failure life is not observed; (instead we find an increase). On the other hand, in the compression of strain waves (d) and (b) there is no clear difference in the failure lives at a total strain range of 1%, but a greater decrease in wave (b) was observed at a total strain range of 0·5%. Therefore we concluded that the low-cycle fatigue failure life of the alloy tested is very much influenced by the strain wave shape and the loading sequence.

As to the effect of compressive hold-time on the number of cycles to failure, it has been reported [21, 22] that the compressive hold-time is more dangerous than tensile hold-time. On the contrary, it has also been reported [23, 24] that the compressive hold-time has no effect on the cycles to failure. The experimental data shown in Fig. 6.30 support the latter claim, but in all the tests, except the present experiment, the plastic strain range controlling method was employed. Therefore a generalized

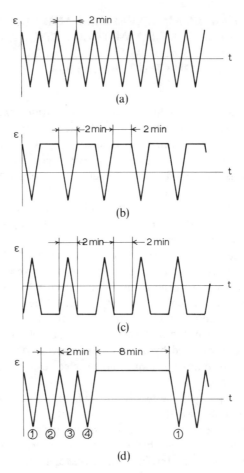

Fig. 6.29. Strain waveforms. (a) Push–pull low-cycle fatigue, (b) fatigue with tensile strain hold in each cycle, (c) fatigue with compressive strain hold in each cycle, and (d) fatigue with tensile strain hold every four cycles.

conclusion concerning the effect of compressive hold-time on failure lives cannot yet be given. Moreover, in the test with compressive hold-time it was observed that the compressive plastic strain range is larger than the tensile plastic strain range in the hysteresis loop. So, when a plastic strain-controlled test with compressive hold-time is conducted, it is reasonable to expect a shift of the hysteresis loop to the tensile side, and consequently the failure life is remarkably decreased by the tensile mean

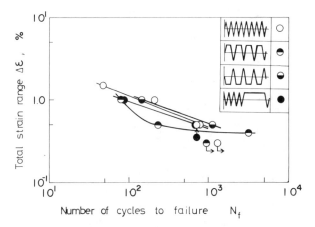

Fig. 6.30. Experimental relation between total strain range, $\Delta\varepsilon$, and the number of cycles to failure, N_f, of superalloy X-40 in four strain waveforms at 1073 K in the atmosphere. (Reprinted from ref. 20 by permission of Japan Soc. Mech. Engrs)

stress as is suggested by Lord and Coffin [22]. Although it seems here that the strain-controlling method is obviously very influential on determination of the failure life in the test with compressive hold-time, it is too early to make any general conclusion about the effect of compressive hold-time on life.

In our opinion the reduction in the failure cycle of the alloy in strain wave (b) as compared to that in strain wave (a) is caused by an increase in the inelastic strain range by tensile hold-time in each cycle, as shown in Fig. 6.31 [20]. In the case of strain wave (d), a reduction of life also seems to result from an increase of the inelastic strain range as in the case of strain wave (b) in the total strain range 1·0%, as shown in the figure. However, in the lower total strain range 0·5%, the inelastic strain range in cycle 1 which increased with the hold-time in tension abruptly decreased in cycles 2, 3 and 4, and the failure cycle in strain wave (d) did not decrease as in strain wave (b). In the test with hold-time in compression, the increase of inelastic strain range with the hold-time is not much larger than that in the test with hold-time in tension. Therefore it is likely that the hold-time in compression was quite influential on the failure life of the alloy tested.

Figures 6.32 (a) and (b) [20] show the crack growth curves of the alloy tested. The faster crack propagation rate in strain waves (b), (c) and (d)

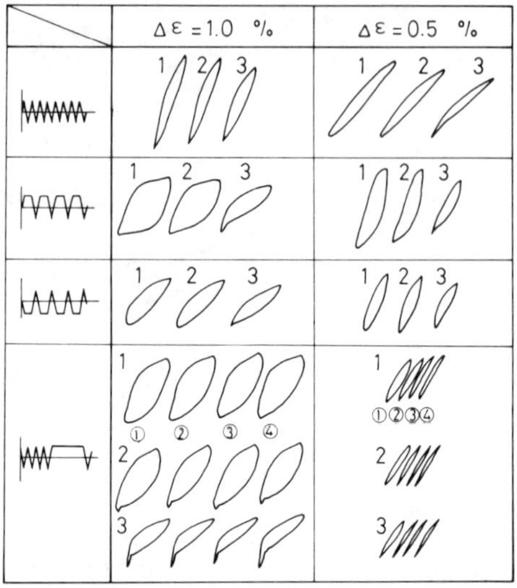

284   PLASTICITY AND HIGH TEMPERATURE STRENGTH OF MATERIALS

Fig. 6.31. Change in the shape of a hysteresis loop of superalloy X-40 in four strain waveforms at 1073 K in the atmosphere. The numbers 1, 2 and 3 show the hysteresis loop at $N = 0.1N_f$, $0.5N_f$ and $0.9N_f$ respectively. (Reprinted from ref. 20 by permission of Japan Soc. Mech. Engrs)

than that in strain wave (a) in the case where $\Delta\varepsilon = 1.0\%$ is a result of the accumulation of creep damage of the material at the crack tip during the hold-time, i.e. *pre-damage of the material*. We can consider that a slower crack propagation rate in strain wave (c) than in strain waves (b) and (d) in the region of long cracks is a result of the decrease of stress concentration at the crack tip due to closing of the crack during the hold-time in compression. On the other hand, in the case where $\Delta\varepsilon = 0.5\%$ the crack propagation rate in strain wave (b) is greatest. The crack propagation rate is faster in strain wave (b) than in strain wave (a) and seems to be caused for the same reasons as mentioned in the case of $\Delta\varepsilon = 1.0\%$. Even the crack propagation rate in strain wave (d) is as fast as strain wave (a). The disappearance of the effect of hold-time from the crack propagation rate in strain wave (d) is also caused by the same mechanism mentioned that changes the hysteresis loop in Fig. 6.31. That is, in the total strain range of 0.5%, since the strain is comparatively small, the hold-time effect which appears in cycle 1 disappears sooner in the following cycles and the accumulation of damage in the material does not become so influential. These phenomena will be interpreted from the following mechanical model.

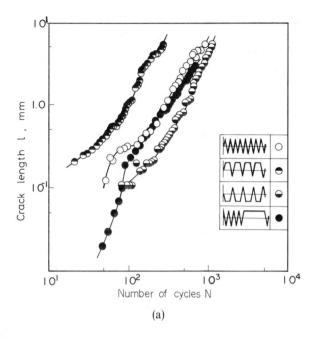

(a)

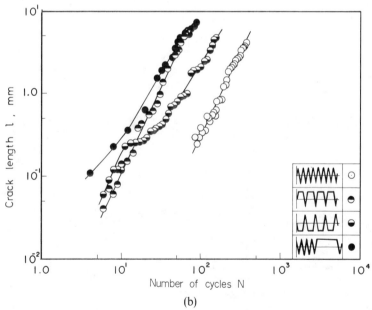

(b)

Fig. 6.32. Crack growth curves of superalloy X-40 in four strain waveforms at 1073 K in the atmosphere: (a) $\Delta\varepsilon = 0.5\%$; (b) $\Delta\varepsilon = 1.0\%$. (Reprinted from ref. 20 by permission of Japan Soc. Mech. Engrs)

6.3.2 Mechanical Model of Crack Propagation in Material under Creep–Fatigue Interaction Conditions

Considering the influence of stress–strain history on the material, we applied the following concept of pre-damage at the crack tip and the discrete computation method to the fatigue crack. The following equation was used as the constitutive equation of the material:

$$\varepsilon = (\dot{\varepsilon}/n)^{n/(n-1)}[1/(a\sigma^{\alpha})]^{1/(n-1)} \tag{6.12}$$

where the material constants α and n are represented as

$$\varepsilon = a\sigma^{\alpha}t^{n} \tag{6.13}$$

and ε, σ and t are strain, stress and time, respectively. We also employed the following equation [25] for the stress distribution at the front of the crack tip, as shown in Fig. 6.33:

$$x = 1/[\sigma F(\sigma)] + \int_{0}^{\sigma} d\sigma / [F(\sigma)\sigma^{2}] \tag{6.14}$$

and the equation can be applied to an arbitrary constitutive equation $\varepsilon = F(\sigma)$. Substituting eqn (6.12) into eqn (6.14), we obtain the following relation:

$$\sigma = \Phi[\dot{\varepsilon}^{n/(n-1)}x]^{(n-1)/(\alpha-n+1)}l^{1/2} \tag{6.15}$$

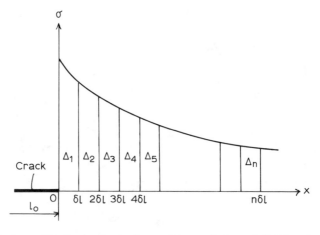

Fig. 6.33. Stress distribution in the front of the crack tip and dividing the front of the crack tip into infinitesimal lengths dl.

where $\Phi = [(1/n)^{n/(n-1)} a^{-1/(n-1)} \{(\alpha-n+1)/\alpha\}]^{(n-1)/(\alpha-n+1)}$. In this relation the term for crack length l is included from the analogy with fracture mechanics. Crack propagation in the fatigue test with hold-time is evaluated as follows.

Dividing the front of the crack tip into infinitesimal lengths dl, as shown in Fig. 6.33, we calculate the inelastic strain energy in the distance by partitioning the stress–strain relation into two parts, namely, the plastic part due to the fatigue component and the creep component during hold-time, as shown in Fig. 6.34. This strain partitioning method is the same as that proposed by Manson and Halford [26]. The plastic strain and creep strain are evaluated from eqn (6.12) and eqn (6.13) respectively. The creep–fatigue interaction condition is represented by stress hold-time in both tension and compression, and it is assumed that the compressive stress hold is subject to the same damage as that in tension. A criterion under which the crack extends to the infinitesimal distance nearest to the crack tip is written as

$$\sum_k (E_c/EC)_k + \sum_k (E_f/EF)_k = 1 \qquad (6.16)$$

where E_c and E_f are the inelastic strain energy in each cycle due to creep and fatigue components, respectively, and EC and EF are the respective critical accumulated inelastic strain energies. The calculation procedure is as follows.

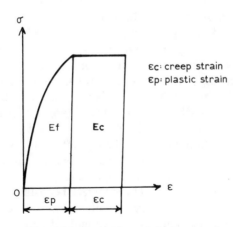

Fig. 6.34. Evaluation of the damage.

At the first step we calculate the number of cycles, N_1, in which the crack extends to the division, $\Delta 1$, by eqn (6.16) giving EC and EF and then evaluate the plastic strain energy, E_{fi1}, and creep strain energy, E_{ci1}, accumulated in the division from $\Delta 2$ to Δn during N_1. In the same way we calculate N_2 in the division, $\Delta 2$, by eqn (6.16) using the critical strain energies $(EF - E_{f21})$ and $(EC - E_{c21})$, where subtraction of E_{f21} and E_{c21} from EF and EC, respectively, is due to the pre-damages E_{f21} and E_{c21} in the front of the crack during N_1. After calculating N_2 we can calculate N_3 by using the critical energies $[EF - (E_{f31} + E_{f32})]$ and $[EC - (E_{c31} + E_{c32})]$. It is difficult to obtain an explicit generalized representation in the test with hold-time, but in the test without hold-time we can formulate the number of cycles N_i in an arbitrary division Δi as

$$N_i = \frac{EF - \sum_{j=1}^{i-1} E_{fij}}{2\varepsilon_p(\sigma) \cdot \sigma[dl, l_0 + (i-1)dl]} \quad (6.17)$$

where $E_{fij} = 2\varepsilon_p(\sigma) \cdot \sigma\{[(i-j)+1]dl, l_0 + (j-1)dl\}$ and ε_p is the plastic strain range; the subscript i in E_{fij} shows the division number and j the step of crack propagation during the division j. Also, for pure fatigue, the crack propagation law is derived from the similar method by omitting the first term of the left side of eqn (6.16). This model is discussed below.

Figure 6.35 [27] shows the analysis of crack propagation curves in the test with hold-time. In the figure the material constants n, α and a are 0.3, 3.0 and 4.33×10^{-6} mm^6/kgf^3 s$^{0.3}$, and EF and EC are 5000 kgf/mm^2 and 4000 kgf/mm^2, respectively. The dashed curve shows the crack propagation curve in the test with a hold-time of 10 min every ten cycles. It is seen from the figure that the crack propagation rate in the test with a hold-time of 10 min every ten cycles is much smaller than that with a hold-time of 1 min each cycle. In this regard we can refer to the experiments mentioned above. The effect of the loading sequence shown in Fig. 6.32(a) is well interpreted by the present model, but in the data of Fig. 6.32(b) at a larger total strain range there is no apparent effect and the data are inadequate for such a comparison.

Next we discuss the influence of the material constants n and α. In the present study, the number of cycles to failure N_f is defined as the instant at which unstable crack growth occurs. Figures 6.36 (a) and (b) [27] are analytical representations of the influence of n and α on the failure life

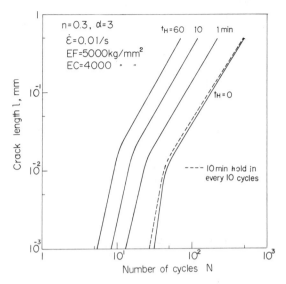

Fig. 6.35. Analytical representation of crack growth curves in the tests with hold-times, and the load sequence effect. (Reprinted from ref. 27).

ratio N_f/N_{f0}, where N_f and N_{f0} are the failure life in the test with and without hold-time respectively. It is seen from these figures that the life ratio decreases as hold-time increases, and the decrease of N_f/N_{f0} is influenced by the material constants n and α. Therefore we can estimate that a small n value and a large α value are adequate for the material used under creep–fatigue interaction conditions. It seems that a large n value is favorable for the material, but a remarkable decrease in N_f with hold-time is analyzed in the case of a large n value, and this is even more dangerous.

Figures 6.37 (a) and (b) [27] also show the influence of the material constants n and α on the normalized failure life $N_f/N_{f(n=0.1)}$ and $N_f/N_{f(\alpha=2.0)}$, where $N_{f(n=0.1)}$ and $N_{f(\alpha=2.0)}$ are the failure lives of the material having the material constants with $n=0.1$ and $\alpha=2.0$ respectively. We can see from these figures that the failure life decreases as n and α increase, and the reduction of life due to the increase in α is remarkable. It was previously noted in Figs 6.36 that a small n value and a large α value are adequate for the material but, since material with a large α value has a short failure life, small n and α values are in the end the most favorable for designing the material. Therefore we can conclude

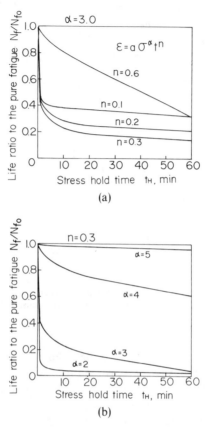

Fig. 6.36. Analytical representation of (a) the influence of the material constant, n, on failure life ratio, N_f/N_{f0}, and (b) that of the material constant, α, on N_f/N_{f0}. (Reprinted from ref. 27)

that material having both small n and α values, i.e. small stress dependence in the constitutive equation, is the most effective material which resists failure life reduction under creep–fatigue interaction conditions at elevated temperatures. We can also derive a similar conclusion for the strain-controlled tests.

In these sections (6.3.1 and 6.3.2), in order to clarify the effect of wave shapes on the creep–fatigue interaction of metallic materials, total strain range controlled experiments under four strain waveforms were performed by using a cobalt-based superalloy X-40 at 1073 K in the atmosphere. In order to discuss the effects of hold-time, wave shape, and

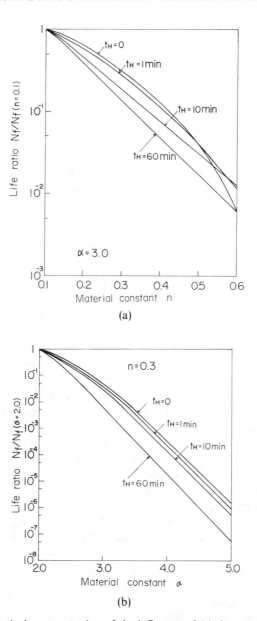

Fig. 6.37. Analytical representation of the influence of (a) the material constant, n, on failure life ratio, $N_f/N_{f0}(n=0.1)$, and (b) that of the material constant α on $N_f/N_{f0}(\alpha=2.0)$. (Reprinted from ref. 27)

material constants in constitutive equations, a simple mechanical model was proposed. From these discussions we concluded the following.

(1) The number of cycles to failure for the alloy in the tests with tensile hold-time in each cycle and every four cycles was remarkably smaller than that in the test without hold-time, but in the test with hold-time in compression no reduction of failure life without hold-time was observed. The difference between the effect of hold-time in tension and that in compression on failure lives was interpreted by presuming that in the case of hold-time in compression the plastic strain energy dissipated in a cycle is smaller than that in tension. This difference was also explained from the delay of crack initiation and a slower crack propagation rate in the case of hold-time in compression compared to the case of hold-time in tension. Also, a slow crack propagation rate in a region of long cracks in the test with hold-time in compression seems to be caused by the decrease of strain concentration due to closing of the crack during the hold-time.

(2) In the comparison of the failure life in the test with tensile hold-time every four cycles and that in the test with tensile hold-time each cycle, no greater difference was observed in a larger total strain range of 1.0%. However, in a lower strain range of 0.5% the former life was greatly extended compared to the latter and was, in fact, as long as the life in the test without hold-time. These facts were also explained from the plastic strain energy dissipated in a cycle. That is, the plastic strain energy dissipated in a cycle was almost the same in two tests with a larger total strain range of 1.0%, but with the tests performed under a smaller total strain range of 0.5% the matter is quite different. In this smaller range, the magnitude of the plastic strain range in cycle 1 abruptly decreased in the following cycles in the test with hold-time every four cycles, and the plastic strain energy dissipated in a cycle was smaller than that in the test with hold-time in each cycle. Moreover, a delay of crack initiation and a slower crack propagation rate in the test with hold-time in every four cycles resulted in a longer failure life than in other tests.

(3) The effects of hold-times and wave shapes were well explained by a mechanical model in which the pre-damage at the front of the crack tip was considered. It is necessary to know under which type of stress wave shape the material has the shortest failure life. In regard to the influence of the material constants in the constitutive equation (6.12), it was found that a small stress dependence is favorable for the material used under the creep–fatigue interaction condition from the viewpoint of resisting the reduction in failure life.

6.4 EFFECT OF STRAIN WAVE SHAPES AND VARYING PRINCIPAL STRESS AXES ON HIGH-TEMPERATURE BIAXIAL LOW-CYCLE FATIGUE

In order to extend the study of creep–fatigue interactions, we will examine the effect of biaxial stresses on the crack behavior and failure life of solution heat treated type 304 stainless steel under combined axial and torsional straining with hold-times. We will also discuss the effect of varying principal stress axes on creep–fatigue interaction. The electro–hydraulic servo-type test machine and the centrally notched hollow cylindrical specimen used are identical to those employed in the previous sections.

6.4.1 Effect of Strain Hold-times on Biaxial Low-cycle Fatigue Lives at Elevated Temperatures

Figure 6.38 [28] shows the experimental relationship between N_f/N_{fo} in

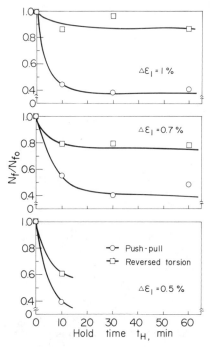

Fig. 6.38. Experimental relationship between lifetimes and strain hold-time for type 304 stainless steel at 923 K in the atmosphere. (Reprinted from ref. 28 by permission of Fatigue of Engineering Materials Ltd)

both the push-pull and reversed torsional low-cycle fatigue tests and strain hold-time at 923 K in the atmosphere, where N_f is the number of cycles to failure in hold-time tests and N_{f0} the number in tests without hold-time. It is apparent that the torsional fatigue lives are greater than uniaxial lives in the low-cycle fatigue test with hold-time. In order to discuss the difference of the strain hold-time effect due to the loading mode, Fig. 6.39 [28] represents an experimental relation between the crack propagation rate and the half crack length. The crack propagation rate in the push-pull test without strain hold-time is almost the same as that in the torsion test, but when the hold-times are introduced the crack propagation rate in the former test is larger than that in the latter. This is interpreted by a method similar to that previously mentioned in Section 6.2.

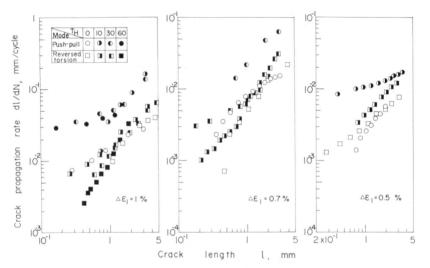

Fig. 6.39. Fatigue crack propagation rate versus crack length for type 304 stainless steel at 923 K in air. (Reprinted from ref. 28 by permission of Fatigue of Engineering Materials Ltd)

Figure 6.40 [28] shows the correlation of the biaxial low-cycle fatigue data in hold-time and no hold-time tests with the equivalent stress range $\Delta\sigma^*$ based on eqn (6.8). A good correlation is found in the figure, and we also find that the push-pull and the torsional fatigue data with and without strain hold-times are classified into groups according to the length of the hold-time but the difference in the fatigue data due to the

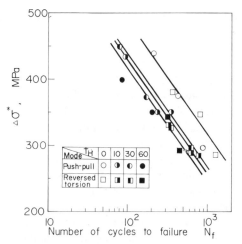

Fig. 6.40. Experimental relationship between COD stress range, $\Delta\sigma^*$, and the number of cycles to failure, N_f, of type 304 stainless steel at 923 K in the atmosphere. (Reprinted from ref. 28 by permission of Fatigue of Engineering Materials Ltd)

loading mode has vanished. Therefore we can conclude that the equivalent stress range $\Delta\sigma^*$ can be successfully applied to biaxial low-cycle fatigue data in tests with strain hold-times as well as in tests without hold-times.

In the present section, the following conclusions were obtained from biaxial low-cycle fatigue tests with and without strain hold-times using a centrally notched hollow cylindrical specimen of type 304 stainless steel at 923 K in air. The ratio of the minimum principal strain to maximum principal strain, $\varepsilon_3/\varepsilon_1$, is within the range -1 to -0.5.

(1) Both the maximum principal strain range and the von Mises equivalent strain range are a nearly adequate comparative basis for biaxial low-cycle fatigue data in tests without hold-time but are inadequate for strain hold-time tests. These are in agreement with the conclusions in Section 6.2.1.

(2) Strain hold-times in the push-pull tests are more detrimental to fatigue lives than those in the torsional tests.

(3) In the tests without strain hold-time, crack initiation and propagation behave in the push-pull tests almost the same as those in the torsion tests. However, earlier crack initiation and faster crack propagation in the push-pull test result in lower fatigue lives in the hold-time test.

(4) An equivalent stress range $\Delta\sigma^*$ based on eqn (6.8), including the effect of the principal stress parallel to the fatigue crack, can be successfully applied to the low-cycle fatigue data even in the hold-time tests.

6.4.2 Effect of Changing Principal Stress Axes on Low-cycle Fatigue Lives under Various Strain Wave Shapes

A study of the effect of changing principal stress axes on low-cycle fatigue lives of a given material is an important aspect of the strength and fracture of such materials under nonproportional loading. Nonproportional cyclic loading appears in structural materials subjected to both cyclic mechanical and thermal stresses with an arbitrary phase. Historically, out-of-phase loading is employed as a means of nonproportional loading. Shorter fatigue lives were reported [29] compared with those under proportional loading. However, a careful discussion of the results of out-of-phase tests is necessary. Usually the mean stress in out-of-phase tests is higher than that in in-phase tests. Therefore it is necessary that the pure effect of nonproportional loading be separated from that of mean stress.

In the present section, biaxial low-cycle fatigue data for solution heat treated type 304 austenitic stainless steel at 823 K and 923 K in air are described for five strain waveforms with and without the alternation of principal stress axes, as shown in Table 6.2. In the table the solid lines denote axial loading and the dashed lines torsional loading. In three waveforms out of four, an extended test program was employed in which the push-pull and the reversed torsional load were applied alternately. In this case a push-pull cycle followed by a torsional cycle was counted as two separate cycles. A push-pull cycle was interchanged with a torsional cycle at zero strain. This test program permits the principal stress axes to change while maintaining the equivalent total strain range of the von Mises type strain constant during the test (see Fig. 4.31(c) in Chapter 4). The net angular change of the principal stress axes is $\pi/4$ radians per cycle. Total strain range and strain rate are also listed in Table 6.2.

Figure 6.41 [30] shows the experimental results at 823 K in air under five strain waveforms, the so-called fast-fast, fast-slow, slow-fast, slow-slow, and trapezoid, with and without the alternation of principal stress axes. The equivalent total strain range of the von Mises type $\Delta\bar{\varepsilon}$ was used as the unit of the ordinate. No significant difference in fatigue lives is observed between push-pull and reversed torsion in the same strain waveforms tested, as previously mentioned in Section 6.2. A noticeable characteristic in the figure is the reduction in the fatigue lives caused by

Table 6.2
Strain Waveforms and Symbols

Wave form	Loading mode	Key	T=923K Total strain range(mm/mm)	$\bar{\varepsilon}$ [Ks^{-1}]	$\bar{\varepsilon}$ [Ks^{-1}]	T=823K Total strain range(mm/mm)	$\bar{\varepsilon}$ [Ks^{-1}]	$\bar{\varepsilon}$ [Ks^{-1}]
(fast-fast)	Not alternated	○	0.01	2.000	2.000	0.02	4.000	4.000
			0.007	1.400	1.400			
		●	0.005	1.000	1.000	0.01	2.000	2.000
	Alternated	⊕	0.003	0.600	0.600	0.005	1.000	1.000
(slow-fast)	Not alternated	□	0.01	0.020	2.000			
		■	0.007	0.014	1.400	0.01	0.200	2.000
	Alternated	⊞	0.005	0.010	1.000			
(fast-slow)	Not alternated	◇	....	....	....	0.01	2.000	0.200
		◆						
(hold)	Not alternated	○	0.01	2.000	2.000	0.01	2.000	2.000
t_H=600s		●	0.007	1.400	1.400	0.005	1.000	1.000
	Alternated	⊕	0.005	1.000	1.000			
(slow-slow)	Not alternated	△	0.01	0.020	0.020	0.01	0.200	0.200
$2\tau_t$		▲	0.007	0.014	0.014	0.005	0.100	0.100

Solid lines denote push-pull loading, and dashed lines torsional loading; τ_t, tension going time; τ_c, compression going time; τ_0, time for a cycle in the fast-fast strain wave shape.

the change of principal stress axes in all the strain waveforms tested. Also, observation of the fracture surface by both scanning electron microscopy (SEM) and optical microscopy revealed that the fracture mode was transgranular in all the strain waveforms. Especially in the fast-fast strain waveform, clear striation marks were observed. Figure 6.42 [30] is similar to Fig. 6.41 but the test temperature is higher at 923 K. No effect from the alternation of principal stress axes was observed on the fatigue lives. Comparing this result to that at 823 K, we know that the effect of the change of principal stress axes depends on the

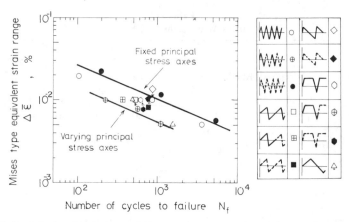

Fig. 6.41. Relationship between Mises strain range, $\Delta\bar{\varepsilon}$, and the number of cycles to failure, N_f, for type 304 stainless steel at 823 K in air. (Reprinted from ref. 30 by permission of Amer. Soc. Test. Mater.)

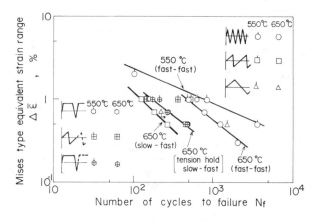

Fig. 6.42. Relationship between $\Delta\bar{\varepsilon}$ and N_f at 923 K in air. (Reprinted from ref. 30 by permission of Amer. Soc. Test. Mater.)

temperature level. This temperature level dependence can be explained as follows.

It is well known that type 304 stainless steel is a typical cyclic hardening material, and this fact is also observed in the present study. About a 35% increase in the stress amplitude is found in cyclic push-pull tests when compared with monotonic tests. Further cyclic hardening is discovered in alternated tests. About a 40% increase in stress amplitude is caused by changing principal stress axes at both temperatures. In this regard, in Chapter 4 we pointed out the increase in stress amplitude and observed this by means of electron transmission microscopy; we found that most of the dislocation structure in the push-pull test is ladder while in the reversed torsion test it is maze and that dislocations are on the main slip plane {111} where the maximum shear stress occurs in both modes. Only the alternation test yields a cell structure because of the complex interaction of the slip systems, which results in a greater degree of cyclic work-hardening. Further evidence was provided by means of the SEM and the optical microscope from which the transgranular and intergranular fracture modes were found at 823 K and 923 K, respectively, in all strain waveforms. Therefore we can conclude that, in the case of transgranular fracture at 823 K, the alternation of principal stress axes affects the fatigue lives, but in the case of intergranular fracture at 923 K it does not; this phenomenon is not affected by the strain waveforms.

There is one exception. In the test of the non-alternated fast-fast strain waveform at 823 K, clear striation marks were observed on the fracture surface, and this specimen fractured transgranularly. However, we could not obtain clear fracture surfaces for the same strain waveform with a change of principal stress axes because of damage to the surfaces by rubbing of the surfaces during the alternation of principal stress axes. Hence a definite conclusion could not be obtained.

In addition, the deviation of the principal stress axes from the strain axes will be mentioned briefly. In proportional loading tests, the direction of the principal stress axes usually agrees with that of the principal strain axes [19], but in nonproportional loading tests this is not always guaranteed [19]. Sometimes a disagreement was observed in the present tests where the direction of the principal stress axes almost coincides with that of the principal strain axes programmed in the push-pull cycle, but in the torsional cycle it does not.

At the end of this section we will check the applicability of three life prediction methods, that is, Ostergren's method [31], the frequency separation method [32], and Tomkin's method [33], for the experimental data in various strain waveforms with and without a change of the principal stress axes.

Ostergren [31] modified the frequency modified fatigue life equation [34] by adding the tensile stress amplitude and obtained

$$\Delta\varepsilon_p \sigma_t v^{\beta(k'-1)} N_f^{\beta} = C \tag{6.18}$$

where $\Delta\varepsilon_p$ is the inelastic strain range, σ_t the tensile stress amplitude in a hysteresis loop at $N = N_f/2$, and v the frequency. In this equation, v is defined as $1/\tau_0$ for the case of the fast-fast test and as $v = 1/(\tau_c + \tau_t)$ for slow-fast, fast-slow and slow-slow tests. Here τ_0 is the cycle time in the fast-fast test while τ_t and τ_c are the tension and the compression times (see Table 6.2). In addition, for the tension-hold tests, v is defined as $1/(\tau_0 + t_H)$, where t_H is the tension hold-time. The exponents β and k' in eqn (6.18) are determined from $\Delta\varepsilon_p$ versus N_f curves in the push-pull test at specific frequencies.

Concerning the frequency-modified fatigue life equation, Coffin [32] proposed the following frequency separation method for the prediction of fatigue lives under asymmetrical waveforms:

$$\Delta\varepsilon_p [(v_c/2)^{k_1} + (v_t/2)^{k_1}]^{1/n'} v_0^{-k'_1 n'} N_f^\beta = C \qquad (6.19)$$

where v_c is the frequency of compression, v_t the frequency of tension, and v_0 the frequency in the fast-fast test. The exponents k_1 and n' are determined also from the cyclic constitutive relation of $\Delta\sigma$ versus $\Delta\varepsilon_p$ curves in the push-pull slow-fast test at specific frequencies. The exponents k'_1 and β correspond to those in eqn (6.18).

Tomkins [33] proposed the following life prediction method based on the following crack propagation consideration:

$$\Delta\varepsilon_p (\Delta\sigma/\sigma_u)^2 N_f = C \qquad (6.20)$$

where $\Delta\sigma$ denotes the stress range in a hysteresis loop at $N = N_f/2$ and σ_u is the static tensile strength of the material.

Figure 6.43 [30] shows the experimental data compared with the four prediction methods. Parts (a)–(c) are the results at 923 K and (d)–(f) are those at 823 K in air. Ostergren's parameter is comparatively superior for the arrangement of the experimental data at 923 K in which no reduction in fatigue lives occurred due to stress alternation. All the data fall into a scatter band with a factor of 2. Nearly half of the experimental data drops out of the band with a factor of 2 in the other prediction methods at 923 K. These methods predict the fatigue lives unconservatively. So frequency separation and the Tomkins method cannot cover the variety of fatigue lives with different strain wave shapes, with and without alternation of principal stress axes.

On the other hand, three prediction methods (d)–(f) are adequate for the data at 823 K without the alternation of principal stress axes, but inadequate for the data with alternation. The data in the tests with alternation of the stress axes fall well outside the scatter band. Therefore

FRACTURE LAWS IN A CONTINUUM MECHANICAL APPROACH

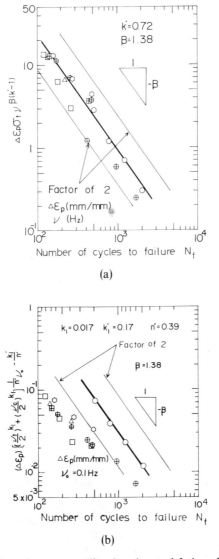

Fig. 6.43. Comparison between predicted and actual fatigue lives: (a) Ostergen method ($T=923$ K in air); (b) frequency separation method ($T=923$ K in air); (c) Tomkins method ($T=923$ K in air); (d) Ostergren method ($T=823$ K in air); (e) frequency separation method ($T=823$ K in air); (f) Tomkins method ($T=823$ K in air). (Reprinted from ref. 30 by permission of Amer. Soc. Test. Mater)

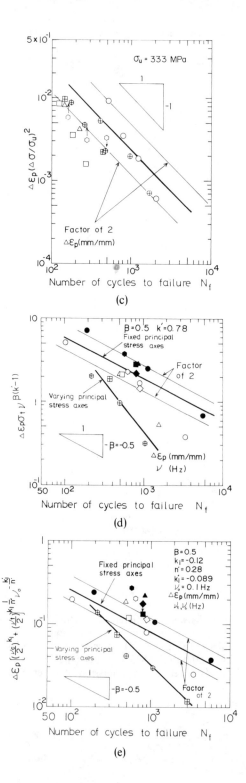

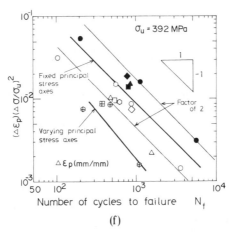

(f)

Fig. 6.43.—contd.

we can conclude that an additional safety factor margin is needed for predicting the low-cycle fatigue life in the case of alternated principal stress axes with transgranular fracture.

In this section, the effect of changing the principal stress axes on low-cycle fatigue lives of type 304 stainless steel was clarified by using four strain waveforms at 823 K and 923 K in air. From these tests we obtained the following conclusions.

(1) Changing principal stress axes reduces the low-cycle fatigue failure life of the material at 823 K in air, but not at 923 K, for the four strain wave shapes tested. Observations of the fracture surfaces by SEM and optical microscopy reveal that the specimens fractured transgranularly at 823 K but intergranularly at 923 K. Therefore the alternation of principal stress axes is detrimental to fatigue in the case of transgranular fracture.

(2) Greater cyclic work-hardening of the material is observed in the tests with the alternation of the principal stress axes at 823 K and 923 K in comparison with tests without such alternation. About a 40% increase in the stress amplitude is observed by changing principal stress axes while at the two temperatures in air. From the observations by SEM and optical microscopy, in addition to those by electron transmission microscopy in Chapter 4, a much greater intensity of deformation in grains occurs in the tests with the change of principal stress axes, which contributes to the increase in cyclic work-hardening at the two temperatures.

(3) Changing principal stress axes causes an earlier initiation of the fatigue crack and enhances the crack propagation rate at 823 K in air, while it has no effect on initiation and propagation at 923 K.

(4) Four prediction methods for low-cycle fatigue lives are applied to the experimental data. The best method is that proposed by Ostergren which can arrange the data within a factor of 2 at 923 K in air, but not 823 K in air. It predicts the low-cycle fatigue life in the alternated tests unconservatively. Therefore an additional safety factor margin is needed for predicting the low-cycle fatigue life under changing stress axes with transgranular fracture.

6.5 NOTCH EFFECT ON LOW-CYCLE FATIGUE UNDER CREEP–FATIGUE INTERACTION CONDITIONS

6.5.1 Mechanical Factors Determining Notch Strengthening and Notch Weakening in Creep Rupture

Figure 6.44 [35] shows the nominal tensile stress σ_n versus rupture time t_r curves for an unnotched and a 60° V-shaped round notched cylindrical specimen of 1 Cr–1 Mo–0·25 V steel (used for steam turbine rotors) at 873 K in the atmosphere. The heat treatment of the material tested was as follows: 1373 K × 20 h → air cooling and 948 K × 24 h → furnace cooling. The specimen was machined by tool cutting following heat treatment, and the root diameter of the notched specimen is 4·9 mm. The elastic stress concentration factor K_t of the round notched bar specimen is 4·7. It was found from the figure that the material tested has the transition from *notch strengthening* to *notch weakening* at an elapsed time of nearly 120 h, and also that the rupture ductility of the unnotched specimen markedly decreases after the elapsed time of this transition, as shown in the parentheses near the data points.

Figure 6.45 [35] shows an analytical representation of the variation of distribution for both axial stress σ_z and the von Mises equivalent stress $\bar{\sigma}$ on the cross-section of the notch root with elapsed time during creep under a nominal stress of 304 MPa, which is the level in the region of notch strengthening as seen from Fig. 6.44. It was found, from the present inelastic finite element analytical results, that the stress near the notch root of the round notched bar specimen markedly relaxes in the early stage of creep and reaches a steady state.

Figure 6.46 [35] shows the variation of the von Mises equivalent stress $\bar{\sigma}$ at the notch root of a round notched bar specimen with elapsed time

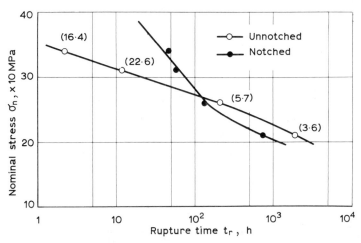

Fig. 6.44. Transition phenomena from notch-strengthening to notch-weakening in the tensile creep rupture of a round notched bar specimen of 1Cr–1Mo–1/4V steel at 873 K in the atmosphere. Numerical values in parentheses show the rupture elongation, %. (Reprinted from ref. 35 by permission of Japan Soc. Mech. Engrs)

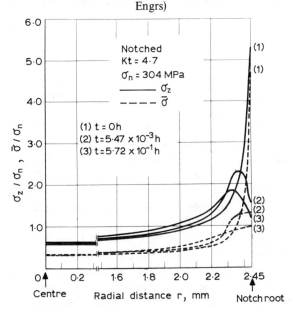

Fig. 6.45. Analytical representation of the variation in both distributions of axial stress, σ_z, and Mises equivalent stress, $\bar{\sigma}$, on the cross-section of the notch root with the elapse of creep time, by inelastic finite element analysis. Test conditions as for Fig. 6.44. (Reprinted from ref. 35 by permission of Japan Soc. Mech. Engrs)

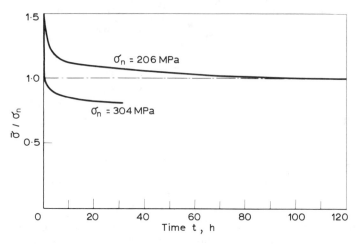

Fig. 6.46. Analytical representation of the variation of Mises stress, $\bar{\sigma}$, at the notch root with time elapse, by inelastic finite element analysis ($\sigma_n = 206$ and 304 MPa). Test conditions as for Fig. 6.44. (Reprinted from ref. 35 by permission of Japan Soc. Mech. Engrs)

during creep under a nominal stress of 206 MPa and 304 MPa. The level of 206 MPa is in the notch weakening region as seen from Fig. 6.44. It was found that under comparatively higher nominal stress level for a notch strengthening the equivalent stress at the notch root rapidly decreases to less than the nominal stress, but does not decrease in the case of lower nominal stress level in notch weakening. Therefore we can conclude that the decrease in the level of the von Mises equivalent stress $\bar{\sigma}$ at the notch root of the notched specimen to less than the nominal stress σ_n in the majority of the rupture life of the material tested is one of the important mechanical factors determining the notch strengthening of the material. On the contrary, the qualitative relation mentioned above is the converse in notch weakening.

Similarly we found that the positive value of hydrostatic stress σ_m at the notch root of the notched specimen is smaller than that for the unnotched specimen under nominal stress in notch strengthening. On the contrary, under nominal stress in notch weakening the former value is larger than that of the latter. It can be estimated that the time to crack initiation in metallic creep at elevated temperatures becomes longer and the crack propagation rate also becomes smaller in accordance with a smaller positive value of hydrostatic stress under a constant nominal stress. This is well estimated from the experimental evidence mentioned in Section 6.1. Therefore we can conclude that the hydrostatic stress is also an influential factor for both notch strengthening and notch weakening.

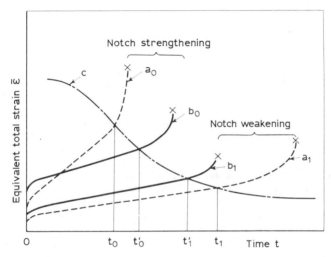

Fig. 6.47. Schematic representation of both notch-strengthening and notch-weakening of the material on the common basis of Mises strain, $\bar{\varepsilon}$.

In connection with these conclusions, the schematic representation of both notch strengthening and notch weakening on the basis of the von Mises equivalent total strain $\bar{\varepsilon}$ can be drawn as shown in Fig. 6.47. It is seen from the figure that under applied stress in notch strengthening the time to crack initiation of the notched specimen t'_0 becomes longer than that for the unnotched specimen t_0 because the level of the equivalent total strain b_0 at the notch root of the notched specimen is lower than that for the unnotched specimen a_0 for most of the rupture life of the material. On the contrary, under the stress in notch weakening the former time t'_1 becomes shorter than the latter t_1 because the level of the equivalent total strain b_1 for the notched specimen is higher than that of a_1 for the unnotched specimen. The qualitative relation of the creep curves of both the round notched and unnotched bar specimens can be examined from the inelastic finite element analysis, as shown in Fig. 6.48 [35].

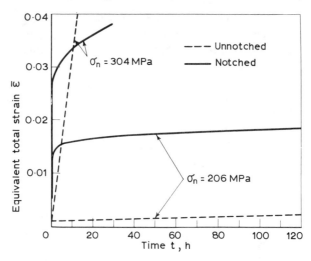

Fig. 6.48. Analytical representation of both creep curves at the notch root of a round notched bar specimen, and of unnotched bar specimens on the basis of Mises strain, $\bar{\varepsilon}$, ($\sigma_n = 206$ and 304 MPa). Test conditions as for Fig. 6.44. (Reprinted from ref. 35 by permission of Japan Soc. Mech. Engrs)

Figure 6.49 [36] also shows the experimental comparison between the nominal tensile stress σ_n versus rupture time t_r curve of a double-edged 60° V-shaped notched ($K_t = 5 \cdot 5$) plate specimen and that of a 60° V-shaped round notched ($K_t = 4 \cdot 7$) bar specimen at 873 K in the atmosphere. The cross-section of the notch root of the notched plate specimen is 6 mm × 3 mm. We found from the figure that the rupture time of the notched plate specimen is shorter than that of the round notched bar specimen in the region of notch strengthening but there is no great difference between the data for the region of notch weakening. In the figure, the other data points for the round bar specimens having a K_t of 5·3 and 5·5 are plotted with reference to the collaboration test data of the 129th Research Committee, Japan Society for the Promotion of Science, and we can see that there is no great difference between the data for the round notched bar specimen having a K_t of 4·7 and those with 5·3 and 5·5. In order to clarify the decrease in the rupture time of the notched plate specimen in comparison with that of the round notched bar specimen, a series of numerical examinations by inelastic finite element analysis was performed. Figure 6.50 [36] shows an example of comparison between the creep curves at the notch root of both types of

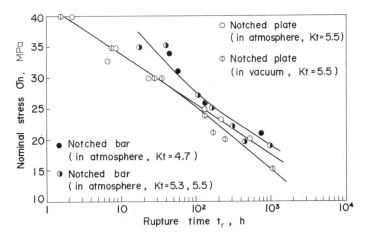

Fig. 6.49. Experimental comparison between the creep rupture time curves of notched plate ($K_t = 4\cdot7$) and round notched bar ($K_t = 5\cdot5$) specimens. Test conditions as for Fig. 6.44. (Reprinted from ref. 36 by permission of Soc. Mater. Sci., Japan)

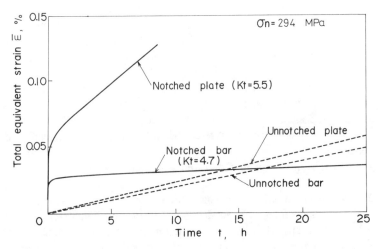

Fig. 6.50. Analytical comparison between the creep curves at the notch root of notched plate and round notched bar specimens, on the basis of Mises strain, $\bar{\varepsilon}$. Test conditions as for Fig. 6.44. (Reprinted from ref. 36 by permission of Soc. Mater. Sci., Japan)

notched specimen under $\sigma_n = 294$ MPa on the basis of the von Mises equivalent total strain $\bar{\varepsilon}$. We can conclude from the present analyses that the difference in the rupture life between the two types of specimen under comparatively higher nominal stress, which is in the region of notch strengthening, is due to the effect of strain constraint at the notch root of the round notched bar specimen.

In this section, the tensile creep rupture tests for an unnotched and a round notched (60° V-shape, $K_t = 4\cdot 7$) bar specimen and a double-edged notched (60° V-shape, $K_t = 5\cdot 5$) plate specimen of 1 Cr–1 Mo–0·25 V steel were performed at 873 K in the atmosphere, and the numerical calculation was carried out by inelastic finite element analysis. From the results the following was concluded.

(1) The material tested shows the transition phenomenon from notch strengthening to notch weakening at an elapsed time of about 120 h. The rupture ductility of the material decreases markedly after the elapsed time of this transition.

(2) Under the same nominal stress in the region of notch strengthening, both the level of the von Mises equivalent stress and that of the equivalent total strain at the notch root of the round notched bar specimen are lower than those for the unnotched bar specimen. This quantitative relation remains true also in the case of hydrostatic stress at the notch root. On the contrary, the quantitative relations mentioned above are reversed in notch weakening. The former is one of the important mechanical factors determining notch strengthening, while the latter is a factor for notch weakening; this is seen from observations of crack initiation by interruption creep tests under nominal stress in the region of notch weakening.

(3) The difference between the rupture life of the round notched bar specimen and that of the double-edged notched plate specimen under comparatively higher nominal stress is due to the effect of strain constraint at the notch root of the round notched bar specimen.

6.5.2 Transition Phenomenon from Notch Weakening to Notch Strenthening under Creep–Fatigue Interaction Conditions

First we will describe the transition phenomenon of notch weakening to notch strengthening under creep–fatigue interaction conditions using the data for one unnotched and three kinds of round notched (60° V-shape; $K_t = 2\cdot 6$, $4\cdot 2$ and $6\cdot 0$) bar specimens of two austenitic stainless steels (types 316 and 304) which were solution heat treated at 1373 K. The test apparatus consisted of conventional static creep test equipment modified to enable a dead weight to lift and lower by means of an air cylinder to perform zero-to-tension tests of triangular stress waveform (stress ratio $R=0.03$) with peak stresses of 314 MPa and 274 MPa at 873 K in the atmosphere. The stress hold-times employed at the tension peak stress in each cycle are: 0, 10, 30, 60 and 1440 min for SUS 316 stainless steel and 0, 10, 60 and 1440 min for SUS 304 stainless steel. The frequency of the stress wave with no hold-time is 0.043 Hz. In this section we will also describe the results of a fully reversed low-cycle fatigue test using the same shape and dimensions of unnotched and round notched bar specimens of type 304 stainless steel as those used above, under a stress range of 441 MPa at 873 K in comparison with the zero-to-tension data.

Figure 6.51 [37] shows the variation of failure life with an increase in the elastic stress concentration at the peak stress of 314 MPa and 274 MPa. For both steels the increase in elastic stress concentration decreases the low-cycle fatigue failure life in the tests with shorter hold-times, while increases in the stress concentration factors do not reduce the fatigue failure life in longer hold-time tests. In longer hold-time tests notch strengthening occurs, in which the fatigue life of the notched specimens is longer than that of unnotched specimens. This transition from notch weakening in the test with shorter hold-times to notch strengthening in the test with longer hold-times was also observed in the stress-controlled fully reversed push-pull test (stress ratio $R=-1$) [38, 39] so that the transition occurs in not only the zero-to-tension test but also the fully reversed push-pull test in relation to notch strengthening behavior in the static creep rupture of the materials tested.

In order to discuss the relation between the low-cycle fatigue data and the static creep rupture data, Fig. 6.52 [37] shows the relation between the failure cycle and the test frequency, defined as the inverse of the time in a cycle for hold-time tests and the inverse of the rupture time for the static creep test. The failure cycle in the static creep data is defined as $N_f = 1/4$ considering that the test was carried out in zero-to-tension conditions. As the test results for both steels exhibit almost the same tendency, we will mainly discuss the results in Fig. 6.52(a), i.e. the case of SUS 316 with a peak stress of 314 MPa, and then discuss the other cases concerning different points derived from the results in Fig. 6.52(a).

In Fig. 6.52(a) the data for the unnotched specimen can be expressed by a straight line with slope 1 from the no hold-time data to the static

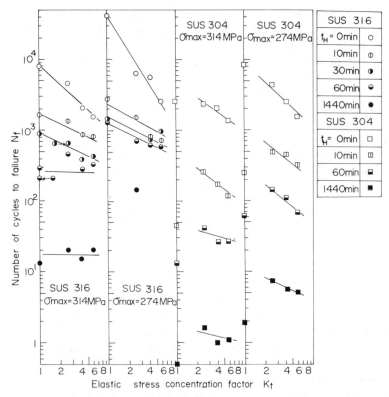

Fig. 6.51. Experimental relation between the number of cycles to failure, N_f, and the elastic stress concentration factor, K_t, in zero-to-tension tests with and without stress hold-times for type 304 and type 316 stainless steels at 873 K in air (σ_{max} = 314 and 274 MPa). (Reprinted from ref. 37 by permission of Soc. Mater. Sci., Japan)

creep data, while for the case of the notched specimens the data can be expressed by two lines; in the frequency range from 4×10^{-4} Hz to 10^{-1} Hz the slope of the data is smaller than that of the unnotched specimen while the same slope occurs in the frequency range smaller than 4×10^{-4} Hz. A slope of unity in N_f versus v plots means that the failure is purely time-dependent while a zero slope means that the failure is purely cycle-dependent. Regarding the fracture mode of the unnotched specimen, all the data can be represented by the line with slope 1, so the fracture mode of the specimen is purely time-dependent in all the frequency ranges. For the notched specimens, combined time- and cycle-

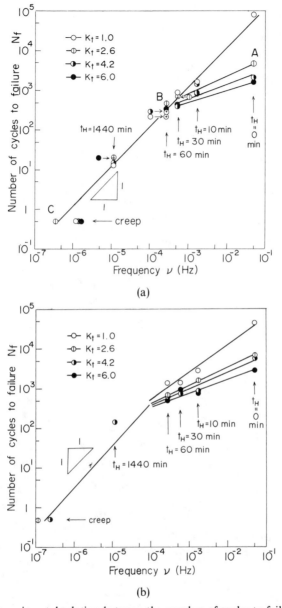

Fig. 6.52. Experimental relation between the number of cycles to failure, N_f, and frequency, v, for (a) type 316 stainless steel, 873 K, $\Delta\sigma = 314$ MPa, (b) type 316 stainless steel, 873 K, $\Delta\sigma = 274$ MPa, (c) type 304 stainless steel, 873 K, $\Delta\sigma = 314$ MPa, and (d) type 304 stainless steel, 873 K, $\Delta\sigma = 274$ MPa. (Reprinted from ref. 37 by permission of Soc. Mater. Sci., Japan)

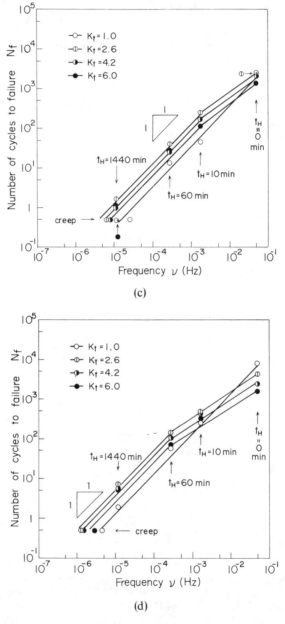

Fig. 6.52.—contd.

dependent fracture occurs in a frequency range larger than 4×10^{-4} Hz while purely time-dependent fracture occurs in a frequency range smaller than 4×10^{-4} Hz.

Comparing the fracture mode, i.e. cycle-dependent or time-dependent, with the notch effect, notch weakening occurs in the cycle- and time-dependent fracture region and the specimen with a larger elastic stress concentration factor exhibits a smaller failure cycle. On the other hand, in the purely time-dependent fracture region, notch strengthening occurs. In that region there is no difference in the failure cycle of the specimen with a different elastic stress concentration factor. These experimental results show that we have to treat the notch effect by separating the fracture region when considering the notch effect in creep–fatigue conditions; in the time- and cycle-dependent fracture region we have to take the reduction in the failure life due to the notch into consideration, while in the purely time-dependent fracture region we do not need to take account of the reduction in the fracture life by the notch.

The above discussion remains, for the most part, true in other cases, i.e. Figs 6.52 (b)–(d), in cases with slightly different points. The different points are as follows. The slope of the data for the unnotched specimen is smaller than unity at $\sigma_{max} = 274$ MPa for type 316 so that the time-dependent fracture region exists for the unnotched specimen at a smaller stress level. For the case of type 304, on the other hand, the data for the unnotched specimen can be expressed by a straight line at the two stress levels and the data for the notched specimens can also be expressed with two lines. For type 304 the notch strengthening occurs at the two stress levels in the test with a hold-time longer than 10 min, which is the same tendency as that of the data for the case of type 316.

Figure 6.53 [38a] shows that notch weakening occurred in the time- and cycle-dependent fracture region while notch strengthening occurred in the purely time-dependent fracture region in push-pull tests with and without stress hold-time for type 304 stainless steel. This transition from notch weakening to notch strengthening was caused by a larger reduction in the failure life of the unnotched specimen with a decreased test frequency compared with that of the unnotched specimen. In the following we will discuss this transition in connection with the testing conditions.

The stress wave was a fully reversed triangular waveform with frequencies ranging from 10^{-4} Hz to 5 Hz and a trapezoidal waveform with frequencies ranging from 10^{-4} Hz to 10^{-2} Hz, where the stress hold-times were introduced at the peak stress in each cycle. The time required

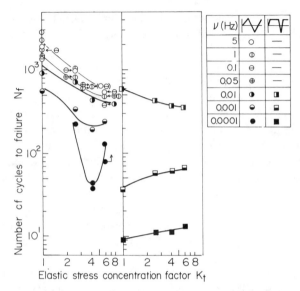

Fig. 6.53. Experimental relation between the number of cycles to failure, N_f, and the elastic concentration factor, K_t, in load-controlled push-pull tests with and without hold-time for type 304 stainless steel at 873 K in the atmosphere, $\Delta\sigma = 441$ MPa. (Reprinted from ref. 38a by permission of Soc. Mater. Sci., Japan)

from peak tension to peak compression in the trapezoidal stress wave was 30 s. Figure 6.54 [38b] shows the relation between the number of cycles to failure N_f of the unnotched and notched cylindrical specimens of type 304 stainless steel and the frequency, v, at 873 K in the atmosphere. Figure 6.55 is a schematic diagram showing the transition in the fracture mode observed by SEM in the push-pull tests with stress hold-time. From these figures, the frequency dependence on failure life can be classified into the following three regions:

(1) $0.05\,\text{Hz} \leqslant v \leqslant 5\,\text{Hz}$ in triangular waveform (region I). In this region, the number of cycles to failure does not depend on frequency, so pure cycle-dependent fracture occurs in this region. Clear striation marks are observed on the fracture surface.

(2) $10^{-4}\,\text{Hz} \leqslant v \leqslant 0.05\,\text{Hz}$ in triangular waveform (region II). In this region, N_f decreases as the frequency decreases. The fracture mode in this region is a mixed type of transgranular and intergranular fracture.

FRACTURE LAWS IN A CONTINUUM MECHANICAL APPROACH 317

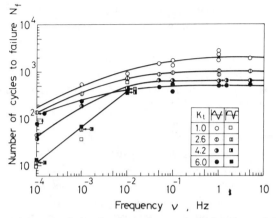

Fig. 6.54. Experimental relation between the number of cycles to failure, N_f, and frequency, v, in the load-controlled push-pull tests for type 304 stainless steel at 873 K in the atmosphere. $\Delta\sigma = 441$ MPa. (Reprinted from ref. 38b by permission of Soc. Mater. Sci., Japan)

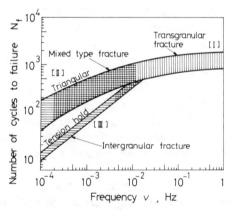

Fig. 6.55. Schematic diagram showing the transition of the fracture mode in load-controlled push-pull tests for type 304 stainless steel at 873 K in air, $\Delta\sigma = 441$ MPa. (Reprinted from ref. 38b by permission of Soc. Mater. Sci., Japan)

(3) 10^{-4} Hz $\leqslant v \leqslant 10^{-2}$ Hz in trapezoidal waveform (region III). In this region, N_f drastically decreases as the frequency decreases. The slope of the data is 0.77, which is slightly smaller compared with the slope in purely time-dependent fracture, but most of the purely

time-dependent fracture occurs in this region. The fracture mode in this region is of the intergranular type.

Transition from notch weakening to notch strengthening occurs in the frequency range from 5×10^{-3} Hz to 5×10^{-2} Hz. This transition frequency agrees with that mentioned above in the zero-to-tension tests, namely, the frequency is 10^{-3} Hz to 5×10^{-2} Hz at 314 MPa and 10^{-3} Hz to 10^{-2} Hz at 274 MPa.

We have also carried out the same kind of push-pull test as above for type 316 stainless steel, but with a stress level of $\Delta\sigma = 392$ MPa and a frequency range of 2×10^{-4} Hz to 5×10^{-2} Hz [39]. For type 316, the transition from notch weakening to notch strengthening also occurred as the frequency decreases. The transition frequency was 5×10^{-4} Hz which agrees with the transition frequency for type 316 mentioned above, i.e. 5×10^{-4} Hz at 314 MPa and 10^{-4} Hz at 274 MPa. The comparison of the transition frequencies from these studies show that the transition frequency does not depend on the stress wave shape, whether it is fully reversed wave or zero-to-tension wave, but depends on the material.

The results of the inelastic finite element analysis, which were used to analyze the cause of notch strengthening in creep rupture, showed that the main cause of strengthening was the lower von Mises equivalent stress at the notch root due to strain constraint compared with the unnotched specimen as mentioned in Section 6.5.1. During the stress hold-time in this study the stress relaxation at the notch root of the notched specimen also occurs and the relaxation behavior is characterized by the creep constitutive equation, i.e. the material constants. These results suggest that the transition from notch weakening to notch strengthening only depends on the material property due to the difference in relaxation behavior.

At the end of this section, we briefly describe the data of a superalloy steel A286 (C 0·07, Si 0·47, Mn 1·31, Cr 14·95, Ni 25·07, Mo 1·13, Ti 2·31, Fe bal, wt%). The material tested was prepared by vacuum melting and the heat treatment was as follows: $1172 \text{ K} \times 5 \text{ h} \rightarrow$ oil cooling and $991 \text{ K} \times 16 \text{ h} \rightarrow$ air cooling. The 60° V-shaped triple round notched specimen had $K_t = 6·4$ and the root diameter was 6 mm. The machining of the specimens was done by tool cutting following heat treatment.

Figure 6.56(a) [40] shows the experimental relation of rupture time t_r to the stress concentration factor, K_t, in the three types of static creep rupture, and a zero-to-tension with and without stress hold-time under peak nominal stress, σ_{max}, of 441 MPa at 923 K in the atmosphere. The

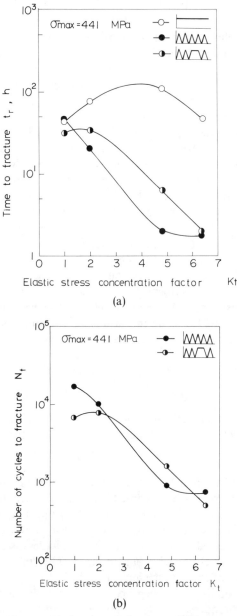

Fig. 6.56. Notch effect on (a) the rupture time, t_r, and (b) the number of cycles to rupture, N_f, in zero-to-tension tests with and without stress hold-time, in the three types of test for alloy A286 at 923 K in the atmosphere. (Reprinted from ref. 40 by permission of Soc. Mater. Sci., Japan)

frequency of cycling in the zero-to-tension low-cycle fatigue test is 8 cpm and the wave of the test with hold-time is composed of cyclic tension for a duration of 15 min with constant tension during the same time interval. We found from the figure that there is no large difference between the rupture lives of the unnotched specimens in the three types of test, but a distinct notch effect on the rupture time of the notched specimen is observed in the three types of test. Figure 6.56(a) [40] also shows the experimental relation between the number of cycles to rupture, N_f, and the stress concentration, K_t, in both the simple fatigue test and the fatigue test with hold-time. The number of cycles to rupture of the material tested in the fatigue with hold-time in tension is larger than that in the simple fatigue in the region of K_t from nearly 2 to nearly 6 in connection with the notch strengthening in the static creep rupture shown in Fig. 6.56(a). This is the converse of the usual effect of hold-time for structural metals at elevated temperatures.

In this section, the zero-to-tension test and the push-pull test with stress hold-time were performed at 873 K in air using both type 316 and 304 stainless steels. The zero-to-tension test with stress hold-time was carried out at 923 K in the atmosphere using a superalloy steel A286. From the results we concluded the following.

(1) In the time- and cycle-dependent fracture region, notch weakening occurred, while notch strengthening occurred in the purely time-dependent fracture region for both types of austenitic stainless steel.

(2) An effect opposite to the usual hold-time effect on the high-temperature low-cycle fatigue is clearly observed for the notched specimen having an eleastic stress concentration factor ranging from nearly 2 to nearly 6 for the superalloy steel A286.

6.5.3 Empirical Formula for a 'Frequency–Elastic Stress Concentration Factor Modified Fatigue Life Equation'

In this section we describe both the frequency and hold-time effects on the low-cycle fatigue failure lives of cylindrical notched specimens of solution heat treated type 316 austenitic stainless steel under fully reversed strain waveform with push-pull at 873 K in the air. The total strain was controlled over a gauge length of 6 mm on the specimen, including the notch for notched specimens. We obtained an empirical formula of a 'frequency–elastic stress concentration factor modified equation' by analyzing the experimental data. We also describe the contribution of fatigue crack initiation life to the notch effect by using the data of solution heat treated SUS 304 austenitic stainless steel. In the

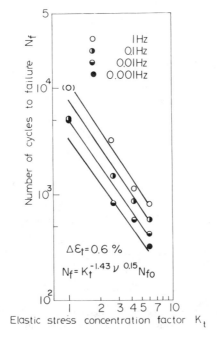

Fig. 6.57. Experimental relation between the number of cycles to failure, N_f, and the elastic stress concentration factor, K_t, for type 316 stainless steel at 873 K in the atmosphere. (Reprinted from ref. 41 by permission of Amer. Soc. Mech. Engrs)

following, the number of cycles to failure N_f is defined as the number of cycles at which the tensile stress amplitude decreases to 3/4 of the maximum value.

Figure 6.57 [41] shows the experimental relationship between the number of cycles to failure N_f and the elastic stress concentration factor K_t. A linear relation on a log-log plot is observed between N_f and K_t at each frequency. A linear relationship was also observed between the number of cycles to failure N_f and frequency v for both unnotched and notched cylindrical specimens at constant K_t, and the following empirical formula is obtained:

$$N_f = N_{f0} K_t^m v^l \quad (m = -1.43, l = 0.15) \tag{6.21}$$

where N_f is the number of cycles to failure for arbitrary K_t and v, and N_{f0} is the reference number of cycles to failure for the unnotched

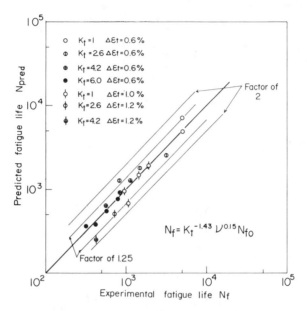

Fig. 6.58. Correlation between the life predicted by eqn (6.21) and the actual life. Test conditions as for Fig. 6.57. (Reprinted from ref. 41 by permission of Amer. Soc. Mech. Engrs)

specimen tested at 1 Hz. Figure 6.58 [41] shows a comparison between the predicted lives N_{fpred} and the actual lives N_f. In the figure, N_{fo} is taken as 10 000 cycles when the total strain range is 0·6% and 3000 and 2000 when it is 1% and 1·2% respectively. A good correlation is found between the two values for a wide range in the values of K_t and v. The scale of the data is almost covered by a factor of 2. Thus we can conclude that eqn (6.21) has sufficient accuracy for predicting the lives of both the unnotched and notched specimens tested at any v. Equation (6.21) is analogous to the frequency-modified fatigue life equation [32], but eqn (6.21) is modified additionally by K_t, so eqn (6.21) shall be called the 'v–K_t modified fatigue life equation'. In the following, the basis of eqn (6.21) will be discussed from two different viewpoints: a numerical model analysis described in Section 6.3.2 and an analytical derivation of eqn (6.21).

Figure 6.59 [41] shows a result of the model calculation. The abscissa represents the stress gradient normalized by the reference stress gradient at the notch root, and the ordinate represents the number of cycles to

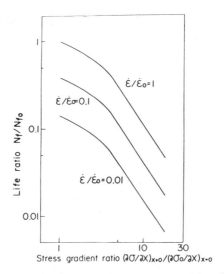

Fig. 6.59. Analytical result of the model of crack propagation in low-cycle fatigue in the creep range. (Reprinted from ref. 41 by permission of Amer. Soc. Mech. Engrs)

failure normalized by the reference number of cycles to failure. As the stress gradient at the notch root corresponds to the elastic stress concentration, Fig. 6.59 is understood as the figure showing the correlation between N_f and K_t by the model analysis. A linear relationship was found between N_f/N_{f0} and $(\partial\sigma/\partial x)_{x=0}/(\partial\sigma_0/\partial x)_{x=0}$ in Fig. 6.59, which is the same result as that found in the effect of frequency on the notched specimen as shown in Fig. 6.57. Thus the result of the model analysis proves that the effect of frequency on the notched specimen, as shown in Fig. 6.59, is due mainly to the strain rate dependence of the material behavior at the notch root and crack tip. It must be pointed out that the model analysis includes the crack propagation process when determining the time to failure.

We now return to the discussion of the analytical derivation of the equation which is analogous to eqn (6.21), combining three well known equations: the constitutive equation (6.12), Neuber's rule [24], and the frequency modified fatigue life equation [32].

A slight transformation of eqn (6.12) yields

$$\sigma = (\varepsilon^{1-n}/\alpha)^{1/\alpha}(\dot{\varepsilon}/n)^{-n/\alpha} \qquad (6.22)$$

If subscripts 0 and m are used to indicate the nominal stress or strain and local stress or strain at the notch root, Neuber's rule, $K_\sigma K_\varepsilon = K_t^2$ [25], can be written as follows:

$$(\sigma_m/\sigma_0)/(\varepsilon_m/\varepsilon_0) = K_t^2 \tag{6.23}$$

Substitution of eqn (6.22) into eqn (6.23), after inserting the subscripts 0 and m, gives

$$\varepsilon_m = A\varepsilon_0 K_t^{2\alpha/(\alpha-n+1)} \tag{6.24}$$

assuming that the strain rate at the notch root is proportional to the nominal strain rate, i.e. $\dot{\varepsilon}_m = s\dot{\varepsilon}_0$. In eqn (6.24), A is the material constant and equals $s^{-n/(\alpha-n+1)}$. Combination of eqn (6.24) with the frequency-modified fatigue life equation [32], $(N_f v^{\kappa-1})^\beta \Delta\varepsilon_p = C_1$, gives the following equation for the notched specimens:

$$A(N_f v^{\kappa-1})^\beta \Delta\varepsilon_0 K_t^{2\alpha/(\alpha-n+1)} = C_1 \tag{6.25}$$

For the smooth specimen tested at a frequency v_0, the frequency-modified fatigue life equation [32] is also shown as

$$(N_{f0} v_0^{\kappa-1})^\beta \Delta\varepsilon_0 = C_1 \tag{6.26}$$

Taking the ratio of eqn (6.25) and eqn (6.26), we finally obtain

$$N_f = AN_{f0}(v/v_0)^{1-\kappa} K_t^{-2\alpha/[\beta(\alpha-n+1)]} \tag{6.27}$$

The exponent of v in eqn (6.27) coincides with that in eqn (6.21) and it takes the value 0·15 when v_0 is 1 Hz. There exists, however, a difference between the exponents of K_t in these equations. The exponent in eqn (6.27) is equal to $-2·4$ (β 0·7, α 4, n 0·27), which is smaller than in eqn (6.21), $-1·43$. The cause of the disagreement in the values of the exponents is due to Neuber's rule which overestimates the reduction in the life of the material with an increase in K_t. The overestimation of the strain concentration at the notch root will be discussed in Section 6.6. More directly, we can understand the overestimation of the strain concentration due to that rule by finite element analysis. A modified Neuber equation, $K_t^b = K_\sigma K_\varepsilon$, was proposed [42] based on finite element analysis and it was reported that b in the equation is not equal to 2 but takes a value from 0·9 to 1·5 under creep loading. The results of the creep analysis cannot be applied rigorously to the low-cycle fatigue case, but their application has some validity in the sense that the creep equation (6.22) is being used as the constitutive equation. So the results of this analysis are considered as a base when discussing the strain concentration in low-cycle fatigue tests. As b in the modified Neuber rule reaches a value of 1·2 when the value of the exponent of K_t in eqn (6.27) is $-2\cdot4$, there is a possibility of using eqn (6.27) instead of eqn (6.21) if we use the modified Neuber rule. Neuber's rule in low-cycle fatigue in the creep range will be described in Section 6.6.

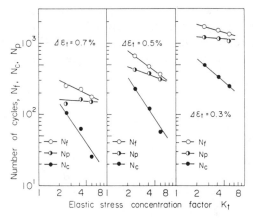

Fig. 6.60. Effect of the elastic stress concentration factor, K_t, on the number of cycles to failure, N_f, number of cycles to crack initiation, N_c, and number of cycles for crack propagation, N_p, for type 304 stainless steel at 873 K in the atmosphere, $v = 0\cdot1$ Hz. (Reprinted from ref. 43 by permission of Amer. Soc. Mech. Engrs)

At the end of this section, we describe the contribution of crack initiation life to notch effect on low-cycle fatigue failure life. Figure 6.60 [43] shows the notch effect on the number of cycles to failure N_f, the number of cycles to crack initiation N_c and the number of cycles for crack propagation N_p of a round notched bar specimen of solution heat treated type 304 stainless steel in strain-controlled fully reversed push-pull low-cycle fatigue at 873 K in air. The shape and dimensions of the unnotched and notched cylindrical specimens, and the machining and the test methods employed are identical to those above. The number of cycles to crack initiation N_c of the notched specimens was determined from the direct observation of striations on the fracture surface of the specimens, and N_c was defined as the cycle at which the crack length reaches 100 μm. The crack propagated in depth from the notch root in a concentric circular manner. So crack propagation life N_p is calculated from $N_p = N_f - N_c$.

We found from the figure that N_c depends on K_t but N_p only has slight dependence. Therefore we can conclude that the notch effect on the failure life results mainly from the effect on initiation life, N_c, and the life prediction using Neuber's rule estimates conservatively the failure lives, but gives predictions for crack initiation (see following Section 6.6). On the other hand, an applied stress/strain criterion can estimate proper crack propagation periods, N_p, of notched specimens. Predicted failure lives, N_f, obtained by adding the predicted N_c and N_p, agree well with the failure lives obtained for notched specimens [43]. We also found that non-detection of crack initiation by d.c. potential and notch opening

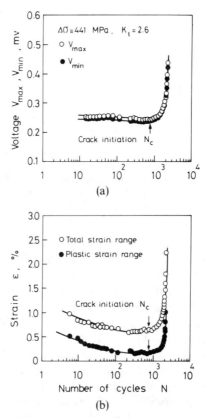

Fig. 6.61. Variation of (a) d.c. potential and (b) notch-opening displacement (NOD), with the number of strain cycles, in load-controlled tests under $\Delta\sigma = 441$ MPa, for a type 304 stainless steel at 873 K in air. (Reprinted from ref. 44 by permission of Soc. Mater. Sci., Japan)

displacement (NOD) measurements are effective. Figure 6.61 [44] shows the non-destructive measurement of N_c of a round notched bar specimen in low-cycle fatigue at 873 K in the atmosphere. Comparing the striation detection, both the d.c. potential and NOD methods are adequate for non-detection of N_c.

In this section, fully reversed low-cycle fatigue tests with tensile strain hold-time were performed at 873 K in air using round notched ($K_t = 2.6$, 4.2 and 6.0) bar specimens of solution heat treated type 316 stainless steel. From the results we concluded the following.

(1) An empirical formula for a 'v–K_t modified equation', $N_f = N_{f0} K_t^m v^l$, is proposed. The applicability of the equation has been confirmed by model analysis. Also, the same type of equation has been derived from the combination of three already well known equations. The value of the exponents in the empirical formula agrees well with those of the equation derived from a modified Neuber rule.

(2) From direct observations of striation on the fracture surface of the notched specimen, and the d.c. potential and the notch opening displacement (NOD) methods, we found that the number of cycles to crack initiation depends on the elastic stress concentration factor but the number of cycles for crack propagation scarcely does. The notch effect on failure life under creep–fatigue interaction conditions results mainly from the crack initiation life of the notched specimen.

6.6 LIFE PREDICTION OF NOTCHED SPECIMENS UNDER CREEP–FATIGUE INTERACTION CONDITIONS

6.6.1 Life Prediction by Neuber's Rule and by the v–K_t Modified Fatigue Life Equation

In the following, two life prediction methods are applied to the failure life data for creep–fatigue interactions, and the validity of these prediction rules is discussed. One is the ASME Boiler and Pressure Vessel Code Case N-47 and the other is the v–K_t modified fatigue life equation (6.21).

Following the ASME Code case, the strain at the notch root other than thermal strain can be estimated by the equation

$$\varepsilon_T = (S^*/\bar{S})K_t^2 \varepsilon_n + K_t \varepsilon_c \tag{6.28}$$

where ε_n and ε_T are the nominal strain and the strain at the notch root, respectively, while S^* and $\bar{S}$ are the nominal stress and the stress at the notch root, respectively, and ε_c is the nominal strain observed during the

Table 6.3
Comparison between Predicted Life and Actual Life (Frequency Effect)
(Reprinted from ref. 41 by permission of Amer. Soc. Mech. Engrs)

	Predicted life			Actual life
K_t	N_{fA} ASME	$N_{f0\cdot3}$ $\Delta\varepsilon=0\cdot3\%$	N_{fE} Elastic	N_f $v=1-0\cdot01\ Hz$
2·6	170	2 300	11 000	32 000–830
4·2	—	270	1 050	11 20–560
6·0	—	<200	230	790–320[a]

[a] 0·001 Hz.

hold-time. So in the tests without hold-time only the first term of the right-hand side of the equation is used.

Table 6.3 [41] shows the comparison between the predicted lives at 1% without hold-time and the experimental lives in the frequency range from 0·01 Hz to 1 Hz without hold-time. For calculating the predicted life, the strain at the notch root was first calculated by combining eqn (6.28) with the cyclic stress–strain diagram of the materials. The cyclic stress–strain relation can be written as $\sigma = K\varepsilon^{n'}$ (K 220 MPa, n' 0·53). The predicted life is then obtained using this strain value in the $\Delta\varepsilon$ versus N_f diagram at 1 Hz. The three predicted values in the table correspond to different nominal strains. Here N_f was obtained according to the ASME Code Case N-47; $N_{f0\cdot3}$ was calculated by taking the nominal strain as equal to 0·3%. The nominal strain in the column for N_{fE} was calculated by dividing the nominal stress amplitude by the elastic modulus. We can see that life predicted by the ASME method gives an unnecessarily large margin in which a life for K_t more than 2·6 could not be calculated. The predicted method for $N_{f0\cdot3}$ seems to be appropriate for smaller K_t, but for large K_t it gives a very conservative estimate. The predicted life in the column for N_{fE} goes from non-conservative to conservative with increasing K_t. From the results in Table 6.3, we can conclude that the Neuber rule, as used in the ASME Code Case N-47, overestimates the reduction in the life of the material with increase in K_t.

The same is found from the results of zero hold-time tests, in Table 6.4 [41] and Fig. 6.62 [41], which show the correlation between the predicted lives and the actual lives, including the results of hold-time tests. In the table N_{fTn} is derived from the total strain range ε_{Tn} ($= (S^*/\bar{S})K_t^2\varepsilon_n$) and N_{fn} is derived from ε_n ($= \varepsilon_{Tn} + K_t\varepsilon_c$) including the

Table 6.4
Comparison between Predicted Life and Actual Life (Hold-time Effect)
(Reprinted from ref. 41 by permission of Amer. Soc. Mech. Engrs)

t_H	K_t	ε_{Tn}	$K_t\varepsilon_c$	N_{fTn}	N_{fn}	ε (%) N_f
0	2·6	1·97	—	5·60	—	1 200
	4·2	3·60	—	135	—	610
	6·0	7·20	—	27	—	580
10	2·6	1·85	0·060	760	700	230
	4·2	3·50	0·130	170	150	180
	6·0	5·10	0·096	70	68	170
30	2·6	1·65	0·072	990	880	290
	4·2	3·30	0·134	195	175	220
	6·0	5·50	0·174	60	54	180
60	2·6	1·90	0·047	750	650	453
	4·2	3·45	0·105	160	145	227
	6·0	6·40	0·144	40	39	190

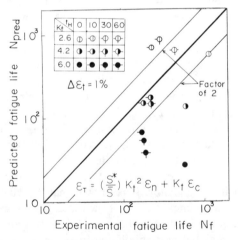

Fig. 6.62. Correlation between the life predicted by eqn. (6.28) and the actual life. Test conditions as for Fig. 6.57. (Reprinted from ref. 41 by permission of Amer. Soc. Mech. Engrs)

second form of eqn (6.28). A rather poor correlation is found between the two calculated and measured lives. In the case of zero hold-time tests, eqn (6.28) predicts too long a life for the material when K_t is 2·6, but lives predicted by eqn (6.28) are very much shorter than the acual lives for all the elastic stress concentration factors tested, which is a similar result to that shown in Table 6.3. On the other hand, for the test with hold-time, eqn (6.28) predicts too long a life for the material when K_t is 2·6, but life prediction by eqn (6.28) changes to the safe side as K_t increases; even for K_t 6·0, eqn (6.28) predicts the life of the material very conservatively.

Figure 6.62 shows the comparison between the predicted lives, N_{fpred}, and the experimental lives, N_f, for the specimen in the test with and without hold-time. A noticeable characteristic in Fig. 6.62 is that the predicted lives have a greater dependence on K_t than the actual lives for both zero and non-zero hold-time tests. This greater dependence on K_t is the same as that mentioned in Section 6.5.3, and it is closely related to the estimate of the exponent of K_t in eqn (6.28). Another characteristic is the cancellation of the overestimation of the strain concentration in the first term of eqn (6.28) and the underestimation in the second one. This is evident from Table 6.4. For example, in the case of K_t of 2·6, the predicted life in the zero hold-time test is 560 cycles, which is smaller than that of actual tests of 1200 cycles. This prediction is a conservative one. When the 10 min hold-time is introduced, the reduction in the predicted life is only 60 ($=760-700$), although the reduction of the actual life is much greater, about 1000 ($=1200-230$). Thus, as for the prediction of the reduction in the fatigue life due to the hold-time, the second term of eqn (6.28) underestimates the reduction of the life by the hold-time, in contrast with the overestimation by the first term. Therefore, since the overestimation in the first term of eqn (6.28) cancels the underestimation in the second one, the prediction of the total life yields the results shown in Fig. 6.62.

We will now examine the validity of eqn (6.21) for predicting the life of the material in hold-time tests. Figure 6.63 [41] shows the correlation between the lives predicted by eqn (6.21) and the actual lives. When applying eqn (6.21) to the hold-time test, the durations of hold-times were converted into frequency. It is known that, by changing the wave shape for a given frequency, the life of the material changed significantly, but in this section the hold-time effect was taken as a frequency-delaying factor and it is included as a modifying frequency. A better correlation was obtained in Fig. 6.63 than in Fig. 6.62. Therefore we concluded that the frequency-modified fatigue life equation for the notched specimen can be extended to the hold-time test.

The above discussions were conducted concerning the applicability of Neuber's rule and the frequency-modified fatigue life equation to the estimate of the total failure life of notched specimens. The term 'total' means that it includes both the crack initiation and propagation periods.

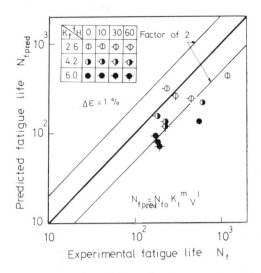

Fig. 6.63. Correlation between the life predicted by eqn (6.21) and the actual life. Test conditions as for Fig. 6.57. (Reprinted from ref. 41 by permission of Amer. Soc. Mech. Engrs)

So, even when Neuber's rule can accurately estimate the strain concentration at the notch root, it may not necessarily predict the total life of the material because of the existence of a crack propagation period. To check this, we must study the notch effect by dividing the total life into the crack initiation and propagation periods, as shown in Fig. 6.60.

At the end of this section, we describe the application of the v–K_t modified fatigue life equation to long-term low-cycle fatigue data. Figures 6.64 (a) and (b) [37] compare the life predicted by eqn (6.21) with the experimental failure life of type 316 and 304 stainless steels mentioned in Section 6.5.2. Material constants used to perform the prediction are written in the captions of the figures. The figures show that the predicted life agrees with the experimental failure life within a scatter band with a factor of 2 under widely varied test conditions including the static creep data, so that eqn (6.21) is available for predicting the failure lives of notched specimens. As for the material constants in eqn (6.21), values of m and l were -1.45 and 0.15 for a fully reversed test of type 316 steel, while in the time- and cycle-dependent fracture region of the zero-to-tension tests the material constants were $m = -1$ and $l = 0.5$ for type 316 steel and they were $m = -0.7$ and $l = 0.7$ for type 304 steel, so that the frequency effect is larger but the effect of elastic stress concentration is smaller in zero-to-tension tests.

In this section, frequency and hold-time effects on the low-cycle fatigue

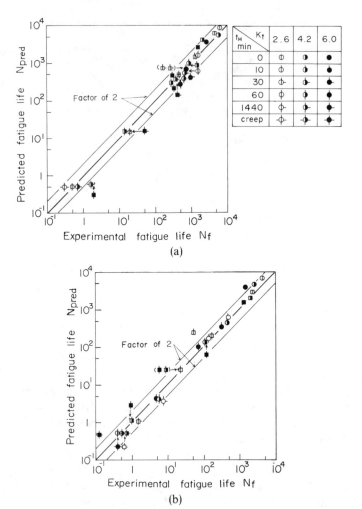

Fig. 6.64. Comparison between the life predicted by eqn (6.21) and the actual one. Circular marks denote the data under $\Delta\sigma = 274$ MPa and square marks the data under $\Delta\sigma = 314$ MPa. (Reprinted from ref. 37 by permission of Soc. Mater. Sci., Japan)

(a) Type 316 stainless steel at 873 K in air (under $\Delta\sigma = 314$ MPa, $N_f = 75\,000\,K_t^{-1} v^{0.5}$ for $t_H = 0$–60 min; $N_f = 1\,300\,000\,v$ for $t_H = 1440$ min and creep; under $\Delta\sigma = 274$ MPa, $N_f = 100\,000\,K_t^{-1} v^{0.5}$ for $t_H = 0$–60 min; $N_f = 280\,000\,v$ for $t_H = 1440$ min and creep).

(b) Type 304 stainless steel at 873 K in air (under $\Delta\sigma = 314$ MPa, $N_f = 44\,000\,K_t^{-0.7} v^{0.7}$ for $t_H = 0, 10$ min; $N_f = 93\,600\,v$ for $t_H = 60, 1440$ min and creep; under $\Delta\sigma = 274$ MPa, $N_f = 110\,000\,K_t^{-0.7} v^{0.7}$ for $t_H = 0, 10, 60$ min; $N_f = 360\,000\,v$ for $t_H = 1440$ min and creep).

lives of cylindrical notched specimens of type 316 stainless steel were studied at 873 K in the atmosphere. From these tests we made the following conclusions.

(1) Neuber's rule, as used in the ASME N-47 Code, very conservatively predicts the failure life of the material in the tests without hold-time, but it gives a non-conservative estimate for the reduction in the life of the material by the introduction of hold-time.

(2) The frequency–elastic stress concentration factor modified fatigue life equation can predict the failure life of the notched specimen in long-term tests within a factor of 2.

6.6.2 Collaborative Research Efforts Concerning the Notch Effect on Low-cycle Fatigue under Creep–Fatigue Interaction Conditions at High Temperatures

In this section we will briefly describe the collaborative research effort concerning the notch effect on low-cycle fatigue under creep–fatigue interaction conditions by experimentation and inelastic finite element analyses. The Notch Effects Test Subcommittee of the Iron and Steel Institute of Japan (chairman the author) planned to coordinate research efforts on the notch effect on cree–fatigue interactions [45, 46].

First we will describe the inelastic cyclic stress–strain response at the notch root based on an inelastic finite element analysis for a trapezoidal strain wave up to five cycles. Three computer programs shown in Table 6.5 [45, 46] were used to examine how the cyclic hardening rule affects the inelastic stress–strain response at the notch root. They are kinematic hardening, MARC type combined hardening and ORNL kinematic hardening rules. The material used was solution heat treated type 304 stainless steel, which has typical cyclic strain hardening qualities, as mentioned in Section 6.4.2, and the ORNL kinematic hardening rule takes this cyclic hardening behavior into account. All the finite elements are isoparametric elements with either 4 or 9 integration points, and the three finite element analysis programs used were HINAPS made by Hitachi, Tepicc-4 made by Babcock–Hitachi and MARC-J2. Figure 6.65 shows the shape and dimensions for the specimen analyzed ($K_t = 1.5$), where only the shaded part is modeled to finite element analysis. The strain wave used in the analysis is a fully reversed triangular wave of the push-pull mode (strain ratio $R = -1$) with a 30 min strain hold-time at the tension peak in each cycle.

Figure 6.66 [45, 46] compares the measured hysteresis loop at the notch root with that of the finite element analysis. Special high-

Table 6.5
Constitutive Equation and Inelastic Finite Element Analysis Program for Cyclic Calculation (Reprinted from ref. 45 by permission of Iron and Steel Inst. Japan)

	Program	HINAPS	MARC-J2	TEPICC-4
Elastic-plastic characteristic	Hardening rule	Kinematic hardening	MARC type combined hardening	ORNL Kinematic hardening
	σ_{yo} (MPa)	132.4	Multilinear approximation	132.4
	σ_y (MPa)	132.4		$\sigma_y = 230.5$
	E (MPa)	1.499×10^5	1.499×10^5	1.499×10^5
	ν	0.30	0.31	0.31
	E_p (MPa)	2530		2530
Creep equation		$\dot{\varepsilon}_c$ (1/h) $= 2.036 \times 10^{-14} \sigma^{4.674}$		(σ : MPa)

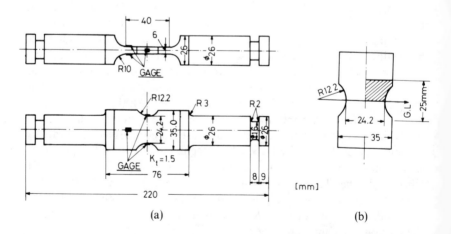

Fig. 6.65. Shape and dimensions of (a) the test specimen used for the measurement of strain, and (b) the portion for finite element mesh.

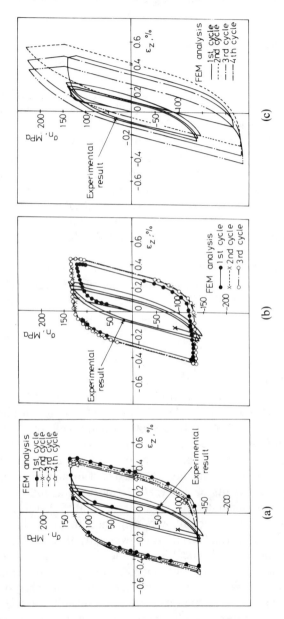

Fig. 6.66. Experimental and finite element analytical results of cyclic stress/strain, based on the nominal stress, σ_n, and the axial strain, ε_z, at the notch root, for type 304 stainless steel at 873 K: (a) kinematic hardening rule; (b) MARC type combined hardening rule; (c) ORNL kinematic hardening rule. (Reprinted from ref. 45 by permission of Iron and Steel Inst., Japan)

temperature resistant strain gauges were attached to the bottom of the notch grooves. Regarding the comparison between the analytical results, the kinematic hardening and the MARC type hardening rules exhibit almost a closed hysteresis loop with a slight movement to the compression side. This movement is caused by the strain hold-time because in the analysis without hold-time completely closed hysteresis loops were found. Stress relaxation during the 30 min hold-time is small and leads to a slight increase in the axial strain during the hold-time; nevertheless, the strain along the gauge length maintains a constant value. For the case of ORNL kinematic hardening, on the other hand, the hysteresis loop exhibits different behavior; at the first cycle smaller cyclic peak stress in tension is found but the second cycle yields significantly larger cyclic stress. After the second cycle there is no large difference in peak stress but the hysteresis loop moves to the compression side after the first cycle. Stress relaxation in the ORNL kinematic hardening rule during the hold-time is larger compared with those in the other two hardening rules.

Measured nominal stress agrees well with the finite element analysis, with the exception of the ORNL hardening rule, but the infinite element analysis overestimates the total strain range by about 1·6 times. The main cause of this disagreement may be attributed to improper modeling of the finite element mesh. As shown in Fig. 6.65 only the notched part, with a height of 1·5 mm, was modeled for finite element analysis, so the finite element model does not precisely simulate the actual manner of deformation. The finite element analysis evaluates the excessive notch opening so that a larger inelastic strain concentration occurs.

In order to check the way in which this deformation manner affects the strain concentration at the notch root, an additional finite element analysis was made for a model with a height of 30 mm, i.e. an element with a height of 17·5 mm was added to the original model shown in Fig. 6.65, in order to improve the constant axial deformation near the notched portion. The results of the recalculation were as follows. After unloading from a nominal tensile peak strain of 0·15%, the residual axial strain was 0·26% in the original model at the notch root, but in the additional model the residual axial strain was only 0·16%. The strain concentration in the original model is 1·6 times larger than that in the additional model, so if we use proper finite element analysis the calculated strain agrees well with the strain measured at the notch root.

Here we describe a life prediction method applicable for notched plate and notched round bar specimens. Three prediction methods are applied to the data of fatigue tests with and without stress hold-time: the inelastic finite element analysis method, the method following the ASME Code Case N-47 and the $v-K_t$ modified fatigue life equation. All the life prediction methods are based on the local strain theory. Note that only the $v-K_t$ equation does not appear to be based on local strain theory, but the equation implicitly includes the concept of local strain and its derivation of the equation as mentioned in Section 6.5.3. From the present study the conclusions are as follows.

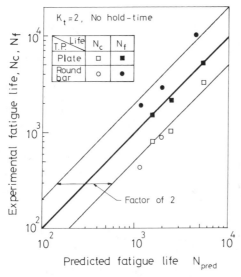

Fig. 6.67. Comparison between the predicted fatigue life, by cyclic finite element analysis, and the actual one, for the load-controlled push-pull test without hold-time. Test conditions as for Fig. 6.66. (Reprinted from ref. 45 by permission of The Iron and Steel Inst., Japan).

(i) A prediction method based on cyclic finite element analysis gives a proper estimate for failure life in tests without hold-time for $K_t = 2 \cdot 0$ while it gives a slightly unconservative estimate for crack initiation life, as shown in Fig. 6.67 [45, 46]. Since a bilinear stress–strain relation was used in the finite element analysis, the deviation of the stress–strain relation used for the analysis from the actual material behavior might result in the unconservative estimation of the crack initiation life. To examine this, we recalculated the local strain by using a multilinear stress–strain relation for the plate specimen with $K_t = 2 \cdot 0$. The result of the life prediction obtained was $N_{pred} = (1/3)N_f$ and $N_{pred} = (1/2)N_c$, where N_{pred} is the predicted failure life, N_f is the actual failure life, and N_c is the actual crack initiation life. Comparing these results with those using the bilinear relation, the predicted values decrease to 3/10 of those for the bilinear type, so we know that an accurate constitutive relation is essential in predicting the failure or crack initiation life of the notched specimen.

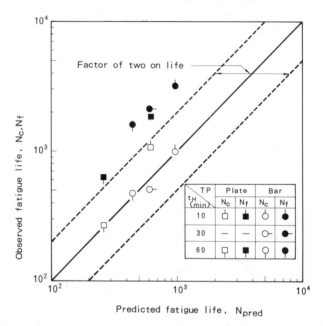

Fig. 6.68. Comparison between the life predicted by cyclic finite element analysis and the observed life, for the load-controlled push-pull test with hold-time. Test conditions as for Fig. 6.66. (Reprinted from ref. 45 by permission of Iron and Steel Inst. Japan).

Figure 6.68 [45, 46] shows the results of life prediction in a stress-hold test for $K_t = 2 \cdot 0$ using finite element analysis. The prediction procedure was as follows. The fatigue damage, ϕ_f, was evaluated by applying the calculated strain to the $\Delta \varepsilon$ versus N_f curve of the unnotched specimen; also the creep damage was calculated from the equation, $\phi_c = \Sigma(dt/t_r)$, from the stress relaxation curve, and then the predicted life is obtained from the ϕ_c versus ϕ_f diagram in the ASME Code Case N-47. The predicted life clearly agrees well with the crack initiation life, indicating that finite element analysis was fairly accurate for predicting the crack initiation life. However, the finite element analysis gave a slightly conservative estimate for the failure life. We previously described that the stress concentrations due to the notches mainly had an effect on crack initiation and had almost no effect on crack propagation behavior. As N_f includes both crack initiation and propagation periods, the finite element analysis estimates a conservative life span for the failure life.

(ii) The $v-K_t$ modified fatigue life equation can properly predict the failure lives of the round bar specimen and the plate specimen in the stress-hold tests, as shown in Fig. 6.69 [45, 46] as well as Section 6.5.3, where the material constants are $m = -0.59$ and $l = 0.31$. Almost all the predicted data are within a scatter band with a factor of 2, so this prediction method can be valid for predicting not only the failure lives in tests without hold-time but also the lives in hold-time tests.

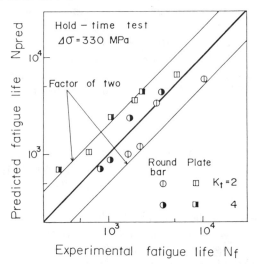

Fig. 6.69. Comparison between the life predicted by $v-K_t$ modified fatigue life equation and the actual life, for the load-controlled push-pull test with hold-time. Test conditions as for Fig. 6.66. (Reprinted from ref. 45 by permission of Iron and Steel Inst. Japan).

(iii) Regarding the results of life prediction by the ASME Code Case N-47, the method gives a conservative estimate especially for $K_t = 6.0$. This estimation was made as a load-controlled test because the tests were made under load-controlled conditions. Besides, this estimation was based on the data in the Code Case which is far too conservative and has an unnecessarily large safety factor margin. The value of the safety factor margin is 22 on average, and the value in some cases becomes larger.

In this section, we arrived at the following conclusions from the collaborative efforts.

(1) Plastic-creep cyclic finite element analysis is made for the trape-

zoidal strain wave in five cycles using three hardening rules, i.e. the kinematic hardening rule, the MARC type combined hardening rule and the ORNL kinematic hardening rule. The stress–strain response at the notch root exhibits an almost closed hysteresis loop, even in creep–fatigue conditions, so we can replace cyclic analysis with monotonic analysis. There exists almost no difference between the cyclic stress–strain concentrations with the exception of the ORNL kinematic hardening rule. The strains calculated by finite element analysis agree well with the experimental measurement if we use the proper finite element mesh and the cyclic constitutive equation.

(2) Three life prediction methods, i.e. the inelastic finite element analysis method, the v–K_t modified fatigue life equation method and the ASME Code Case N-47 method, are applied to the failure life in the load-controlled tests with and without hold-time. Among them the v–K_t modified fatigue life equation method and finite element analysis are quite accurate while the ASME N-47 method has an excessive safety factor margin.

6.7 THERMAL FATIGUE OF A CIRCULAR DISK SPECIMEN IN CONNECTION WITH THE STRENGTH OF GAS TURBINE DISKS

Before describing the thermal fatigue behavior of a circular disk specimen, we first discuss the unsteady state radial temperature distribution of a circular disk whose periphery is cyclically heated and cooled. We will also examine the influence of the history of heat transfer on the radial temperature distribution of the disk.

The heat conduction equation of a circular disk having a uniform thickness and a small circular hole in the center, and of which the periphery is rapidly heated by hot gas (gas temperature T_g) as shown in Fig. 6.70, was solved under the following conditions. The boundary conditions for heating are given by $k(\partial T/\partial r) = h(T_g - T)$ and $k(\partial T/\partial r) = 0$ at $r = r_o$ and $r = r_i$, respectively, and the initial condition is given by $T = 0(°C)$ at time $t = 0$, where k is the conductivity, h is heat transfer, and r_o and r_i are the outer and inner radius of the disk specimen.

Figures 6.71 (a) and (b) [47] show an example of the analytical results of the radial temperature distribution on the circular disk specimen having a 75 mm radius subjected to a T_g of 773 K and an H of 1·0 in heating and cooling processes, respectively, where $H = hr_i/k$ (nondimen-

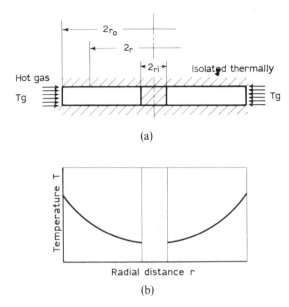

Fig. 6.70. Schematic representation of both the heat transfer (a) and radial temperature distribution (b) of a circular disk whose periphery is rapidly heated.

sionalized heat transfer), $\rho = T/T_g$, $\tau = at/r_i^2$ (nondimensionalized time) and $R = r/r_i$. The analytical results of Fig. 6.71 were examined by the experiments.

Figure 6.72 [47] shows an analytical example of the influence of the history of heat transfer on the radial temperature distribution of the disk specimen. We found that the gradiant of the radial temperature distribution in the circular disk specimen subjected to a thermal load of a linear increasing heat transfer of H to elapsed time τ of 2·04 is smaller than that for an impulsive thermal load (stepwise loading in H). Therefore the thermal stress at the periphery of the circular disk specimen under the former loading conditions becomes smaller than that under the latter. Besides, we found that after an elapsed time τ of 2, the difference between the temperature gradients under two types of thermal loading becomes comparatively small. Since an elastic-plastic tangential thermal stress, σ_θ, at the periphery of the disk specimen under a heat transfer of 0·1 takes the maximum value at an elapsed time τ of 6·75, as shown in the following, we concluded that there is no influence of the transient phenomena of heat transfer on the gradient of the radial

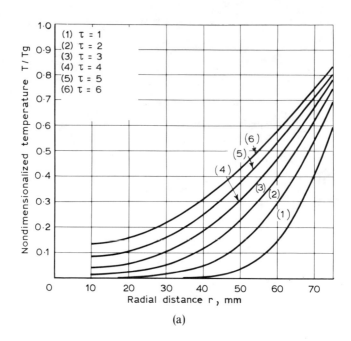

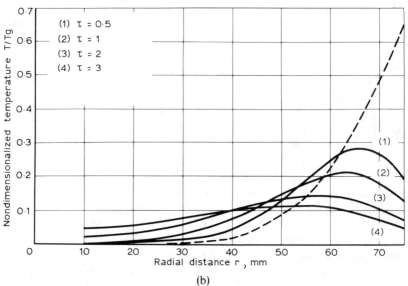

Fig. 6.71. Analytical results of unsteady state radial temperature distributions in a circular disk, for (a) heating and (b) cooling processes, where H is nondimensional heat transfer and c is nondimensional time. (Reprinted from ref. 47 by permission of Japan Soc. Mech. Engrs).

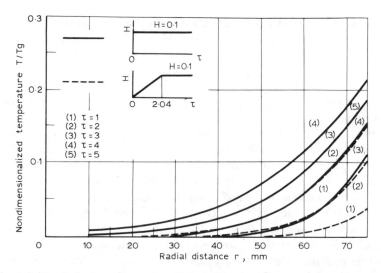

Fig. 6.72. Analytical representation of the influence of heat transfer on temperature distribution in a radial direction for a circular disk specimen during the heating and cooling processes. (Reprinted from ref. 47 by permission of Japan Soc. Mech. Engrs).

temperature distribution of the disk specimen from a practical viewpoint.

Next we describe an analytical consideration of both thermal crack initiation and propagation on the circular disk. We assume that the fatigue damage, $\Delta\phi$, imposed on the material during a half cycle of thermal cycling is given by

$$\Delta\phi = \lambda(\Delta\bar{\varepsilon}_p)^\alpha \tag{6.29}$$

where $\Delta\bar{\varepsilon}_p$ is the equivalent strain range based on the von Mises criterion and λ and α are coefficients which depend upon the temperature cycling [48]. Then we assume that the thermal fatigue crack initiation begins once the accumulation of eqn (6.29) reaches a finite value of ϕ_0, which is a material constant. Formulating it, we have

$$(\Delta\bar{\varepsilon}_p)N_c^{1/\alpha} = (\phi_0/2\lambda)^{1/\alpha} = C \tag{6.30a}$$

When the magnitude of the equivalent total strain range $\Delta\bar{\varepsilon}$ is large enough for the component of plastic strain range to occupy the greater part of the total strain range, eqn (6.30a) is written as

$$(\Delta\bar{\varepsilon})N_c^{1/\alpha} = C \tag{6.30b}$$

Now let us consider the application of the damage rule of eqn (6.29) to the thermal crack tip of the material. Here the crack propagation rate $(\mathrm{d}l/\mathrm{d}N)$ of the material with a half crack length of l is written as

$$(\mathrm{d}l/\mathrm{d}N)_{x=l} = -(\delta\phi/\delta N)_{x=l}/(\delta\phi/\delta x)_{x=l} \qquad (6.31)$$

where x shows the distance from the crack tip for the direction of crack propagation.

In eqn (6.31), $(\partial\phi/\partial N)_{x=l}$ is given by $2\lambda(\Delta\bar{\varepsilon})^\alpha$ and $(\partial\phi/\partial x)_{x=l}$ is written as $-k\alpha\lambda(\bar{\varepsilon}_{x=l})^\alpha l^{-1}$, provided that the gradient of the distribution of the von Mises equivalent total strain $\bar{\varepsilon}$ at the crack tip is given as $(\partial\bar{\varepsilon}/\partial x)_{x=l} = -k\bar{\varepsilon}_{x=l}l^{-1}$, where the coefficient k has a positive value.

This relation can be adopted from the singular behavior in the front of the crack tip which gives $\bar{\varepsilon}_e(x) = C_e K x^{-1/2}$ and $\bar{\varepsilon}_p(x) = C_p K_\varepsilon x^{-1/(n+1)}$ [11], where $\bar{\varepsilon}_e$ and $\bar{\varepsilon}_p$ are the elastic and plastic components of the von Mises equivalent strain, K and K_ε are an elastic stress intensity factor and an inelastic strain intensity factor, n is a strain-hardening exponent of the material, and C_e and C_p are constants. Therefore we obtain the relation $(\partial\bar{\varepsilon}/\partial x)_{x=l} = -(1/2)\bar{\varepsilon} l^{-1}$ in the region of dominant elasticity and $(\partial\bar{\varepsilon}/\partial x)_{x=l} = -[1/(n+1)]\bar{\varepsilon} l^{-1}$ in the region of dominant plasticity. In fact, this is confirmed from the inelastic finite element analysis for the $\bar{\varepsilon}$ distribution in the front of the crack tip, as shown in Fig. 6.73 [47]. This calculation is performed for the material having a bilinear constitutive equation under the conditions of stress ratio σ_x/σ_y of 0 (uniaxial) and 0·1 and 0·2 (biaxial), by taking into consideration the distribution of elastic-plastic thermal stress at the periphery of the circular disk specimen. We found from the present calculation that there is no large difference between the distribution of $\bar{\varepsilon}$ under uniaxial tension and that under biaxial tension, but there is a tendency for the slope in the strain distribution to become a gentle slope as the σ_x/σ_y value increases.

Therefore the crack propagation rate $(\mathrm{d}l/\mathrm{d}N)$ of eqn (6.31) is given by

$$(1/l)(\mathrm{d}l/\mathrm{d}N) = B, \qquad B = C_1(\Delta\bar{\varepsilon}/K_\varepsilon)^\alpha \quad \text{or} \quad C_2(\Delta\bar{\varepsilon}/K_\sigma^{1/n})^\alpha \qquad (6.32)$$

In eqn (6.32), $\bar{\varepsilon}_{x=l}$ is the equivalent total strain at the crack tip and is rewritten in terms of K_ε or K_σ, where K_σ is an inelastic stress intensity factor at the peak bulk stress during temperature cycling, n is a cyclic work-hardening exponent and C_1 and C_2 are constants. This equation indicates that the nondimensional crack propagation rate, $(1/l)(\mathrm{d}l/\mathrm{d}N)$, is proportional to $(\Delta\bar{\varepsilon}/K_\varepsilon)^\alpha$ or $[(\Delta\bar{\varepsilon})/K_\sigma^{1/n}]^\alpha$. This means that $\mathrm{d}l/\mathrm{d}N$ is explicitly represented by not only the equivalent strain range, $\Delta\bar{\varepsilon}$, but also the singularity field of the crack tip. Strictly speaking, K_ε or K_σ

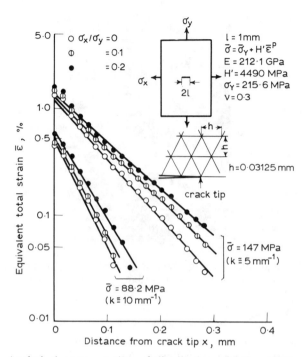

Fig. 6.73. Analytical representation of distribution of Mises strain, $\bar{\varepsilon}$, on the cross-section of the crack tip of a plate specimen of 0·21% carbon steel with a central crack subjected to uniaxial and biaxial tensions, by the finite element method. (Reprinted from ref. 47 by Japan Soc. Mech. Engrs).

depends on multiaxiality of stress [8], but it is supposed that the effect is comparatively small and the term of K_ε or K_σ in eqn (6.32) has a constant value irrespective of multiaxial stress state in thermal fatigue of the circular disk specimen. Historically, this type of crack propagation law has been discussed by many researchers, and recently Marsh et al. [49] reported the validity of this type of equation in multiaxial thermal fatigue of type 316 stainless steel. Therefore it is worthy of note here that the thermal crack propagation rate of the circular disk specimen at an arbitrary crack length is predicted from the coefficient B in eqn (6.32) which is obtained from the uniaxial thermal fatigue data.

Figure 6.74 [47] shows the thermal fatigue test apparatus for the circular disk specimen. Cyclic heating and cooling of the periphery of the disk specimen are performed by using both a high-frequency coil set up around the circle of the periphery of the disk specimen and a cooling air pipe set up to surround the heat coils, respectively. To avoid excessively

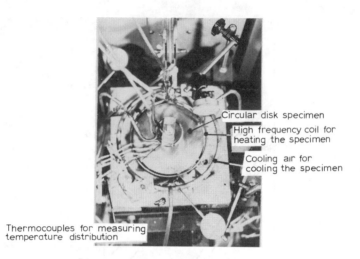

Fig. 6.74. Thermal fatigue test apparatus for a circular disk specimen. (Reprinted from ref. 47 by permission of Japan Soc. Mech. Engrs).

heating the center of the disk, both the cooling water pipe and the air pipe are set and a large gradient of radial temperature on the disk specimen is created. Radial displacement of the disk specimen subjected to thermal cycling is measured by the displacement of a quartz glass bar slightly touching the periphery of the disk specimen through the magnification mechanism. The temperature difference on the periphery of the disk specimen is kept within 10°C.

In addition, the uniaxial thermal fatigue test is performed by using a test apparatus in which the thermal strain constraint of the specimen is varied arbitrarily. Cyclic heating and cooling of the specimen are also performed by use of both a heat coil surrounding the gauge length of the specimen and inner cooling of the hollow specimen. The circular specimen has a 150 mm diameter and a 3 mm thickness and was taken from a structural carbon steel SS41 (0·21% C) bar with diameter 160 mm. All the hollow cylindrical specimens are taken from the roll direction. The specimen surfaces are polished after machining. The inner surface of the hollow cylindrical specimen (10 mm ID, 12 mm OD, 30 mm gauge length) used for the uniaxial thermal fatigue test is specially polished by emery paper No. 400. The circular disk specimen for the measurement of X-ray residual stress is annealed at 873 K for 2 h after machining, and the surface of the disk specimen is also removed by electrolytic polishing.

All the disk specimens, except the one used for the X-ray measurement, are used as machined for the thermal fatigue tests. The X-ray residual stress measurement is made from a diffraction profile plane of the (310) plane based on a $CoK\alpha_1$ line using the $\sin^2\psi$ method, where ψ is the angle between the incident X-rays and the specimen surface normal.

The lower temperature of thermal cycling is kept at a constant 473 K for both types of test; the upper temperatures selected were 723 K, 773 K, 823 K and 873 K for the circular disk specimen, and the upper temperature limit for the hollow cylindrical specimen was also kept at a constant 773 K. Several total thermal strain ranges of the cylindrical specimen are adopted by several choices of the strain constraint $\eta = 1 - (\delta_s/\delta)$, where δ is the free linear expansion in the gauge length of the specimen and δ_s is the thermal displacement of the specimen constrained at both ends. The frequency of thermal cycling was 1–2 cpm for both types of tests.

Figure 6.75 [47] shows the analytical result of the cyclic tangential stress, σ_θ, versus the tangential strain, ε_θ, relation at the periphery of a disk specimen subjected to thermal cycling, based on the experimental results of Fig. 6.76 by using a successive approximation method [50]. In order to examine the validity of the analytical results of Fig. 6.75, a comparison between the experimental results of the diametral displacement of the disk specimen and the analytical results is made in Fig. 6.76 [47], and Fig. 6.77 [47] also shows a comparison between the analytical results of the radial distribution of tangential thermal stress, σ_θ, and the experimental results of X-ray residual stress. From these figures, we found that the results of inelastic thermal stress and strain analyses agree well with those arrived at experimentally.

Figure 6.78 [47] shows an example of the experimental variation of residual tangential stress, $\sigma_{Re\theta}$, and the ratio of the X-ray half-value width b/B at both the periphery and the crack tip of the disk specimen with the number of thermal cycles. The figure also shows the variation of micro-Vickers hardness H_v at the periphery of the disk specimen and crack length l of the disk specimen with number of thermal cycles. We found that the residual stress at the periphery of the disk specimen has the maximum value at a life fraction of about 5% of N_f, and a thermal microcrack 0·1 mm in length was observed with this life fraction. We also found that after this life fraction the residual stress decreases gradually but the thermal crack grows with thermal cycling. In this case, the thermal fatigue life of the disk specimen is defined as the number of cycles to growth of a crack 20 mm in length. Moreover, it is found that variation of both the ratio of the half-value width and the hardness at the

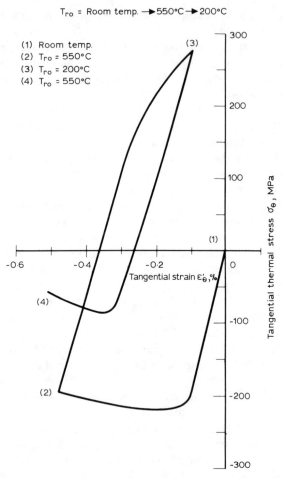

Fig. 6.75. Analytical cyclic thermal stress/strain curves in tangential direction for a circular disk specimen at the first thermal cycle. (Reprinted from ref. 47 by permission of Japan Soc. Mech. Engrs)

periphery of the disk specimen are similar to those in the case of residual stress. It seems that an increase of the residual stress at the crack tip of the disk specimen results in an increase in the mean tensile thermal stress with thermal cycling, and this has the effect of accelerating crack propagation. We can also suppose that the cyclic hardening of the material occurs in the early stages of the fatigue process, from the good

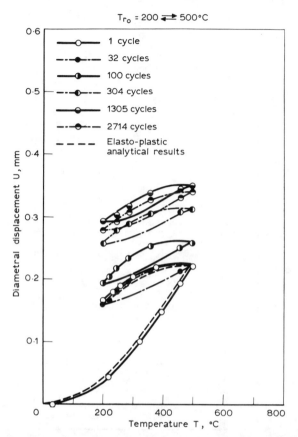

Fig. 6.76. Comparison between the analytical result of diametral displacement of a circular disk specimen and the experimental result. (Reprinted from ref. 47 by permission of Japan Soc. Mech. Engrs)

correlation of the variation in the hardness with the half-value width.

Figure 6.79 [47] shows an experimental comparison between the von Mises equivalent total strain range, $\Delta \bar{\varepsilon}$, versus number of cycles to crack initiation, N_c, relation of the circular disk specimen and that of the hollow cylindrical specimen with and without a small circular hole 1 mm in diameter. As seen from the figure, there is not a large difference between the slopes of the curves for both types of specimen, and the value of α in eqn (6.30b) for both types of specimen is about 5. We also found that the number of cycles to crack initiation in the disk specimen

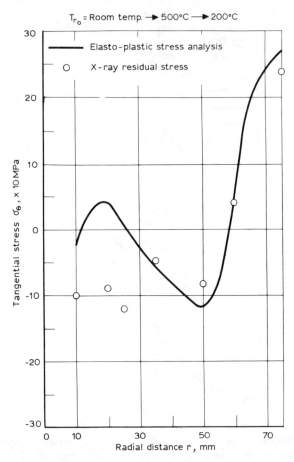

Fig. 6.77. Comparison between the elasto-plastic analysis and the X-ray residual stress measurement, in tangential thermal stress, σ_θ, for a circular disc specimen of 0·21% carbon steel. (Reprinted from ref. 47 by permission of Japan Soc. Mech. Engrs)

as machined is larger than that in the annealed specimen, and the compressive residual stress of -196 MPa according to the X-ray measurement has the effect of retarding thermal crack initiation. Moreover, the crack initiation life of the disk specimen is slightly shorter than that of the hollow cylindrical smooth specimen but, in the range of the stress ratio of the minimum principal stress to the maximum principal stress tested, it seems that there is no remarkable effect of principal

○ Tangential residual stress at the periphery of the disk specimen
● Tangential residual stress at the crack tip
◐ Half-value width at the periphery of the disk specimen
◐ Half-value width at the crack tip
⊖ Micro vickers hardness
⊗ Thermal crack length

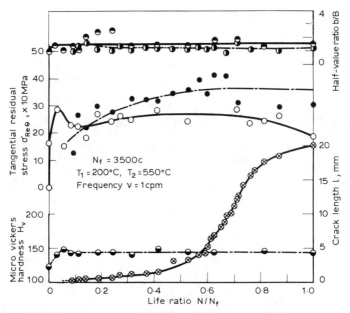

Fig. 6.78. Experimental variation in tangential residual stress, $\sigma_{Re\theta}$, micro-Vickers hardness, H_v, half-value breadth, b/B, and thermal fatigue crack length, l, in a circular disk specimen, with the number of thermal cycles, N. (Reprinted from ref. 47 by permission of Japan Soc. Mech. Engrs)

stress parallel to the crack on the thermal fatigue crack propagation rate in the circular disk.

Figure 6.80 [47] shows an experimental examination of eqn (6.32) for the disk specimen and the hollow cylindrical specimen, both with and without a small circular notch. As seen from the figure, the non-dimensional crack propagation rate, $(1/l)(dl/dN)$, is primarily represented by a power function of $\Delta\bar{\varepsilon}$, as shown in eqn (6.32), for both specimens. Therefore we can conclude that the thermal fatigue crack

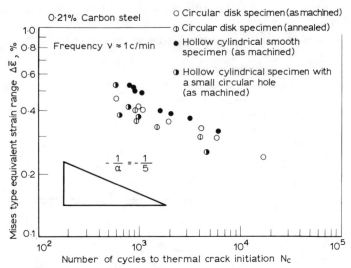

Fig. 6.79. Experimental comparison between the number of cycles to thermal crack initiation, N_c, of a circular disk specimen, and that of a hollow cylindrical specimen, on the comparison basis of Mises strain range, $\Delta\bar{\varepsilon}$. (Reprinted from ref. 47 by permission of Japan Soc. Mech. Engrs)

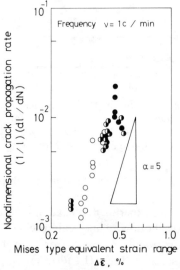

Fig. 6.80. Experimental comparison between the nondimensional crack propagation rate, $(1/l)(dl/dN)$, of the circular disk specimen, and that of the hollow cylindrical smooth specimen of 0·21% carbon steel. (Reprinted from ref. 47 by permission of Japan Soc. Mech. Engrs)

propagation rate of the circular disk specimen can be predicted from that of the hollow cylindrical smooth specimen on the bases of eqn (6.32).

In this section, the analytical and experimental results of thermal fatigue on a structural carbon steel circular disk specimen were presented. The periphery of the disk was cycled between two programmed temperature levels, while a water-cooled sink at the inner radius was maintained at a constant lower temperature level. We made the following conclusions.

(1) Analytical results of an unsteady inelastic thermal stress and the strain of a circular disk specimen of which the periphery is subjected to cyclic heating and cooling agree well with the experimental results of the diametral displacement measurements and those of the X-ray measurement of radial distribution of tangential residual stress.

(2) The maximum residual stress at the periphery of the disk specimen is observed at the early stage of thermal cycling, and crack initiation is also observed at this stage. Continuous growth of the macrocrack, which originated at the periphery and propagated in a direction toward the center of the disk specimen, is observed during thermal cycling, and the residual stress at the crack tip is larger than that at the periphery.

(3) The number of cycles to crack initiation of the disk specimen is predicted from a hollow cylindrical specimen of the material under uniaxial thermal stress on the common basis of the von Mises equivalent total strain range $\Delta \bar{\varepsilon}$.

(4) The thermal macrocrack propagation rate in the disk specimen is predicted from that in the hollow cylindrical smooth specimen on the common basis of eqn (6.32).

REFERENCES

1. Y. S. Garud, Multiaxial fatigue: a survey of the state of the art, *Journal of Testing and Evaluation*, **19** (1981), 165, Amer. Soc. Test. Mater.
2. M. W. Brown and K. J. Miller, Two decades of progress in the assessment of multiaxial low-cycle fatigue life, *ASTM Special Technical Publication*, **770** (1982), 482, Amer. Soc. Test. Mater.
3. M. Ohnami and M. Sakane, A study on creep-fatigue interaction of polycrystalline metal at elevated temperatures (especially on the effect of hydrostatic pressure on crack behavior), *Bulletin of the Japan Society of Mechanical Engineers*, **20** (1977), 1.
4. F. C. Monkman and N. J. Grant, An experimental relationship between rupture life and minimum creep rate in creep-rupture tests, *Proceedings of the American Society for Testing and Materials*, **56** (1963), 593.

5. J. R. Rice, A path-independent integral and the approximate analysis of strain concentration by notches and crack, *Transactions of the American Society of Mechanical Engineers*, **E35** (1968), 379.
6. M. Ohnami, The influence of hydrostatic pressure on fracture toughness of polycrystalline metallic materials, *Journal of the Japan Society for Technology of Plasticity*, **15** (1974), 769 (in Japanese), CORONA-Sha.
7. M. Ohnami and M. Sakane, A study of metallic creep-fatigue interaction at elevated temperatures (especially based on strain history), *Bulletin of the Japan Society of Mechanical Engineers*, **21** (1978), 1057.
8. P. D. Hilton, Plastic intensity factor for cracked plate subjected to biaxial loading, *International Journal of Fracture Mechanics*, **9** (1973), 149, Pergamon Press.
9. N. Hamada, M. Sakane and M. Ohnami, A study of high temperature low cycle fatigue criterion in biaxial stress state, *Bulletin of the Japan Society of Mechanical Engineers*, **28** (1985), 1341.
10. N. Hamada, M. Sakane and M. Ohnami, Effect of temperature on biaxial low cycle fatigue crack propagation and failure life, *Proceedings of 30th Japan Congress on Materials Research* (1987), Soc. Mater. Sci., Japan.
11. J. W. Hutchinson, Singular behavior at the end of a tensile crack in hardening materials, *Journal of Mechanics and Physics of Solids*, **16** (1968), 13, Pergamon Press.
12a. M. Sawada, M. Sakane and M. Ohnami, Mode I–mode II crack transition in high temperature biaxial low cycle fatigue. (Paper of the Japan Society of Mechanical Engineers, No. 864–2, 1986), p. 16 (in Japanese).
12b. M. Sakane, M. Ohnami and M. Sawada, Fracture modes and low cycle biaxial fatigue life at elevated temperature, *Journal of Engineering Materials and Technology, Transactions of the ASME*, **109** (1987), 236.
13. F. A. Kandil, M. W. Brown and K. J. Miller, Biaxial low cycle fatigue failure of 316 stainless steel at elevated temperatures, *Mechanical Behavior of Nuclear Applications for Stainless Steel at Elevated Temperatures* (1982), p. 203.
14. R. D. Lohr and E. G. Ellison, A simple theory for low-cycle multiaxial fatigue, *Fatigue of Engineering Materials and Structures*, **3** (1980), 1, Pergamon Press.
15. M. W. Brown and K. J. Miller, High temperature low cycle biaxial fatigue of two steels, *Fatigue of Engineering Materials and Structures*, **1** (1979), 217.
16. M. Sakane, T. Itsumura and M. Ohnami, High temperature biaxial low cycle fatigue testing machine by using a cruciform specimen and some experiments (Paper of the Japan Society of Mechanical Engineers, No. 864–2, 1986), p. 13 (in Japanese).
17. M. Ohnami, M. Sakane, S. Nishino and T. Itsumura, Physical and mechanical approaches to cyclic constitutive relationships and life evaluation of structural materials at high temperature, in *Advanced Materials for Severe Applications* (Japan-US Seminar, 1986, Tokyo), (ed. K. Iida and A. J. McEvily), Elsevier Applied Science Publishers, 1987.
18. M. W. Brown and K. J. Miller, Initiation and growth of cracks in biaxial fatigue, *Fatigue of Engineering Materials and Structures*, **1** (1979), 231.

19. M. Ohnami and N. Hamada, Biaxial low-cycle fatigue of a SUS 304 stainless steel at elevated temperatures, *Proceedings of 25th Japan Congress on Materials Research* (1982), p. 93, Soc. Mater. Sci., Japan.
20. M. Ohnami and M. Sakane, A study on the creep-fatigue interaction of a cobalt-base super alloy X-40 at high temperatures (especially the effect of strain wave shapes on creep-fatigue interaction), *Bulletin of the Japan Society of Mechanical Engineers*, **21** (1978), 547.
21. C. H. Wells and C. P. Sullivan, Interaction between creep and low cycle fatigue in Udimet 700 at 1400°F, *ASTM Special Technical Publication*, **459** (1969), 59, Amer. Soc. Test. Mater.
22. B. C. Lord and L. F. Coffin, Jr, Low cycle fatigue hold time behavior of cast Rene 80, *Metallurgical Transactions*, **4** (1973), 1647, American Soc. Metals.
23. C. H. Jaske, H. Mindlin and J. S. Perrin, Influence of hold-time and temperature on the low-cycle fatigue of Incoloy 800, Transactions of the American Society of Mechanical Engineers, **94** (1972) 930.
24. C. Y. Cheng and D. R. Diercks, Effects of hold time on low cycle fatigue behavior of AISI 304 stainless steel at 593°C (Communications), *Metallurgical Transactions*, **4** (1973) 615, Amer. Soc. Metals.
25. H. Neuber, Theory of stress concentration for shear-strained prismatical bodies with arbitrary nonlinear stress-strain law, *Transactions of the American Society of Mechanical Engineers*, **E28** (1961), 544.
26. S. S. Manson, G. R. Halford and M. H. Hirschbery, Creep fatigue analysis by strain range partitioning, NASA, TMX **6783** (1971), NASA.
27. M. Ohnami and M. Sakane, Elastic-plastic model of crack propagation under creep-fatigue interacted conditions at elevated temperatures, *Research Report of 123rd Committee on Heat-resisting Materials and Alloys*, **19** (1978), 275 (in Japanese), Japan Society for the Promotion of Science.
28. N. Hamada, M. Sakane and M. Ohnami, Creep-fatigue studies under a biaxial stress state at elevated temperatures, *Fatigue of Engineering Materials and Structures*, **7** (1984), 85.
29. K. Kanazawa, K. J. Miller and M. W. Brown, Low-cycle fatigue under out-of-phase loading conditions, *Journal of Engineering Materials and Technology*, **99** (1977), 222, Amer. Soc. Mech. Engrs.
30. M. Ohnami, M. Sakane and N. Hamada, Effect of changing principal stress axes on low-cycle fatigue life in various strain wave shapes at elevated temperatures, *ASTM Special Technical Publication*, **853** (1985), 622.
31. W. Ostergren, A damage function and associated failure equation for predicting hold-time and frequency effects in elevated temperature, low-cycle fatigue, *Journal of Testing and Evaluation*, **44** (1976), 327.
32. L. F. Coffin, Jr, Fatigue at high temperature, *Fracture 1977*, Vol. **1**, ICF 4, Waterloo, p. 263.
33. B. Tomkins, Fatigue crack propagation—an analysis, *Philosophical Magazine*, **18** (1968), 1041, Taylor and Francis Ltd., London.
34. L. F. Coffin, Jr, Predictive parameters and their application to high temperature low-cycle fatigue, *Proceedings of 2nd International Conference on Fracture* (Brighton, 1969), p. 643.
35. M. Ohnami, K. Umeda, Y. Awaya and M. Takada, Study on creep rupture of notched cylindrical specimens of 1Cr-1Mo-1/4V steel at elevated tem-

peratures, *Bulleting of the Japan Society of Mechanical Engineers,* **19** (1972), 1100.

36. M. Takada, M. Ohnami and Y. Awaya, Creep rupture of notched plate and cylindrical specimens of 1Cr-1Mo-1/4 V steel at elevated temperatures, *Journal of the Society of Materials Science, Japan,* **24** (1975), 747 (in Japanese).
37. M. Sakane and M. Ohnami, Long-term fatigue life prediction in creep-fatigue, *Proceedings of 29th Japan Congress on Materials Research* (1986), p. 69, Soc. Mater. Sci., Japan.
38a. M. Sakane, M. Ohnami, Y. Awaya and N. Shiraishi, A study of notch effect in creep-fatigue, *Proceedings of 21st Symposium of High Temperature Strength of Materials* (1983), p. 102 (in Japanese), Soc. Mater. Sci., Japan.
38b. S. Nishino, M. Sakane and M. Ohnami, Life prediction of high temperature low cycle fatigue for austenitic stainless steel by X-ray fractography, *Proceedings of 29th Japan Congress on Material Research* (1986), p. 61.
39. M. Ohnami, M. Sakane and Y. Awaya, A study of the notch effect of high temperature low cycle fatigue in creep range (especially on the loading mode and crack initiation behaviors), *30th Annual Meeting of the Society of Materials Science, Japan* (1981), p. 626 (in Japanese).
40. M. Ohnami and M. Sakane, A study on the creep-fatigue interaction of super alloy steel A286, *Journal of the Society of Materials Science, Japan,* **24** (1976), 545 (in Japanese).
41. M. Sakane and M. Ohnami, A study of the notch effect on the low cycle fatigue of metals in creep-fatigue interacting conditions at elevated temperatures, *Journal of Engineering Materials and Technology,* **105** (1983), 75, Amer. Soc. Mech. Engrs.
42. R. Ohtani and S. Taira, Effect of nonlinear stress-strain rate relation on the deformation and fracture of materials in the creep range, *Journal of Engineering Materials and Technology,* **101** (1979), 369.
43. M. Sakane and M. Ohnami, Notch effect in low-cycle fatigue at elevated temperatures: life prediction from crack initiation and propagation considerations, *Journal of Engineering Materials and Technology,* **108** (1986), 279.
44. M. Sakane and M. Ohnami, Detection of crack initiation by DC potential and notch opening displacement, *Proceedings of 23rd Symposium of High Temperature Strength of Materials* (1985), p. 137 (in Japanese), Soc. Mater. Sci., Japan; Electrical potential drop and notch opening displacement methods for detecting high temperature low cycle fatigue cracks of circumferential notched specimens, *Journal of Engineering Materials and Technology, Transactions of the ASME,* in press.
45. The Notch Effects Subcommittee of The Iron and Steel Institute of Japan, Collaborative effort of notch effect on low-cycle fatigue in creep-fatigue at high temperatures: experiment and FEM analysis (1985) (in Japanese).
46. M. Ohnami, Y. Asada, M. Sakane, M. Kitagawa and T. Sakon, Notch effect on low-cycle fatigue in creep-fatigue at high temperatures: experiments and FEM analysis, *Symposium on Low Cycle Fatigue—Direction for the Future* (Lake George, 1985), Amer. Soc. Test. Mater.
47. M. Ohnami and H. Nakamura, A study of the thermal fatigue of a circular

disk (in connection with the strength of gas turbine disk), *Bulletin of the Japan Society of Mechanical Engineers*, **19** (1976), 1409.
48. S. Taira, M. Ohnami, H. Minata and T. Shiraishi, Thermal fatigue and mechanical strain fatigue at elevated temperatures of 18–8 Cb stainless steel and 2·25Cr–21Mo steel, *Bulletin of the Japan Society of Mechanical Engineers*, **6** (1963), 169.
49. D. J. Marsh and F. D. W. Charlesworth, The determination and interpretation of thermally promoted crack initiation and growth data and its correlation with current uniaxial design data, *ASTM Special Technical Publication*, **853** (1985), 700.
50. A. Mendelson and S. Manson, Practical Solution of plastic deformation problems in elastic-plastic range, NACA Technical Note **4088** (1957).

Chapter 7

Fracture Mechanics of Solids with Microstructure and the J-Integral

In this chapter we will describe the relationship between the fracture mechanics of heterogeneous materials having microstructure, using the theory of continuously distributed dislocations and the J-integral in continuum mechanics.

First, from the influence of hydrostatic pressure on the fracture toughness of pure metals, we will clarify that the J-integral, one of the so-called path-independent integrals, represents the resultant force, i.e. driving force, in a singularity or an inhomogeneity composed of both the crack tip itself and the neighboring plasticity zone by which the crack grows. We will describe the phenomenon where the driving force changes with the substructure in the plasticity zone ahead of the crack tip. Thus we can suppose that the J-integral is effective for evaluating the fracture toughness of actual material as well as a mathematical tool in continuum mechanics, because the contribution of the characteristic plasticity zone to the driving force is localized in front of the crack tip.

Second, we will describe the interaction between crack, plasticity and vacuum environment in the crack behavior of polycrystalline metals at elevated temperatures. The vacuum environmental effect on crack behavior in the static creep and the cyclic creep of pure metals, heat-resisting metals and superalloys will be discussed from the viewpoint of the interaction between the oxide surface film at the notch root or crack tip and the distributed dislocations beneath the surface layer. We will also show that the crack propagation rate in the metals and alloys is well correlated with the creep J-integral irrespective of the stress wave of static/cyclic creep and the atmospheric/vacuum environment.

Thirdly, we will point out that there is a creep–fatigue interaction in the crack propagation rate in stress-hold tests and that a deceleration of the creep crack propagation rate immediately after the stress variation is interpreted by the decrease of the effective stress on the distributed dislocations in front of the crack tip under a steep slope of stress distribution at reloading. Thus creep–fatigue interaction is observed concerning the crack propagation rate in stress-hold test but the crack propagation rate is controlled mainly by the creep J-integral.

Lastly, we will describe an X-ray study of the damage evaluation for a heat-resisting metal under creep–fatigue interaction conditions by using X-ray profile analysis. Here X-ray profile analysis is a useful nondestructive parameter for estimating the crack initiation and propagation lives irrespective of transgranular and intergranular fracture modes.

7.1 INFLUENCE OF HYDROSTATIC PRESSURE ON THE FRACTURE TOUGHNESS OF POLYCRYSTALLINE METALS AND J-INTEGRAL

We will describe the deformation and energy at the notch root of a single-edged V-notched test plate specimen of commercial pure copper and commercial pure aluminum in tension tear tests under hydrostatic pressure, thus clarifying the mechanism regarding the increase in the toughness of the metal with combined loading of hydrostatic pressures.

Since we will describe the test apparatus under high hydrostatic pressure in Fig. 8.11 of Chapter 8, here we describe only the tear testing method. Figure 7.1 shows the shape and dimensions of the specimen which is similar to that used in the Alcoa Research Laboratory, USA [1]. The pure copper specimen was annealed at 673 K for 1 h and the pure aluminum specimen was annealed at 633 K for 1 h after machining. Tensile loading of the specimen was performed by compression on the yoke shown in Fig. 7.2 in the high-pressure cell within the pressure

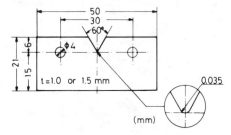

Fig. 7.1. Shape and dimensions of the tear test specimen.

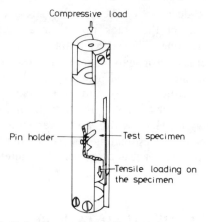

Fig. 7.2. Yoke for tensile loading on the test specimen.

vessel. The tensile load P, and the displacement L, between both end pin holes in the specimen, were measured accurately by the spring 9 and the DTFs 8 and 10 as shown in Fig. 8.11, respectively.

Information obtained from the present test is as follows. First, we can obtain both the energy required to initiate a crack (abbreviated UIE) and that needed to propagate the crack (abbreviated UPE). The former is calculated from $\text{UIE} = \int_0^{L_1} P\delta L/bt$ and the latter from $\text{UPE} = \int_{L_1}^{L_2} P\delta L/bt$ as shown in Fig. 7.3, and these are employed as the resistance of the materials to crack initiation and propagation [1]. Next we can obtain the fracture toughness G by combining both the load P versus displace-

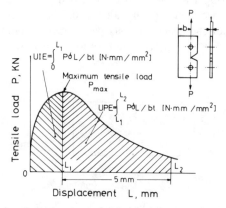

Fig. 7.3. Schematic tensile load versus elongation curve in the tensile tear test.

ment L curve and the crack length l versus displacement L curve as

$$G = (1/t)(\delta U/\delta l) = (1/t)(\delta U/\delta L)/(\delta l/\delta L) \quad (\text{MN m}^{-1}) \tag{7.1}$$

where $\delta U/\delta L$ and $\delta l/\delta L$ are obtained by reading both the increment in strain energy, δU, and that in crack length, δl, in relation to the increment in displacement, δL, as shown in Fig. 7.4.

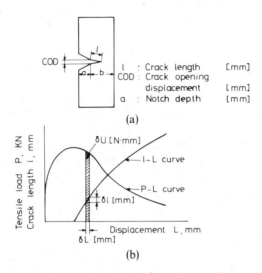

Fig. 7.4. Information obtained from the tensile tear test.

Since G of eqn (7.1) is composed of both the energy required to crack initiation/propagation and the plastic work in the front of the crack tip, we can hardly tell them apart. Therefore we can say that the evaluation of the plasticity zone ahead of the crack tip is very important. This will be discussed from the perspective of the J-integral.

As previously mentioned in Fig. 3.12(c) in Chapter 3, the J-integral value varies according to the respective circuit such as Γ and Γ_1 in Fig. 3.12(c) because there is a difference in singularities, such as fixed dislocations within the circuits. Therefore we can suppose that the J-integral for path Γ_1 is represented by the sum of the resultant force due to the geometrical singularity at the crack tip and that due to the mechanophysical singularity of the fixed pile-up dislocations against the crack tip as

$$J_{\Gamma 1} = J_e + J_d \tag{7.2}$$

where J_e and J_d are the integral for paths which surround only the crack tip (the so-called J-integral in LFM) and the fixed dislocations, respectively. Thus the J_d-integral varies with the path, and we should note that the path independence does not hold true in the case of a heterogeneous material with a dislocation cloud ahead of the crack tip. In other words, it is notable that the resultant force in the dislocation cloud ahead of the crack tip varies with the change in dislocation density, and even if the dislocation density has the same force it is changeable according to distribution.

In connection with the differential geometry, as shown in Table 3.7 in Chapter 3, we can see that the J-integral is related to the Riemann–Christoffel curvature tensor. Therefore it is very important to evaluate the plasticity zone in connection with the substructure (both density and distribution of imperfections) of the material at the crack tip. It is worthy of note that a failure of the path independence in the J-integral, from a micromechanical viewpoint, does not preclude the ineffectiveness of a macromechanical approach. In fact, the J-integral is very useful as a mathematical tool in continuum mechanics. This recognition is very important in order to understand the concept of the so-called path-independent integral. This is because, as mentioned previously, we can consider that the J-integral is the resultant force, i.e. the driving force, on the singular field composed of both the crack tip itself and the neighboring plasticity zone by which the crack grows, and that the contribution of the characteristic plasticity zone to the driving force is localized in the front of the crack tip.

Figure 7.5 [2] shows the pressure effect on the maximum tensile load P_{max}, the UIE and the UPE obtained from the P versus L curve in the following figure for commercial pure copper at an oil temperature of 296 K, and we found that these values increase as the pressure increases. Figure 7.6 [2] shows the experimental results of the P versus L curve, the l versus L curve and the COD versus L curve under atmospheric and hydrostatic pressures of 0·98 and 1·96 kbar, respectively. The displacement speed of the plunger under pressure was 0·5 mm/min and the COD was measured outside the pressure vessel by a micrometer. The P versus L curves shown as dashed curves in Fig. 7.6 are obtained from six specimens, one for each respective hydrostatic pressure. No influence of pressure on both the l versus L curve and the COD versus L curve was observed, and Fig. 7.7 [2] shows a geometrical relationship between the crack length l and the COD irrespective of hydrostatic pressure. Figure 7.8 [2] shows the pressure effect on the fracture toughness G of

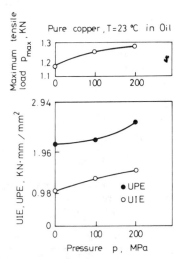

Fig. 7.5. Influence of pressure on the maximum tensile load P_{max}, UIE and UPE of commercial pure copper. (Reprinted from ref. 2 by permission of Japan Soc. Tech. Plasticity)

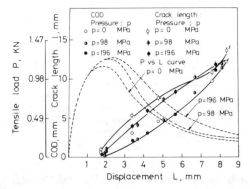

Fig. 7.6. Tensile load, P, versus elongation, L, curve, crack opening displacement, COD, versus L curve, and crack length, l, versus L curve, in the tensile tear test of commercial pure copper at 296 K in oil. (Reprinted from ref. 2 by permission of Japan Soc. Tech. Plasticity)

commercial pure copper. We found from the figure that G increases as pressure increases and that G also increases with the displacement L of the specimen up to P_{max}, but afterwards it decreases with L.

Under atmospheric pressure the fracture surface demonstrates an

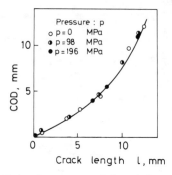

Fig. 7.7. Geometrical relation between crack length, l, and COD. Test conditions as for Fig. 7.6. (Reprinted from ref. 2 by permission of Japan Soc. Tech. Plasticity)

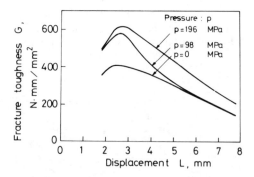

Fig. 7.8. Fracture toughness, G, of commercial pure copper under several pressures. Test conditions as for Fig. 7.6. (Reprinted from ref. 2 by permission of Japan Soc. Tech. Plasticity)

elongated dimple as shown in Fig. 2.2(a) in Chapter 2, but under hydrostatic pressure it shows serpentine glide, shown in Fig. 2.2(b), which inhibits the nucleation of minute voids. Therefore we can suppose that the plastic work γ_p at the crack tip increases under pressure; in fact this is clear from Fig. 7.9 [2]. The figure shows the pressure effect on the plastic work γ_p in front of the crack tip and the fracture toughness K_c in LFM, where γ_p was calculated by $\gamma_p = \sigma_B \varepsilon_f t/2$ (KN/mm), and σ_B and ε_f are respectively the tensile strength and fracture strain of the metal tested. K_c values were evaluated from the three types of calculation shown in the firgure. As seen from Fig. 7.9(b), the K_c value based on

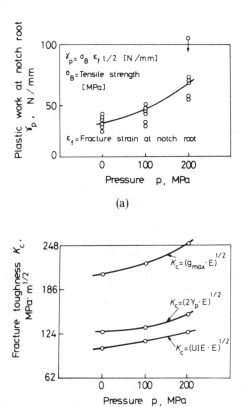

Fig. 7.9. Influence of pressure on (a) the plastic work, γ_p, at the crack tip and (b) the fracture toughness, K_c, of commercial pure copper at 296 K in oil. (Reprinted from ref. 2 by permission of Japan Soc. Tech. Plasticity)

$K_c = (G_{max} E)^{1/2}$ is larger than those based on UIE and $2\gamma_p$, and this suggests that the G value is composed of both the energy required to propagate the crack and the plastic work at the crack tip.

Figure 7.10 [2] is a similar representation for commercial pure aluminum to that of Fig. 7.6, and Fig. 7.11 [2] shows the pressure effect on G for that metal. We found that the pressure effect is small for commercial pure aluminum compared with commercial pure copper. This difference between both pure metals is related well with that in the pressure effect on the necking strain ε_u and the fracture strain ε_f in a static tensile test,

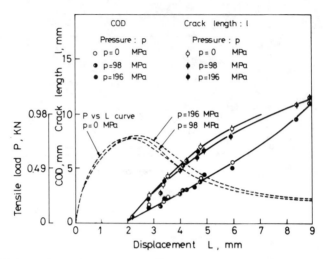

Fig. 7.10. Experimental P versus L, l versus L, and COD versus L curves of commercial pure aluminum at 296 K in oil. (Reprinted from ref. 2 by permission of Japan Soc. Tech. Plasticity)

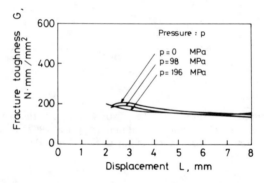

Fig. 7.11. Fracture toughness, G, of commercial pure aluminum under several pressures at 296 K in oil. (Reprinted from ref. 2 by permission of Japan Soc. Tech. Plasticity)

as shown in Fig. 7.12 [2]. We briefly describe here the pressure effect on necking strain as follows.

We can obtain the following condition of unstable deformation in a static tensile test under combined hydrostatic pressure p, using the yield condition on eqn (4.12) which includes the influence of hydrostatic stress

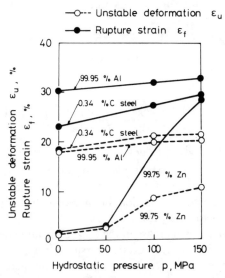

Fig. 7.12. Influence of hydrostatic pressure on necking strain, ε_u, and rupture strain, ε_f, in the simple tensile test for polycrystalline metals. (Reprinted from ref. 2 by permission of Japan Soc. Tech. Plasticity)

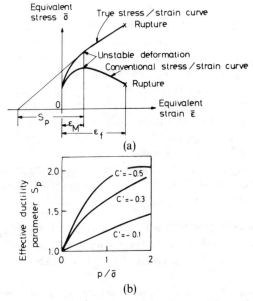

Fig. 7.13. Condition of unstable deformation of materials, and increase of the parameter of effective ductility, S_p, with elevation of hydrostatic pressure, p. (Reprinted from ref. 3 by permission of Soc. Mater. Sci., Japan)

on yielding (see Fig. 7.13(a)):

$$d\bar{\sigma}/d\bar{\varepsilon} = \bar{\sigma}/S_p$$

where

$$S_p = \frac{-(C'/2\lambda) + (1/\lambda^2)\Phi^{-1/2}\{C'^2[1-3(p/\sigma)] + (4/3)\}}{-(C'/2\lambda)[1-3(p/\sigma)] + (1/2)\Phi^{1/2}}$$

(7.3)

In the equation, Φ is given as $\Phi = (1/\lambda^2)\{C'^2[1-3(p/\sigma)]^2 + (4/3)\}$, where $\lambda = (1/2)[-C' + \sqrt{C'^2 + (4/3)}]$. If $p/\sigma = 0$ (in the atmosphere) or $C' = 0$ (von Mises condition), S_p reaches unity and eqn (7.3) is written as $(d\sigma/d\varepsilon) = 0$ for the usual condition of unstable deformation in simple tension under atmospheric pressure. Figure 7.13(b) [3] shows an analytical representation of the increase in the tensile unstable deformation with the combined loading of hydrostatic pressures.

From the results of the tensile tear test on commercial pure copper and commercial pure aluminum under hydrostatic pressures, we made the following conclusions.

(1) Fracture toughness in the tear test for commercial pure copper increases as combined hydrostatic pressure increases, but it does not do so for commercial pure aluminum.

(2) Pressure dependence in the fracture toughness correlates well with the pressure effect on the fracture ductility of the metals.

(3) Fracture toughness G or the J-integral for the resistance of notched material to crack initiation and propagation represents the resultant force in a singularity or an inhomogeneity (lattice defects and minute voids) at the crack tip. It varies with the substructure (density of lattice defects and the distribution) in the plastic zone ahead of the crack tip. Therefore we can suppose that the G or the J-integral is the resultant force, i.e. driving force, on a singular field composed of both the crack tip itself and the neighboring plasticity zone by which the crack grows, and that the latter component of the force increases with pressure.

7.2 INTERACTION BETWEEN CRACK, PLASTICITY AND ENVIRONMENT IN THE CRACK BEHAVIOR OF POLYCRYSTALLINE METALS

7.2.1 Micro and Macro Combined Fracture Mechanics and the Theory of Distributed Dislocations

In this section, we first describe the combined micro and macro fracture mechanics (abbreviated CMMFM; see Table 3.8 in Chapter 3) and macro fracture mechanics from a viewpoint of the theory of distributed dislocations.

Let q_1 be the stress concentration factor due to the crack or notch for a macro defect and q_2 be that due to pile-up of distributed dislocations for a micro defect, as shown in Fig. 7.14; the local stresses are written as $\sigma_{ln} = q_1 \sigma$ and $\sigma_{ld} = q_2 \tau$. If both types of defect are in close proximity to each other these stresses are given as $\sigma_{ln} = \alpha_1 q_1 \sigma$ and $\sigma_{ld} = \alpha_2 q_2 \tau$ respectively, where α_1 and α_2 are the interaction coefficient derived from the distributed dislocations and that from the crack or notch respectively.

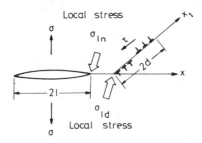

Fig. 7.14. Interaction between crack/notch and distributed slip dislocations array. (Reprinted from ref. 4 by permission of IWANAMI-Shoten).

Yokobori derived both criteria of local tensile stress and energy, and the former is given by [4]

$$\alpha_1 q_1 \sigma_{\phi(m)} + \alpha_2 q_2 \tau_{\phi(m+\theta)} = \sigma_{th} \qquad (7.4)$$

where $\phi(m)$ is the inclination angle between the crack plane and the reference plane, $\phi(m+\theta)$ that between the slip plane and the reference plane, and σ_{th} is the atomic cohesion stress.

Thus CMMFM results in determining q_1, q_2, α_1 and α_2 from the physical, chemical and mechanical models. In particular, in the fracture of metals at high temperatures, a non-conservative motion in dislocations such as 'climb' occurs in addition to a nonlinearity in cracked material, and this complicates matters.

On the other hand, macro fracture mechanics has a physical orientation since it is based on energy criteria. For instance, the J criterion, i.e. $J_{critical} = \gamma$ (γ is the energy required to create a crack plane in unit area), is derived from the critical condition $dE_T/dl = 0$ (see eqns (2.4b) and (2.5) in Chapter 2), where l is the crack half-length and E_T is the total energy which includes the following: self-energy in crack/slip dislocations, total interacted energy in the dislocations, surface energy in crack growth, work done by external stress for crack/slip dislocations, additional surface energy in crack tip opening which occurs due to the generation of slip dislocations under the concentrated stress at the crack tip, and work done by slip dislocations in climbing with an excess atom vacancy. In the J_c criterion no local stress condition of eqn (7.4) is taken into consideration and it is different from CMMFM, but the *theory of distributed dislocations* will be a useful analytical method with which to establish fracture criteria such as crack tip opening displacement (CTOD) criteria.

Next we describe the application of the theory of distributed dislocations to the influence of a vacuum environment on the creep crack behavior of metals at high temperatures.

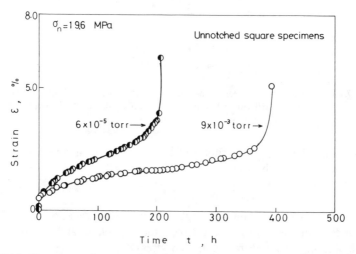

Fig. 7.15. Experimental comparison between the creep curves of a smooth bar specimen, having square cross-section, of commercial pure nickel, under 6×10^{-5} torr and 9×10^{-3} torr at 923 K. (Reprinted from ref. 5 by permission of Japan Soc. Mech. Engrs)

Figure 7.15 [5] shows the drastic influence of a vacuum environment on the creep curve of pure metals. The test was performed under both vacuum pressures of 6×10^{-5} torr (8×10^{-3} Pa) and 9×10^{-3} torr (1·2 Pa) at 923 K using a smooth square bar specimen (cross-sectional area 10 × 10 mm) of commercial pure nickel (Nickel 270, supplied by Huntington Co., USA). Figures 7.16 (a) and (b) [5] show the notch effect on time to rupture t_r and that to crack initiation t_c of the notched specimen (cross-sectional area at the notch root is 16 × 10 mm) in both

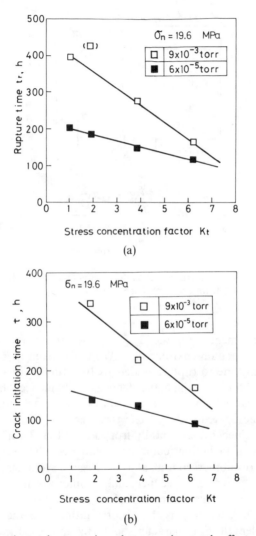

Fig. 7.16. Experimental comparison between the notch effect on (a) time to rupture, t_r, and (b) time to crack initiation, t_c, for pure nickel, under two vacuum levels at 923 K. (Reprinted from ref. 5 by permission of Japan Soc. Mech. Engrs)

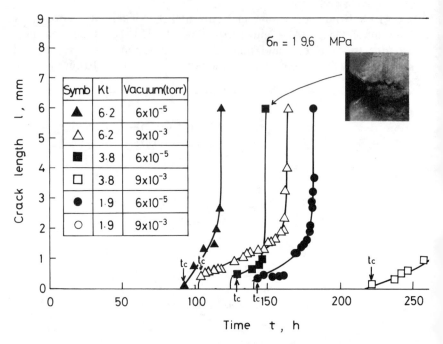

Fig. 7.17. Creep crack growth curves under two vacuum levels for pure nickel at 923 K. (Reprinted from ref. 5 by permission of Japan Soc. Mech. Engrs)

vacuum environments. Figure 7.17 [5] shows the creep crack growth curves in both vacuum environments. We found from these results that the decrease in time to rupture under 6×10^{-5} torr compared with that under 9×10^{-3} torr, in which the decrease is influenced by the elastic stress concentration factor K_t, is a result of early crack initiation under 6×10^{-5} torr compared with 9×10^{-3} torr.

This is examined experimentally from Fig. 7.18 [5] which shows the Hall relationship in both vacuum environments. The examination was made by interruption creep tests at a macroscopic total strain, $\bar{\varepsilon}$, of 2·2% for the smooth square bar specimen, and the microstrain, ε, was obtained from the slope in the experimental representation of $\beta \cos(\theta/\lambda) = (1/\eta) + 2\bar{\varepsilon} (\sin \theta/\lambda)$, where β is the misorientation, θ is the Bragg angle, λ is the wavelength in a characteristic X-ray and η is the particle size. We found that $\bar{\varepsilon}$ is 0.64×10^{-2} under 9×10^{-3} torr and 1.05×10^{-2} under 6×10^{-5} torr, and the former is almost twice as large as the latter.

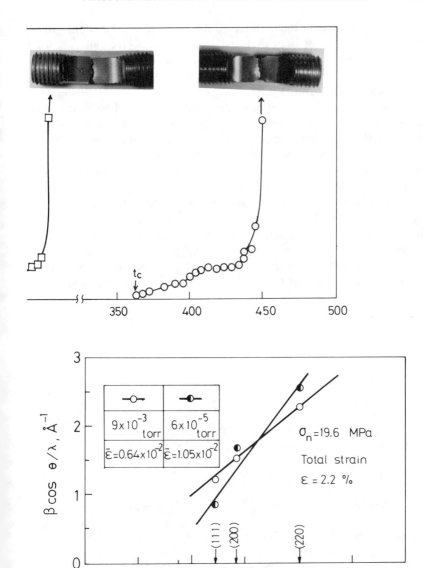

Fig. 7.18. Representation by Hall's equation. Test conditions as for Fig. 7.17. (Reprinted from ref. 5 by permission of Japan Soc. Mech. Engrs)

This suggests that local microstrain due to dislocations under 6×10^{-5} torr is larger than that under 9×10^{-3} torr. We can thus make the analysis that the crack initiates in the early creep stage under the former vacuum pressure provided that the crack occurs when the local strain reaches the critical level.

In the study of the model of continuously distributed dislocations the following assumptions were made. (i) The vacuum environmental effect is interpreted by an oxide surface film, i.e. an *elastic constant effect*. (ii) The analytical result for a mode III crack (anti-plane shear mode) is applicable to the usual mode I crack (opening mode). (iii) The constitutive equation for creep is written by a Norton type of equation, i.e. $\dot{\gamma} = \alpha \tau^n$.

Here the first assumption is based on the experimental result as shown in an example in Fig. 7.19 [6]. The figure shows the influence of change in vacuum level on creep rate in a smooth cylindrical specimen (diameter 8 mm, gauge length 50 mm) of commercial pure copper (Cu 99·5%, O_2 0·04 wt%) at 593 K. We found from the figure that the creep rate in 6×10^{-5} torr changes to that in 9×10^{-3} torr when the vacuum level is

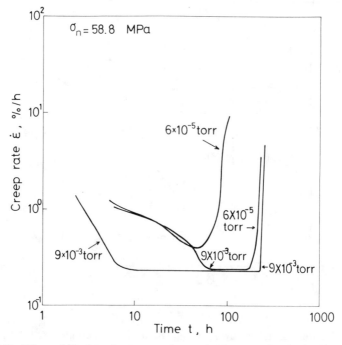

Fig. 7.19. Effect of change of vacuum pressure level on creep rate of commercial pure copper under $\sigma_n = 58 \cdot 8$ MPa at 593 K. (Reprinted from ref. 6)

changed to 9×10^{-3} torr from 6×10^{-5} torr, but the rate does not change when the level is changed in the opposite direction. Since the same protective oxide film of Cu_2O forms on the surface of the specimen at both vacuum pressures, we can consider that the creep rate is mainly controlled by the thickness of the oxide surface film in the vacuum pressure change test.

The following are the numerical results of the continuous distribution of screw dislocations in front of the crack tip in which a thin oxide film, having a thickness h and a larger elastic constant G_1 compared with G_2 of a matrix, exists by using Riedel's method [7].

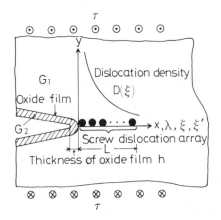

Fig. 7.20. Model for distributed screw dislocations in front of crack tip.

Shear stress τ_D due to a pile-up dislocation array having N screw dislocations within a distance L in a semi-infinite matrix of a notched body as shown in Fig. 7.20 is given as

$$\tau_D(x,t) = Ab \left[\int_0^L \frac{f(x',t)}{x-x'} dx' + K \int_0^L \frac{f(x',t)}{x+x'} dx' \right.$$

$$\left. -(1-K^2) \sum_{m=1}^{\infty} \int_0^L \frac{K^{m-1} f(x',t)}{x+x'+2hm} dx' \right] \quad (7.5)$$

where x, x' are coordinates, t is the creep time, b is the Burgers vector, $A = G_1/2\pi$ and $K = (G_2 - G_1)/(G_2 + G_1)$. The first term in eqn (7.5) is an interacted stress in screw dislocations, the second is stress due to the elastic constant effect and the final one is additional stress to satisfy the free boundary (stress free) condition on the surface of the oxide film based on an *image dislocation model*.† The symbol ⨍ denotes a Cauchy principal value integral.

Local creep strain, γ, ahead of the notch root is written as

$$\gamma(x,t) = w(x,t)/l = b \int_x^L f(x',t)\,\mathrm{d}x'/l = \int_x^L D(x',t)\,\mathrm{d}x'/l \qquad (7.6)$$

where w is the local displacement, $bf(x',t) \equiv D(x',t)$ is the screw dislocation density function (nondimensional), and l is the distance in the creep region. Putting eqns (7.5) and (7.6) into the creep constitutive equation, $\dot{\gamma} = \alpha \tau^n$, we obtain

$$\frac{\partial}{\partial t}\int_x^L D(x',t)\,\mathrm{d}x' = \alpha l \left[A \left\{ \unicode{x2A0D}_0^L \frac{D(x',t)}{x-x'}\,\mathrm{d}x' + K \int_0^L \frac{D(x',t)}{x+x'}\,\mathrm{d}x' \right.\right.$$
$$\left.\left. -(1-K^2)\sum_{m=1}^{\infty}\int_0^L \frac{K^{m-1}D(x',t)}{x+x'+2hm}\,\mathrm{d}x' \right\} + \frac{C\tau}{2\pi x} \right]^n \qquad (7.7)$$

where $C(\mathrm{mm}^{1/2})$ is a constant. We assumed, for the sake of simplicity, that the stress distribution at the notch root is represented by $C\cdot\tau/\sqrt{2\pi x}$ on the analogy of LFM. We should use the relation $C\cdot\tau^{1/(n+1)}$, but the following analytical result is not essentially influenced by this replacement, where n is a stress exponent in the creep constitutive equation.

We can solve eqn (7.7) under an initial condition $D(x,t)=0$ by using the following nondimensional quantities:

Nondimensional time: $t' = t/t_0$
Nondimensional distance: $\xi \equiv x/x_0(t')$ and $\xi' \equiv x'/x_0(t')$ (7.8)
$D(x',t) \equiv \Delta(\xi')f(t')$

where

$$t_0 = (C\tau/A)^2/[(1+n)A^n\alpha l]$$
$$x_0 = [(C\tau)/A]^2 t'^{2/(1+n)}$$
$$f(t') = t'^{-1/(1+n)}$$

Putting eqn (7.8) into eqn (7.7) we obtain

---

†Regarding the method satisfying the stress-free condition, both an image dislocation model (Head [8]) and a surface dislocation model (Marcinkowski *et al.* [9]) have been studied. The latter model has wide applicability for the discrete distribution of dislocations compared with the former [10].

$$\int_0^L \frac{\Delta(\xi')}{\xi-\xi'} d\xi' + K \int_0^L \frac{\Delta(\xi')}{\xi+\xi'} d\xi' - (1-K^2) \sum_{m=1}^{\infty} \int_0^L \frac{K^{m-1}\Delta(\xi')}{\xi+\xi'+(2hm/x_0)} d\xi'$$

$$+ \left[2\xi\Delta(\xi) + \int_{\xi}^{L'} \Delta(\xi') d\xi' \right]^{1/n} = -(2\pi\xi)^{-1/2} \qquad (7.9)$$

where $L' = L/x_0(t')$. Thus the problem ends by solving the nonlinear integral equation of eqns (7.8) for a normalized dislocation density $\Delta(\xi)$.

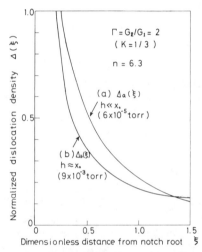

Fig. 7.21. Analytical comparison between the normalized dislocation density, $\Delta(\xi)$, in two vacuum levels. (Reprinted from ref. 5 by permission of Japan Soc. Mech. Engrs)

Figure 7.21 [5] shows the numerical result of the dislocation distribution $\Delta(\xi)$ versus ξ, in the case of both (a) an extremely thin oxide film ($h \ll x_0$) and (b) a thick oxide film ($h \approx x_0$) for $\Gamma = G_2/G_1 = 2$ (or $K = 1/3$). We can assume here that both cases correspond to vacuum environments in 8×10^{-3} Pa and 1.2 Pa respectively, and the oxide films in both vacuums have much the same character due to the fact that the oxide film of pure nickel in both vacuums is protective NiO. We found that the dislocation density distributions are different in both vacuums. We can obtain the time to crack initiation, t_c, as

$$t_c = \left[\hat{\gamma}_c \bigg/ \int_0^{\infty} \Delta(\xi) d\xi \right]^{1+n} \{ A^n/[(1+n)\alpha l(C\tau)^{2n}] \} \qquad (7.10)$$

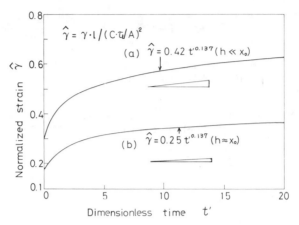

Fig. 7.22. Analytical comparison between the normalized creep curves under two vacuum levels. (Reprinted from ref. 5 by permission of Japan Soc. Mech. Engrs)

We assumed here that the crack initiates when the local creep strain

$$\hat{\gamma}(\xi,t') = t'^{1/(1+n)} \int_\xi^L \Delta(\xi')\,d\xi'$$

reaches the critical strain, $\hat{\gamma}_c$, where $\hat{\gamma}(\xi',t') = \gamma \cdot l/(C \cdot \tau/A)^2$. Figure 7.22 [5] shows the normalized creep curves, $\hat{\gamma}$ versus t', for the non-dimensional distance ξ of 0·3. We found that the creep rate for case (a) is about double that of (b), and this corresponds well to the experimental result of microstrain in Fig. 7.18. The ratio of the t_c values in both vacuums is given as

$$t_c(h \ll x_0)/t_c(h \approx x_0) = \left[\left(\hat{\gamma}_c \bigg/ \int_0^\infty \Delta_a(\xi)\,d\xi \right) \bigg/ \left(\hat{\gamma}_c \bigg/ \int_0^\infty \Delta_b(\xi)\,d\xi \right) \right]^{1+n} \quad (7.11)$$

where n is 6·3 for commercial pure nickel at 923 K. We found that the time to crack initiation under 6×10^{-5} torr ($h \ll x_0$) is almost 1/5 that under 9×10^{-3} torr ($h \approx x_0$), and this agrees well with the experimental results shown in Fig. 7.16.

From the experimental result of the vacuum environmental effect on the static creep rupture behavior of unnotched and notched square bar specimens of commercial pure nickel at 923 K, we concluded the following.

(1) The creep rupture time of the notched specimen in 6×10^{-5} torr was 1/10 to 1/2 that in the atmosphere, and it decreases as the elastic

stress concentration factor increases. This mainly resulted from early crack initiation.

(2) The early initiation of the crack in 6×10^{-5} torr compared with 9×10^{-3} torr was explained from the large microstrain beneath the oxide film in the former vacuum compared with that in the latter, and also from the application of the theory of continuously distributed dislocations beneath the surface film to the creep field.

7.2.2 Vacuum Environmental Effect on the Creep Rupture Behavior of Pure Metals

We call the influence of chemical environment on the strength of materials an *environmental effect*. Chemical change in the neighborhood of the metal surface is influenced by many types of reaction and material geometry and it affects mechanical behavior. Since the reaction rate is controlled by the stress/strain distribution in the reaction region, the interaction between the metal and the environment complicates matters. Table 7.1 [11] shows several types of metal–environment interaction.

In retrospect of the past 30 years we can find several remarkable features in a modern study of the environmental effect. The first is by use of the updated analysis instruments such as EPMA, SEM, ESCA, AES and others. The second is a study of hot corrosion and of the mechanism due to impurities in gas turbine oil or jet engine oil and the developments of new alloys since the 1950s. Since the 1970s the following are features of the study: R & D on a combustion chamber for automobile exhaust regulation applications; the study of hot corrosion due to impurities in helium gas coolant in a helium-to-helium heat exchanger for nuclear steelmaking and high-temperature strength research (see Chapter 8); the study of creep–fatigue–vacuum environment interaction (see Chapter 6).

In this section we describe the influence of oxide film on the static creep and cyclic creep of metals shown in (2) and (4) of Table 7.1, in addition to the influence of decarburization.

Figure 7.23 [12] shows the experimental rupture strength curves of a smooth cylindrical bar specimen (diameter 8 mm, gauge length 50 mm) of commercial pure copper under several vacuum levels at 593 K. After being machine finished, the specimens were annealed for 1 h at 693 K in vacuum. Figure 7.24 [12] shows the creep curves at various vacuum levels under a nominal tensile stress of $\sigma_n = 58.8$ MPa. We found from the figure that each creep curve shows a somewhat different shape with each vacuum level. However, there is no large difference in the creep rupture strain in each. Figure 7.25 [12] represents the variation of both rupture

Table 7.1
Possible Types of Metal–Environment Interaction [11]

Direction of mass transport	Examples of effect
None	Atmosphere inert to metal
From atmosphere to metal	Formation of a thin adsorbed layer on the specimen
	Formation of an adherent surface layer of second phase (e.g. oxidation[a])
	Formation of second phase particles or grain boundary films in the subsurface layer (e.g. internal carbide[b])
	Formation of a second phase in surface cracks in the specimen (e.g. oxide film[c])
From metal to atmosphere	Evaporative loss or chemical dissolution of elements from the specimen to the environment (e.g. carburization)

[a] Oxide is formed by $2(x/y)M + O_2 = (2/y)M_xO_y$ and the oxidation rate is controlled by both oxygen partial pressure and diffusion in carbide.
[b] Internal oxide (SiO_2) in Cu–Si alloy and internal carbide ($Cr_{23}C_6$) in Fe–Ni–Cr alloy.
[c] Remarkable decarburization occurs in mild steel at 750 K and the following decarburizations are observed: 2·25 Cr–1 Mo steel at 925 K and Inconel 617 at 1723 K in helium. Also, sodium environment brings about mass transport of Ni, Cr, Nb and Mo in steels.

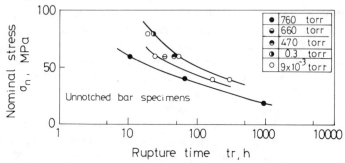

Fig. 7.23. Nominal stress, σ_n, versus rupture time, t_r, curves of commercial pure copper in both a vacuum and the atmosphere at 593 K. (Reprinted from ref. 12 by permission of Soc. Mater. Sci., Japan)

time, t_r, and minimum creep rate, $\dot{\varepsilon}$, under $\sigma_n = 58·8$ MPa with pressure. It is obvious from the figure that the rupture time increases monotonously with increase in vacuum levels from atmospheric pressure to 0·3 torr (40 Pa), but decreases with increase in the vacuum level from

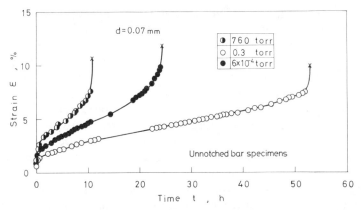

Fig. 7.24. Tensile creep curve of commercial pure copper under $\sigma_n = 58 \cdot 8$ MPa in a vacuum and the atmosphere at 593 K. (Reprinted from ref. 12 by permission of Soc. Mater. Sci., Japan)

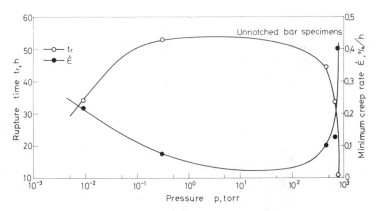

Fig. 7.25. Variation of rupture time, t_r, and minimum creep rate, $\dot{\varepsilon}$, of commercial pure copper under $\sigma_n = 58 \cdot 8$ MPa at 593 K, with pressure. (Reprinted from ref. 12 by permission of Soc. Mater. Sci., Japan)

0·3 torr to 9×10^{-3} torr (1·2 MPa). It is also seen that the product of t_r and $\dot{\varepsilon}$ is not always influenced by the vacuum level.

In this study, the vacuum level is employed as the measure of air pressure, but it does not show the concentration of residual element gases. However, it shows the degree of oxidation potential such as humidity. The humidity measured in the rupture tests at 0·3 torr and 9×10^{-3} torr was 31·1 and 6·1 ppm, respectively, by the SHAW dewpoint meter.

At the end of the 1950s Shahinian and his colleagues [11] performed a series of creep rupture tests on pure metals and they found the inverse phenomenon due to vacuum level concerning rupture time for commercial pure nickel (Nichrome V) similar to that shown in Fig. 7.25. Regarding Shahinian's findings and also the author's findings, it is very important to point out the characteristic influence of vacuum environments on the high-temperature creep rupture of pure metals.

In order to discuss these variations of the rupture time and minimum creep rate with pressure, the following considerations were made from the viewpoint of the change in the structure of the specimen under the creep process at various vacuum levels.

Figure 7.26 [12] shows the scanning profiles of the vicinity of the oxidized surface layer of the specimen ruptured under $\sigma_n = 58.8$ MPa at 0.3 torr ($t_r = 53$ h) by use of an optical microscope, SEM and EPMA. Arrows on the right of the scanning profile of the EPMA indicate the

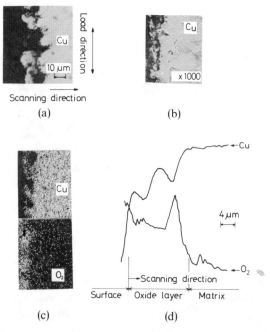

Fig. 7.26. Observations by optical microscope, SEM, and EPMA for a specimen of pure copper ruptured under $\sigma_n = 58.8$ MPa in 0.3 torr ($t_r = 53$ h) at 593 K. (a) Optical microscope; (b) SEM; (c) X-ray image by EPMA; (d) scanning profile of EPMA. (Reprinted from ref. 12 by permission of Soc. Mater. Sci., Japan)

concentration levels of Cu and O_2, respectively. We can suppose from these profiles that the region where the concentration of both Cu and O_2 varies remarkably indicates the oxidized surface layer. We can see from the observations, by both optical microscope and SEM, that the oxidized layer appears to be porous. This can also be confirmed from the EPMA profiles of both Cu and its oxide within the oxidized layer. It was also determined, after identification by Co-K_α X-rays, that the oxidized layer was composed of only CuO_2. Figure 7.27 [12] shows similar observations by both optical microscope and EPMA for the specimen ruptured under $\sigma_n = 58 \cdot 8$ MPa ($t_r = 23$ h) at 9×10^{-3} torr.

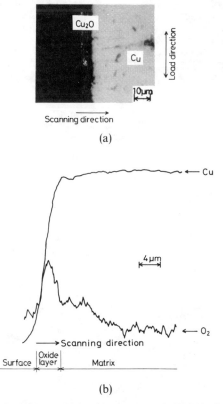

Fig. 7.27. Observations by optical microscope and EPMA for a specimen of pure copper ruptured under $\sigma_n = 58 \cdot 8$ MPa in 9×10^{-3} torr ($t_r = 23$ h) at 593 K. (a) Optical microscope; (b) scanning profile of EPMA. (Reprinted from ref. 12 by permission of Soc. Mater. Sci., Japan)

Figure 7.28 [12] shows the data under $\sigma_n = 58{\cdot}8$ MPa ($t_r = 995$ h) in the atmosphere. Many intergranular cracks were observed in the specimen. The fact that internal cracks become more predominant in the atmosphere than in a vacuum can be interpreted from the following factors: (i) the formation of an unprotective oxidized layer of CuO outside the Cu_2O layer hinders the growth of a protective Cu_2O layer; (ii) the exfoliation of CuO accelerates selective oxidation in the specimen. From the identification by Co-K_α X-rays, it was confirmed that the oxidized layer

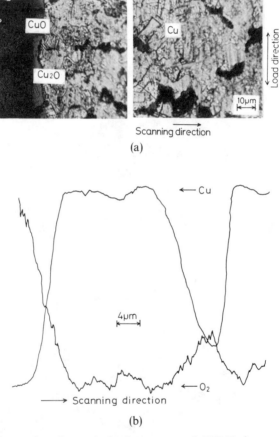

Fig. 7.28. Observations by optical microscope and EPMA for a specimen of pure copper ruptured under $\sigma_n = 19{\cdot}6$ MPa in the atmosphere ($t_r = 955$ h) at 593 K. (a) Optical microscope; (b) scanning profile of EPMA. (Reprinted from ref. 12 by permission of Soc. Mater. Sci., Japan)

of CuO was formed preferentially in comparison with that of Cu_2O. Since the formation of the external layer of CuO in the atmosphere does not protect the matrix from oxidation, the oxidant can diffuse into the matrix through the scale. Such diffusion will cause the subsequent improvement of the metal in the atmosphere. Consequently the contamination of the metal by the oxidant extends more deeply from the surface of the metal into the grain boundaries, and the internal cracks initiated in the grain boundaries, as shown in Fig. 7.28(a). This must be one of the reasons for the remarkable decrease in the rupture time of the metal in the atmosphere. On the other hand, the formation of only a layer of Cu_2O in vacuum will protect the matrix from O_2 oxidation.

We can consider from Fig. 7.24 that the effect of the vacuum level on the creep rupture time of the specimen is a result of the variation of the creep rate with pressure. The substructure beneath the oxidized surface layer of the crept specimen was studied by using the X-ray diffraction technique mentioned in Fig. 7.18. Interest was placed on vacuum levels 9×10^{-3} torr (1·2 Pa) and 0·3 torr (40 Pa) which were distinctly different from one another in the creep rupture times under the same nominal stress of 58·8 MPa. We found that the microstrain, $\bar{\varepsilon}$, of $2·15 \times 10^{-3}$ in 9×10^{-3} torr at macroscopic strain, $\bar{\varepsilon}$, of 7% is almost twice as large as that of $1·21 \times 10^{-3}$ in 0·3 torr at the same total strain. This suggests that, in spite of the same macroscopic total strain, the microstrain near the surface of the specimen in 9×10^{-3} torr is larger than that in 0·3 torr.

There are two main reasons why the microstrain under 9×10^{-3} torr is larger than that under 0·3 torr. First, we observed from the X-ray identification that both oxidized surface layers are composed of the same protective Cu_2O. Secondly, we can assume that the thickness of the oxidized surface film, which directly affects the continuous distribution of the dislocations beneath the film in the specimen, is thicker under 0·3 torr than under 9×10^{-3} torr. The amount of humidity or residual oxygen at 9×10^{-3} torr is naturally less than at 0·3 torr, and consequently the oxidation does not progress so far. So the effect of the thickness of the oxidized film on the microstrain, $\bar{\varepsilon}$, of the specimen beneath the film is an effective means of studying the vacuum environmental effect as mentioned in the preceding Section 7.2.1.

Referring to Fig. 7.29, we can obtain an equation similar to eqn (7.10) for the shear stress τ_D due to N discrete distributed screw dislocations piling up against the surface film within a distance L. Therefore the static force equilibrium equation for the discrete distributed dislocations is given by replacing τ_D by the external shear stress τ. In order to analyze

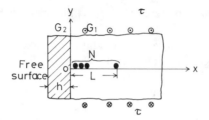

Fig. 7.29. Screw dislocations array piling up against a surface film.

this equation when $K(=(G_2-G_1)/(G_2+G_1))$ is small, the first approximate solution [13], neglecting the second and third terms of the right-hand side of eqn (7.5), is used:

$$f(x') = \frac{\tau}{\pi A} \sqrt{\frac{L-x'}{x'}} \qquad (7.12)$$

Putting eqn (7.12) into the second and the third terms of eqn (7.5), the following equation is obtained:

$$\oint_0^L \frac{f(x')}{x-x'} dx' = \frac{P(x)}{A}$$

where

$$P(x) \equiv \tau \left[(1+K) - K\sqrt{\frac{x+L}{x}} \right.$$

$$\left. + (1-K^2) \sum_{m=1}^{\infty} \left(\sqrt{\frac{x+2hm+L}{x+2hm}} - 1 \right) K^{m-1} \right] \qquad (7.13)$$

The solution of eqn (7.13) can be obtained by the inversion theorem formulated by Muskhelishvili [14]. The number of dislocations, N, which can be arrayed in equilibrium within L is given by $N = \int_0^L f(x')dx'$, and in the limiting case of $h \to \infty$ it becomes $N=(\tau L/2A)[1-(2K/\pi)]$. This agrees with the analytical result discovered by Smith [15] for the screw dislocation array in heterogeneous materials composed of soft and hard semi-infinite phases.

Figure 7.30 [12] shows the analytical result for the effect of the thickness of the hard surface film on the screw dislocation density piling up against the film. The ordinate shows an N/L value which corresponds to the mean screw dislocation density in the case where $\Gamma = G_2/G_1 = 2$. It

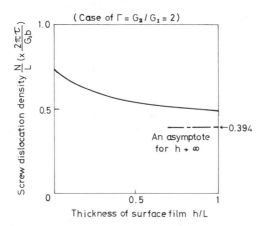

Fig. 7.30. Effect of thickness of surface oxide film on screw dislocation density piling up against the surface film. (Reprinted from ref. 12 by permission of Soc. Mater. Sci., Japan)

is worthy of note that these results are applicable for interpreting the similar tendency in the edge dislocation array [8].

We also found a drastic environmental effect from the viewpoint of both the grain size effect on the creep rupture behavior and the fracture mode for commercial pure copper mentioned above. Figures 7.31 (a) and (b) [16] show an experimental relation between the grain diameter, d, and the minimum creep rate, $\dot{\varepsilon}$, or the creep rupture time, t_r, respectively, under $\sigma_n = 39 \cdot 2$ MPa. In the following, the discussions are separate for the cases in the atmosphere and in a vacuum. The symbols T and I denote transgranular and intergranular fracture of the material, respectively.

The minimum creep rate, $\dot{\varepsilon}$, reaches the minimum value when the grain diameter of the specimen is in a range between 0·2 and 0·5 mm. When the grain diameter is less than 0·2 mm it tends to show that $\dot{\varepsilon}$ is inversely proportional to the grain diameter d, i.e. $\dot{\varepsilon}$ increases as the diameter decreases. Within the range of this inverse proportionality of $\dot{\varepsilon}$ to d, the test specimen always exhibits intergranular fracture. When the grain diameter exceeds the range 0·2–0·5 mm, the contrary is seen, $\dot{\varepsilon}$ shows a tendency to increase in proportion to the grain diameter d. In this case transgranular fracture was observed. The relations mentioned above were also maintained at other stress levels.

In a vacuum, contrary to the data in the atmosphere, the minimum

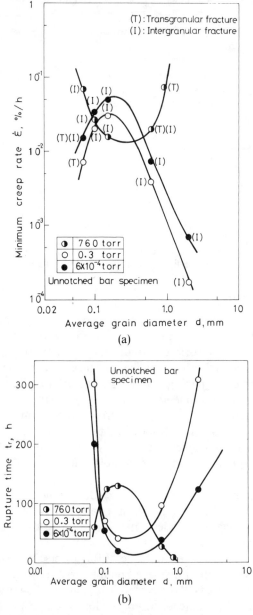

Fig. 7.31. Variation of (a) minimum creep rate, $\dot{\varepsilon}$, and (b) creep rupture time, t_r, with increase of average grain diameter, d, for pure copper in both a vacuum and the atmosphere under $\sigma_n = 39\cdot2$ MPa at 593 K. (Reprinted from ref. 16 by permission of Soc. Mater. Sci., Japan)

creep rate $\dot{\varepsilon}$ takes the maximum value when the grain diameter of the specimen is about 0·15 mm; when the grain diameter is less than that, $\dot{\varepsilon}$ decreases in proportion to the grain diameter d. In this relation the fracture was transgranular. When the grain diameter exceeds the range 0·15–0·2 mm, $\dot{\varepsilon}$ decreases in an inverse proportion to d. In this region the fracture was intergranular. This was also observed at other stress levels, except for the finer grain specimens tested under $\sigma_n = 29\cdot4$ MPa. It is noted that, for all the grain diameters tested, the minimum creep rate of the specimen in a vacuum of 0·3 torr (40 Pa) is about one-half that in a vacuum of 6×10^{-4} torr (0·08 Pa).

Regarding the rupture time t_r shown in Fig. 7.31(b), the contrary tendency to that seen in Fig. 7.31(a) is exhibited. From the present experiment, the empirical equation of t_r (h) to $\dot{\varepsilon}$ (%/h) was shown in connection with d (mm) as follows:

$$\log(t_r/\varepsilon_f) = \log b_r - m_r \cdot \log \dot{\varepsilon} \tag{7.14}$$

where ε_f is the percentage of rupture elongation, and m_r and b_r are represented by $m_r = 0\cdot64 d^{-0\cdot10}$ and $b_r = -0\cdot53 d^{0\cdot13}$.

Figure 7.32 [16] shows the experimental correlation of the number of

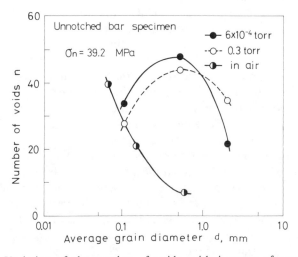

Fig. 7.32. Variation of the number of voids, with increase of average grain diameter, d, for pure copper in both a vacuum and the atmosphere under $\sigma_n = 39\cdot2$ MPa at 593 K. (Reprinted from ref. 16 by permission of Soc. Mater. Sci., Japan)

voids to the grain size in the area of 1 mm² parallel to the specimen axis, which is located at a distance of 20 mm from the fracture surface and 3 mm from the specimen surface. The number of voids larger than 15 μm were plotted in the figure. The voids were the r-type observed in creep rupture tests at elevated temperatures, especially at lower stress levels. In the atmosphere the number of voids increased with decreasing grain diameter, and in the range of grain diameter exceeding 1 mm, voids were scarce along the grain boundaries. Contrary to the data in the atmosphere, in a vacuum the largest number of voids was observed in the specimen with a 0·6 mm grain size. We confirmed, by observation with an optical microscope, that the large number of voids formed on the grain boundary resulted in intergranular fracture. It appears that the formation of voids is closely connected with the minimum creep rate since the data of Fig. 7.32 show the same tendency as those of Fig. 31(a).

From the experimental results of the vacuum environmental effect on the static creep rupture behavior of a smooth cylindrical bar specimen of commercial pure copper, we concluded the following.

(1) The creep rupture lives in both a vacuum of 0·3 torr and 9×10^{-3} torr were 5 and 2 times larger than that in the atmosphere, respectively. In the range between 9×10^{-3} torr and 0·3 torr, a coherent oxidized film of Cu_2O was observed on the surface of the crept specimen. In addition, an exfoliating CuO film was observed to be predominant in the range 0·3–760 torr. This is one of the reasons for the remarkable decrease in the rupture time of the material in the atmosphere.

(2) The remarkable increase in the minimum creep rate in 9×10^{-3} torr was a result of the increase in the dislocation density or the associated lattice strain on the basis of the measurement of microstrain by X-ray diffraction and from the model of distributed dislocations beneath the oxide surface film under stress.

(3) The minimum creep rate reaches the minimum value when the average grain diameter d is about 0·6 mm. The rate decreases in inverse proportion to d due to intergranular fracture when d is larger than 0·6 mm. On the contrary, in a vacuum the minimum creep rate reaches the maximum value under $\sigma_n = 39·2$ MPa when d is nearly 0·15 mm. The creep rate increases in proportion to d due to transgranular fracture, when d is larger than 0·15 mm.

(4) The main causes of the reversal phenomenon mentioned above are the retardation of the creep crack initiation in the specimen with a larger grain size in a vacuum, because of the formation of the protective surface

oxide film and the formation of a comparatively large number of voids along the grain boundaries. On the other hand, in the atmosphere fewer voids are initiated with increasing grain diameter and this results not in intergranular fracture but in transgranular fracture.

(5) The minimum creep rate of the specimen tested in a low vacuum of 0·3 torr is approximately one-half that in a comparatively higher vacuum of 6×10^{-4} torr, because of lower dislocation densities in the base material caused by the thicker oxide film.

7.2.3 Vacuum Environmental Effect on Static Creep and Cyclic Creep Rupture Behavior of Heat-resisting Metals and Superalloys

Figure 7.33 [17] shows the variation in the creep rupture time of a cylindrical bar specimen composed of solution heat treated type 316 stainless steel with the elastic stress concentration factor K_t in both air and 9×10^{-3} torr (1·2 Pa) under $\sigma_n = 255$ MPa at 873 K. We found that the time to rupture of the notched cylindrical bar specimen (60° V-notch, outer diameter 8 mm, notch root diameter 6 mm) in a vacuum is also shorter than that in air in all the ranges of K_t tested, but the effect decreases as K_t increases. For the notched plate specimen (60° V-notch, width 20 mm, thickness 1·5 mm), we found that there is almost no notch effect irrespective of either environment.

In order to confirm these results, observation by optical microscope,

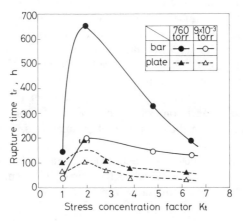

Fig. 7.33. Variation of rupture time, t_r, of type 316 stainless steel, with stress concentration factor, K_t, in both air and a vacuum under $\sigma_n = 255$ MPa at 873 K. (Reprinted from ref. 17 by permission of Soc. Mater. Sci., Japan)

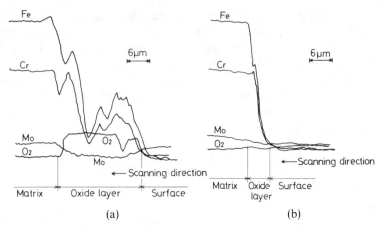

Fig. 7.34. Scanning profiles of EPMA for specimens of type 316 stainless steel, in (a) air and (b) a vacuum (9×10^{-3} torr) at 873 K. (Reprinted from ref. 18)

SEM and EPMA was performed. Figure 7.34 [18] shows an example of scanning profiles of EPMA for cylindrical bar specimens ruptured when $t_r = 1422$ h ($\dot{\varepsilon} = 0.005\%$/h, $\varepsilon_f = 26.2\%$) in the atmosphere and $t_r = 412$ h ($\dot{\varepsilon} = 0.023\%$/h, $\varepsilon_f = 23.1\%$) in 9×10^{-3} torr at 873 K. We found that the profile was wavy in the atmosphere and the 22 μm thick oxide film was porous, but in a vacuum it was coherent with the metal and there was only an 8 μm thin film. We observed from the X-ray diffraction that the product on the specimen surface was an Fe_3O_4 oxide in both the atmosphere and a vacuum. Internal oxidation was also observed in the neighborhood of the intergranular crack tip in the metal, but in a vacuum it was slight. This agrees well with the optical microscope observations in which the creep crack tip was blunt, but in 9×10^{-3} torr it was comparatively sharp. We can suppose from these observations that the surface oxide is one of the controlling factors of the vacuum environmental effect on the creep rate of the metals, as mentioned in the previous sections.

Next we describe the effect of vacuum environment on push-pull cyclic creep fracture of a hollow cylindrical specimen (i.d. 9 mm, o.d. 12 mm, gauge length 20 mm) of the same type 316 stainless steel mentioned above.

Figure 7.35 [19] shows the experimental relation between nominal stress and time to fracture in the total strain-controlled ($\Delta\varepsilon = 2.2\%$) push-pull cyclic creep tests in the atmosphere and 0.1 torr (13.3 Pa) at 873 K.

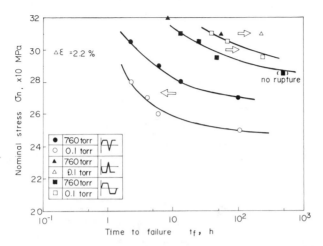

Fig. 7.35. Variation of time to failure, t_f, in push-pull cyclic creep, with environment and the shape of stress wave, for a hollow cylindrical specimen of type 316 stainless steel at 873 K. (Reprinted from ref. 19 by permission of Soc. Mater. Sci., Japan)

We found that the cyclic creep failure time of the material tested was remarkably dependent on the environment and the shape of the stress wave. The time to failure under a 'cp' stress wave with tensile creep hold in 0·1 torr was 6 to 25 times shorter than that in the atmosphere. This is more remarkable compared with the case of static creep rupture in which the rupture time in a vacuum is half that in air at most. On the contrary, the failure time under a 'cc' stress wave with both tension and compression holds in 0·1 torr was 3 to 5 times larger than that in the atmosphere. The latter fact was also revealed under a 'pc' wave with compressive stress-hold.

We found that the above behavior under a cp wave resulted from both early crack initiation and fast crack propagation in the tension cycle in 0·1 torr, and the latter effect under a cc wave resulted from late initiation and slow propagation of the crack in 0·1 torr. The crack propagation rate of the material tested was well represented by a single curve on the basis of the creep J-integral, $J_c = [(n-1)/(n+1)]\sigma_{net}\dot{\delta}$, irrespective of stress levels, shapes of the stress wave and environments, as shown in Fig. 7.36 [19], where σ_{net} is the net section stress, $\dot{\delta}$ is the rate of COD, and n is the stress exponent in the Norton type creep constitutive equation and has a value of 14 for the material tested at 873 K in air.

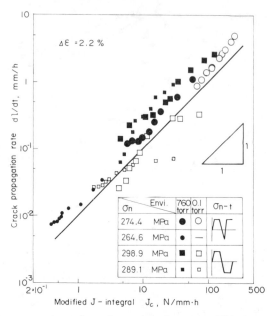

Fig. 7.36. Demonstration of crack propagation rate, dl/dt in push-pull cyclic creep, on the basis of the creep J-integral, J_c. Test conditions as for Fig. 7.35. (Reprinted from ref. 19 by permission of Soc. Mater. Sci., Japan)

This means that, since the J correlates with the local creep rate in front of the crack tip, the effect of both the vacuum environment and the shape of the stress wave on the crack propagation rate is interpreted from the local creep response in the vicinity of the crack tip. We can see from the figure that the crack propagation rate under a cp stress wave in 0·1 torr increases along the upward curve, whereas in the case of a cc stress wave the rate in 0·1 torr decreases along the downward curve.

The variations in both the crack initiation time and the crack propagation rate of the metal in 0·1 torr seemed to result mainly from the variation in the macroscopic creep rate of the whole specimen. We can consider that the increase in the creep rate in 0·1 torr is a result of the increase in the dislocation density, as seen in the case of static creep of the metal in 0·1 torr, and also that the decrease in the creep rate under a cc wave in 0·1 torr is analogous to the deformation and fracture behavior of the high-temperature low-cycle fatigue of the metal.

At the end of this section, we will describe the effect of decarburization on the creep rupture behavior of a notched plate specimen (double-

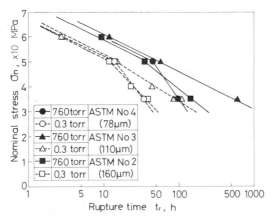

Fig. 7.37. Effect of vacuum environments on creep rupture strength of a notched plate specimen of Inconel 617 at 1273 K, $K_t = 6.4$, in comparison with that in air. (Reprinted from ref. 20 by permission of Japan Soc. Mech. Engrs)

edged 60° V-notch, width 20 mm, thickness 1.5 mm) of Inconel 617, a nickel-base alloy supplied by Huntington Alloy Co., USA.

Figure 7.37 [20] shows the experimental creep rupture strength diagram for the notched plate specimens with a K_t of 6.4 and three kinds of grain size in the atmosphere and a vacuum of 0.3 torr (40 Pa) at 1273 K. The vacuum of 0.3 torr was adopted as the environment of decarburization of Inconel 617 alloy at 1273 K. The H_2O content measured at 18.3 ppm in the vacuum of 0.3 torr was analyzed by the SHAW H_2O meter. We found that time to creep rupture in 0.3 torr decreased remarkably to 1/3 or 1/6 that in air in both ranges of stress and grain size tested. The tests for a grain size of 78 μm showed the shortest rupture time, and all the tests showed grain size dependence for each alloy tested. In the following, we will discuss the test data of the specimens with a grain size of ASTM 4 ($d = 78$ μm).

Figures 7.38 (a) and (b) [20] show the notch effect on both the rupture time t_r and the crack initiation time t_c under $\sigma_n = 34.3$ MPa. We found that t_r in 0.3 torr decreased to one-half that of t_r in air in the range of K_t tested, with an exception when K_t is 2.0. We also found that t_c is 70 or 80% of the t_r and t_c in a vacuum but decreased to half that in air.

Figures 7.39 (a) and (b) [20] show an example of the optical microscopic observation of the notch root of the specimen with a K_t of 2.0 in 0.3 torr and the number of microcracks ahead of the notch root. In order to measure the microcracks, the creep test was interrupted to make

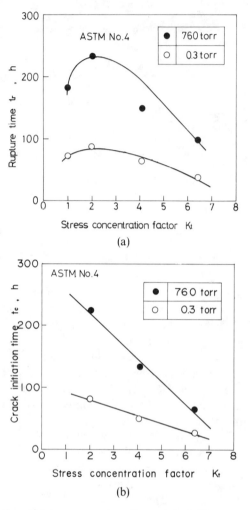

Fig. 7.38. Variation of (a) creep rupture time, t_r, and (b) creep crack initiation time, t_c, of Inconel 617, with stress concentration factor, K_t, in both air and a vacuum under $\sigma_n = 34\cdot 3$ MPa at 1273 K. (Reprinted from ref. 20 by permission of Japan Soc. Mech. Engrs)

observations by optical microscope in the zone of 20×4 mm covering the whole notch cross-section of the specimen. The creep time was interrupted after 82 h. The length of microcrack was smaller than the average grain size of the specimen, and the cracks occurring in a

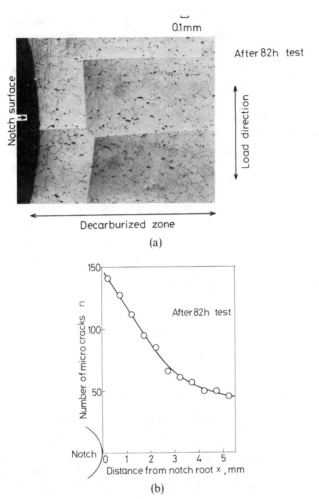

Fig. 7.39. (a) Microcracks under the microscope, and (b) the number of microcracks, N, at the notch root of a notched plate specimen of Inconel 617 (ASTM No. 4) in a vacuum of 0·3 torr under $\sigma_n = 34·3$ MPa at 1273 K, $K_t = 2·0$. (Reprinted from ref. 20 by permission of Japan Soc. Mech. Engrs).

direction perpendicular to the load axis were predominent. We found from the figure that the decarburized zone was localized to the notch root, that the zone was about 1 mm from the notch root on the cross-section of the specimen, and that it was most intense at the notch root and was governed by the stress gradient at the notch root. In this connection, the greatest number of microcracks was found at the notch root and it decreased toward the interior zone of the specimen. It is worthy of note that the zone size is nearly identical to the size of the decarburized zone. On the contrary, we confirmed that there was neither microcrack nor decarburized zone in air at the same interruption time as that with a vacuum. Therefore we can suppose that the decarburization along grain boundaries of the specimen in a vacuum results in a decrease of creep resistance of the alloy tested, and causes intergranular microcracks. We can also suppose that the appearance of macrocracks at the notch root results in an increase in the local creep strain of the specimen and in an increase of the notch opening displacement.

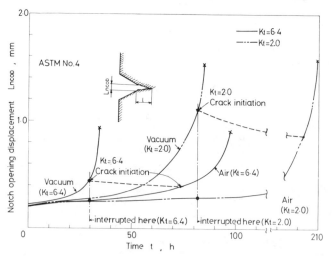

Fig. 7.40. Notch opening displacement curves of notched plate specimen of Inconel 617, in both air and a vacuum under $\sigma_n = 34.3$ MPa at 1273 K. (Reprinted from ref. 20 by permission of Japan Soc. Mech. Engrs)

Figure 7.40 [20] shows the experimental notch opening displacement (NOD) versus time curves of the notched plate specimens with a K_t of 2·0 and 6·4, respectively, in both air and 0·3 torr. We can suppose that the reason for early crack initiation in 0·3 torr compared to that in air is found in the higher rate of NOD in 0·3 torr compared with that in air. In other words, it is logical to assume that the crack initiation is controlled by the local rate at the notch root where microcracking occurs. We also found that the critical magnitude of NOD, at which a creep crack initiated, was nearly independent of environment but dependent on K_t. This was also confirmed in the propagation rate of macroscopic through

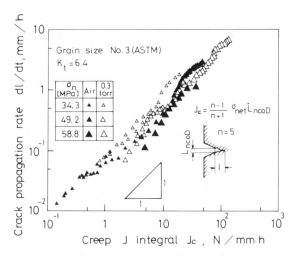

Fig. 7.41. Creep crack growth rate, dl/dt, plotted against the creep J-integral, J_c, of Inconel 617, in both the atmosphere and a vacuum of 0·3 torr at 1273 K. (Reprinted from ref. 20 by permission of Japan Soc. Mech. Engrs)

creep crack, as shown in Fig. 7.41 [20], in which the propagation rate is represented by a single curve on the basis of the creep J-integral, mentioned above, independent of both the environment and the notch shaped K_t.

From the experimental results of the vacuum environmental effect on static and cyclic creep rupture behavior of type 316 stainless steel at 873 K, and the static creep rupture behavior of the Inconel 617 superalloy at 1273 K, we made the following conclusions.

(1) The static rupture lives of both unnotched and round notched cylindrical bar specimens of type 316 stainless steel in 10^{-3} torr at 873 K are smaller than that in the atmosphere. In particular, the largest environmental effect is observed in the unnotched bar specimen. However, for the notched plate specimen no vacuum environmental effect was observed in either the unnotched or the notched specimens.

(2) In 0·1 torr the oxide film is thinner and adheres coherently to the metal when compared with the porous film in the atmosphere. The oxide is the protective Fe_3O_4 or Fe_2O_3 layer in both the atmosphere and a vacuum. From the model of oxide surface film and distributed dislocations beneath the surface film under shear stress, it was suggested that decrease in the dislocation density in 0·1 torr compared with that in the atmosphere is influenced by the stress gradient at the notch root, and it is influential in increasing the local creep rate at the notch root and in bringing about early crack initiation.

(3) The total strain-controlled push-pull cyclic creep failure life of a hollow cylindrical specimen of SUS 316 stainless steel at 600°C is remarkably dependent on the vacuum environment and the shape of the stress wave. The time to failure under a cp stress wave, with tension-hold in 0·1 torr, is markedly smaller than that in air. On the contrary, the failure time under a cc stress wave, with both tension and compression holds in 0·1 torr, is larger than that in the atmosphere. The latter fact is also revealed under a pc stress wave with compression hold time.

(4) The above characteristic behavior under a cp stress wave resulted from both early crack initiation and fast crack propagation, in the tension cycle, in 0·1 torr, and the latter effect under a cc stress wave was a result of the later initiation and slow propagation of the crack in 0·1 torr in connection with the cyclic deformation behavior.

(5) The creep rupture time of both the unnotched and the notched plate specimens of Inconel 617, in 0·3 torr, was shorter than that in air, and the life of the specimen with a grain size of ASTM No. 4 (78 μm) was the shortest. The rupture time of the unnotched and notched plate specimens, was a grain size of No. 4 tested in 0·3 torr, decreased to nearly half that in air. This also was a result of both the early initiation and fast propagation of the creep crack.

(6) It was estimated that the effect mentioned above resulted from the increase in the local creep rate of the specimen at the notch root or the creep crack tip. This was confirmed from the comparison between the notch opening displacement curves and that between the crack propagation rate and the creep J-integral in both the atmosphere and a vacuum. The increase of the local creep rate of the specimen in 0·3 torr resulted from the decarburization of the specimen through the non-protective oxide layer which was produced by the lack of oxygen in the vacuum.

7.3 EFFECT OF LOAD WAVE SHAPE ON THE HIGH-TEMPERATURE LOW-CYCLE FATIGUE OF PURE METALS

7.3.1 Transitional Changes of the Creep Crack Propagation Rate after Reloading in Stress-hold Tests

We previously described the effect of load wave shape and sequence on the high-temperature low-cycle fatigue of pure metals, heat-resisting metals and superalloys in Chapter 6. In this section, we describe this effect in relation to a theory of distributed dislocations and the J-

integral. Static tensile creep rupture tests and load-controlled push-pull tests with and without stress-hold were performed by using a hollow cylindrical specimen (i.d. 4 mm, o.d. 6 mm, gauge length 10 mm), with a circular notch 0·5 mm in diameter, composed of commercial pure copper in diffusion oil at 543 K under atmospheric pressure using the test apparatus shown in Fig. 4.46 in Chapter 4. The specimen was annealed at 673 K for 1 h after machining. Time for one cycle is about 7 min for tests without stress-hold, and about 15 min for stress-hold tests (hold time $t_H = 7·8$ min).

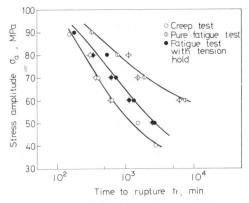

Fig. 7.42. Comparison of rupture time, t_r, of commercial pure copper in three types of test, in diffusion oil of 543 K at atmospheric pressure. (Reprinted from ref. 21 by permission of Japan Soc. Mech. Engrs)

Figure 7.42[21] shows a comparison of the nominal stress versus failure time curve in the three types of test in diffusion oil at 543 K in atmospheric pressure, and Fig. 7.43[21] shows the effect of stress-hold on the number of cycles to failure. We obtained the crack propagation rate at a crack length of 0·1 mm in the three types of test as follows: $dl/dt = 2·4 \times 10^{-3}$ mm/min for the static creep test (at $t = 450$ min); $dl/dN = 6·45$ mm/c for the test without stress-hold (at $N = 580$); $dl/dN = 1·2 \times 10^{-2}$ mm/c for the stress-hold test (at $N = 52$). It is clear that the propagation rate in the stress-hold test is larger than that in the test without hold. These facts were obtained from the experimental relation of the crack propagation rate in the three types of test. On the other hand, in the stress-hold test where $dl/dN = 1·2 \times 10^{-2}$ mm/cycle at $l = 0·1$ mm, we can estimate the partitioning crack length l per cycle; l is 0·021 mm during t_H of 7·8 min, and it is about 6×10^{-4} mm during the

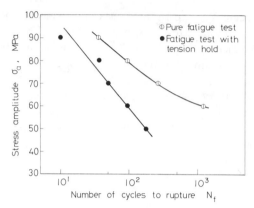

Fig. 7.43. Decrease of low-cycle fatigue failure lives due to stress hold. Test conditions as for Fig. 7.42. (Reprinted from ref. 21 by permission of Japan Soc. Mech. Engrs)

inverse loading and reloading in one cycle, respectively. By using the d.c. potential method, where about 3 μm in crack length is measurable, Kitamura and Ohtani[22] observed in the stress-hold test, for a centrally notched plate specimen of type 316 stainless steel at 873 K in air, that the crack propagated during not only stress-hold but also stress variation, and both crack propagations resulted from the creep damage of the material.

We found from the comparison of partitioning crack lengths, in one cycle in the stress-hold test, that the sum of the crack length estimated from both the static creep crack propagation data during stress-hold and that estimated from the pure fatigue crack propagation data during stress variation is larger than that in the stress-hold test. As a matter of course, the creep crack length during stress-hold is larger than the pure fatigue crack length during stress variation in the stress-hold test.

Figure 7.44 schematically illustrates partitioning of the crack length in the stress-hold test and the comparison of crack length propagated per cycle in the three types of test mentioned above. We found that in the stress-hold test the crack propagation rate during stress-hold decreases markedly due to stress variation compared with that in static creep. Thus we found that there is a *creep–fatigue interaction in crack propagation*. Kitamura and Ohtani[22] also found experimentally that the creep crack propagation stopped in an early stage of stress-hold. The transitional phenomenon at reloading in the stress-hold test can be

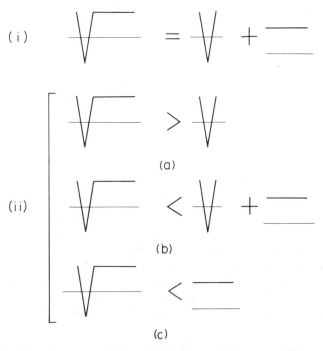

Fig. 7.44. (i) Partitioning crack length in the stress-hold test, and (ii) comparison of crack length per cycle in three types of test; left-hand side shows the one cycle in the stress-hold test, and right-hand side shows the respective components in pure fatigue and static creep.

interpreted from the following factors. The first is the transitional decrease of the local creep rate at the crack tip due to stress variation, and the second is an oxidizing environmental effect. Since the latter will have the effect of crack reshaping due to an oxide film at the crack tip in the atmosphere, it is included in the former. Therefore we describe mainly the first factor using the theory of distributed dislocations as follows.

Since the creep crack propagation rate is proportional to the local strain rate $\dot{\varepsilon}_{loc}$, as mentioned in the previous sections, it is worth noting that $\dot{\varepsilon}_{loc}$ is proportional to $\rho b v$, where ρ is a movable dislocation density, b is the Burgers vector and v is the average velocity. We previously described how ρ increases or decreases due to inverse loading in an early stage of the low-cycle fatigue of pure metals (see 4 and 5 in Fig. 5.34 in Chapter 5), but no remarkable change was observed with advanced

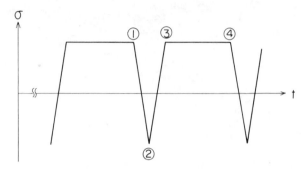

Fig. 7.45. Stress wave in the stress-hold test.

number of cycles (see Fig. 5.35). Therefore we can assume only the time-dependence of v in the following discussion.

First, regarding the velocity of dislocation in static creep (point 1 in Fig. 7.45), the dislocations pile up against an obstacle under monotonic loading and we can consider that the climbing frequency, due to thermal activation, corresponds to the velocity of dislocations. When loading is reversed from tension to compression (point 2 in Fig. 7.45), the dislocations move easily in the inverse direction. At reloading to tension (point 3 in Fig. 7.45), the dislocations tend to return to the original state but time is needed for returning because the stress distribution in front of the crack tip has a steep slope due to inverse loading compared with that in static creep. This will result in the difference in the dislocation distribution at both stages 1 and 3 as follows.

Dislocation distribution under a stress gradient was analyzed by Bilby et al. [23]. We use a discrete distribution of dislocations, $D(x)$, where x is a coordinate on the slip plane. If a fixed dislocation exists in x_0, the stress $\sigma(x)$ at x, due to both an external stress, $\tau(x)$, and the fixed dislocations, is written as

$$\sigma(x) = \tau(x) + \sum_\alpha \frac{A}{x - x_\alpha} \qquad (7.15)$$

where A is $Gb/[2\pi(1-v)]$ for edge dislocations and $Gb/(2\pi)$ for screw dislocations. Since a movable dislocation moves from an arbitrary position x_j to that where $\sigma(x)$ becomes zero, we obtain

$$\sum_{i \neq j} \frac{1}{x_j - x_i} + P(x_j) = 0 \quad (i,j = 1, 2, \ldots, n) \qquad (7.16)$$

where $P = \sigma(x)/A$. Equation (7.16) is the equilibrium equation for movable dislocations. Using a polynomial equation $f(x) = \Pi_{i=1}^{n}(x - x_i)$, eqn (7.16) is rewritten as

$$f''(x) + 2P(x)f'(x) + q(n,x)f(x) = 0 \tag{7.17}$$

where $q(n, x)$ is the x value for $f(x) = 0$ and it has no infinitive value. In order to obtain the dislocation density we use the variable transformation $v(x) = f(x)\exp[\int P(x)\,dx]$, and eqn (7.17) is given as

$$v''(x) + I(x)v(x) = 0 \tag{7.18}$$

where $I(x) \equiv q(n, x) - P^2(x) - P'(x)$. Equation (7.18) has a reasonable solution for only $I(x) > 0$. The solution is approximately written as

$$v(x) = A(x)\sin[(I(x))^{1/2}x + \phi(x)] \tag{7.19}$$

The distance d is between the zero values in eqn (7.19), i.e. the distance between dislocations is given as

$$d = \pi/[I(x)]^{1/2} \tag{7.20}$$

Let us consider the distribution density of n movable dislocations under $\tau(x) = -\alpha x$. We obtain $\sigma(x) = -\alpha x$ from eqn (7.15) and also $P(x) = -\alpha x/A$. Therefore eqn (7.17) is given as

$$f''(x) - 2(\alpha x/A)f'(x) + q(n,x)f(x) = 0 \tag{7.21}$$

and the solution is given by the Hermite polynomial, $H(\sqrt{\alpha/A}x)$, where $q(n, x) = 2\alpha n/A$. Therefore we obtain

$$D(x) = \frac{\alpha}{A\pi}\sqrt{a^2 - x^2} \tag{7.22}$$

where $a \equiv (2nA/\alpha)^{1/2}$.

Figure 7.46 [21] shows an example of the numerical calculation of $D(x)$ (mm^{-1}) for $\alpha = 10, 5, 1$ and 0.1 (kgf/mm^3) under a fixed condition of n. If there is an obstacle ahead of the crack tip, as shown in Fig. 7.47, we can consider that the climbing frequency, $\bar{v}$, is proportional to an effective stress, τ_{eff}, on the distributed dislocations. Under a steep slope of stress the dislocations distribute within only a short distance, and if an obstacle exists at the modified distance from the crack tip, $x' + \sqrt{2/\alpha} \approx 0.5(\sqrt{\text{mm}^3/\text{kgf}})$ in Fig. 7.46, the $D(x)$ is in self-equilibrium under $\alpha = 10(\text{kgf/mm}^3)$ and then $\tau_{\text{eff}} = 0$. On the contrary, under the other gentle slope the τ_{eff} increases due to pile up of dislocations against the obstacle. Therefore, we found from the present calculation that τ_{eff}

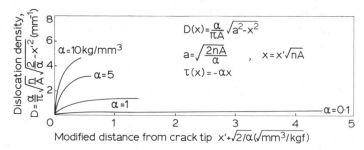

Fig. 7.46. Dislocation distribution under a stress gradient of $\tau = -\alpha x$. (Reprinted from ref. 21 by permission of Japan Soc. Mech. Engrs)

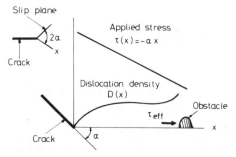

Fig. 7.47. Relation between stress distribution, $\tau(x)$, in front of crack and effective shearing stress, τ_{eff}, imposed on an obstacle.

decreases as the stress gradient increases, and $\bar{v}$ at reloading will decrease due to the transitional steep slope in the stress distribution in front of the crack tip compared with that under static tensile creep.

7.3.2 Crack Initiation and Crack Propagation Damage, and Time-dependent and Cycle-dependent J-Integrals

Figure 7.48 [21] shows an experimental relation between creep crack initiation damage ϕ_{ic} and fatigue crack initiation damage ϕ_{if}, and that between creep crack propagation damage ϕ_{cp} and fatigue crack propagation damage ϕ_{fp}, in the stress-hold test by using a linear life fraction rule, where crack initiation is defined as when the crack reaches $10 \, \mu\text{m}$ in length. We can see from the figure that creep damage is dominant in both crack initiation and propagation damage. If we define crack propagation damage at an instant, a differential form of the crack

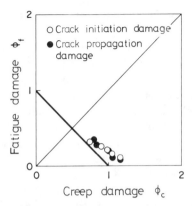

Fig. 7.48. Crack initiation damage and crack propagation damage for pure copper, in the stress-hold test at 543 K in oil, by means of the linear damage rule. (Reprinted from ref. 21 by permission of Japan Soc. Mech. Engrs)

propagation damage by linear damage rule is

$$\phi'_p = \frac{1/\Delta l_{hf}}{1/\Delta l_f} + \frac{1/\Delta l_{hc}}{1/\Delta l_c} \qquad (7.23)$$

where $1/\Delta l_f$ is the number of cycles required to propagate unit pure fatigue crack, $1/\Delta l_c$ is the time required to propagate unit creep crack, $1/l_{hf}$ is the number of cycles required to propagate unit crack in the stress-hold test, and $1/\Delta l_{hc}$ is the time required to propagate unit crack in the stress-hold test. We found that the propagation damage ϕ'_p of eqn (7.23) takes a larger constant value of almost 2 under $\sigma_a = 58 \cdot 8$ MPa in the crack length from 10 μm to 2 mm. This difference in the crack propagation damage based on a linear fraction rule, ϕ_p, and the differential form, ϕ'_p, is caused because ϕ_p includes unstable crack growth prior to failure but ϕ'_p does not.

All in all, we found that a linear damage rule, $\phi_c + \phi_f = D$, applies whatever damage definition we use, but creep damage is dominant in the stress-hold test. Recently, Kitamura and Ohtani [22] reported that crack extension l_h during stress-hold in one cycle and the sum $(\Delta l_f + \Delta l_h)$ during both stress variation and stress-hold in one cycle were well represented by the *time-dependent J-integral range* ΔJ_c per cycle, and that the majority of the data fell into the data band for static creep, where $\Delta J_c = \int_0^t J_c \, dt$ (J_c is the creep J-integral, t is the duration in one cycle). Furthermore, l_f was larger than that estimated from the *cycle-dependent*

J-integral range, i.e. the fatigue *J*-integral range ΔJ_f. We can suppose from these facts that in the stress-hold tests the crack propagates during both stress variation and stress hold, and the propagation rate is mainly controlled by creep damage ahead of the crack tip. Therefore we can say that there is a creep–fatigue interaction in the crack propagation rate, as mentioned above, but there is almost no interaction based on the fracture criterion of the *J*-integral. Life prediction based on crack propagation damage will be described in the next chapter.

In this section we have concluded the following from the low-cycle fatigue with tensile stress-hold for a center-notched plate specimen of commercial pure copper at 543 K.

(1) Partitioning of crack extension during stress variation (Δl_f) and stress hold (Δl_h) per cycle, both crack extensions were estimated from the experimental crack propagation data of the fatigue test without hold and the static creep rupture test, respectively. The crack extension per cycle in the stress-hold test is almost half the sum $\Delta l_\text{f} + \Delta l_\text{c}$. This means that the creep crack propagation rate during stress-hold decelerates due to stress variation compared with that of static creep. Therefore there is a creep–fatigue interaction in the crack propagation rate even if it is temporary.

(2) The deceleration of the creep crack propagation rate at reloading is interpreted by the decrease of an effective stress τ_eff on the distributed dislocations under a steep slope of stress distribution in front of the crack tip using the theory of distributed dislocations under a stress gradient.

(3) By partitioning damage during the time to crack initiation, ϕ_i, and that for crack propagation, ϕ_p, based on a linear life fraction rule, we can see that there is no difference in $\phi_\text{i} = 1$ and $\phi_\text{p} = 1$ and the creep damage component within both ϕ_i and ϕ_p is dominant. The differential form of crack propagation damage, ϕ'_p, of eqn (7.23) is half ϕ_p because ϕ_p includes the unstable crack propagation period prior to failure. The linear damage rule $\phi_\text{c} + \phi_\text{f} = D$ almost holds whatever damage definition we use based on partitioning both life to crack initiation and that for crack propagation, and both the fatigue crack propagation rate and creep crack propagation rate.

(4) Creep–fatigue interaction is observed in the crack propagation rate in the stress-hold test but the crack propagation rate is controlled mainly by the time-dependent *J*-integral range, ΔJ_c, but not by the cycle-dependent *J*-integral range, i.e. the fatigue *J*-integral range, ΔJ_f. This suggests a creep–fatigue interaction in the creep propagation rate, but

FRACTURE MECHANICS OF SOLIDS WITH MICROSTRUCTURE 409

there is almost no interaction on the basis of the fracture criterion of the J-integral.

7.4 X-RAY STUDY OF DAMAGE EVALUATION FOR STRUCTURAL METALS UNDER CREEP–FATIGUE INTERACTION CONDITIONS

7.4.1 Analysis of X-Ray Diffraction Asterism and Evaluation of the Damage of Austenitic Stainless Steel in High-temperature Low-cycle Fatigue

Information obtained from X-ray diffraction is one of the parameters for evaluating the *damage* of materials, but we have not reached a definite conclusion as to whether X-ray parameters are effective or not. This is quite opposite to crack propagation because a through-macrocrack is a concept in which we can measure the crack concretely, but not the damage. In this section, we will describe the X-ray parameters for estimating the damage of the material and the crack initiation/propagation behavior in high-temperature low-cycle fatigue. The total strain controlled push-pull test was performed using a hollow cylindrical specimen (i.d. 9 mm, o.d. 12 mm, gauge length 20 mm), with a circular notch 1 mm in diameter, of solution heat treated type 304 stainless steel (average grain size 990 μm) at 873 K in air. Table 7.2 shows the conditions of the X-ray diffraction by the micro Laue method. A Rigaku Roterflex RU-200 was used to conduct the X-ray diffraction experiment and a JEM100C was used to observe the microstructure of the specimen. Foils for TEM observations were prepared by jet polishing with reagent grade acetic acid containing 6% perchloric acid and finishing with reagent grade acetic acid containing 25% perchloric acid. Misorientation,

Table 7.2
Conditions of Micro Laue X-Ray Diffraction

Target	Mo
Tube voltage	50 kV
Tube current	20 mA
Slit diameter (single pinhole)	ϕ100 μm
Distance from specimen to film	30 mm
Illuminated cross-section	ϕ250 μm
Exposure time	4 h
Film	Fuji 150 type

β, in the X-ray diffraction was calculated by eqn (5.36) in Chapter 5, and the dislocation density, S_{12}^{**i} ($i=1, 2, 3$), and the imperfection density, R_{112}^{***i} ($i=1, 2, 3$), were calculated from eqns (5.34) and (5.35), respectively.

Figure 7.49 [24] shows examples of the variation of misorientation, β, with increase in strain cycles, N. N_c is the number of cycles to crack initiation at the notch root, defined as the stage when the crack length reaches 100 μm, and N_f is the number of cycles to failure defined as the stage when the tensile stress amplitude decreases to 3/4 of the maximum

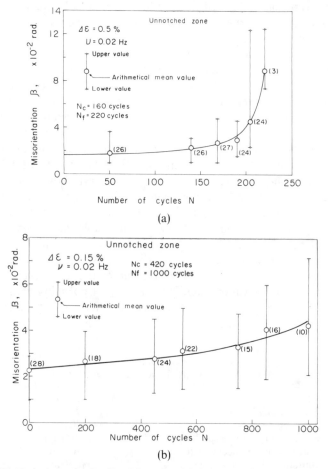

Fig. 7.49. Variation of misorientation, β, with increase of the number of strain cycles, for type 304 stainless steel at 873 K in air; the number in parentheses indicates the number of the asterism: (a) $\Delta\varepsilon=0.5\%$, (b) $\Delta\varepsilon=0.15\%$. (Reprinted from ref. 24 by permission of Soc. Mater. Sci., Japan)

value. Misorientation increased monotonously with increase in cycles, especially in the test at $\Delta\varepsilon=0.5\%$ where misorientation increased significantly at the end of the life, but in the tests at $\Delta\varepsilon=0.15\%$ such an increase at the end of the life was not observed. Therefore the behavior of increasing misorientation can be said to depend on the total strain range. Figure 7.50 [24] shows the experimental relation between the average of misorientations of all the diffraction planes, $\bar{\beta}-\bar{\beta}_0$, and the cumulative plastic strain range, $\Delta\varepsilon_p$, where β_0 is the misorientation at the initial state. The correlation between the two is so good that $\beta-\beta_0$ can be a good parameter for estimating the residual life of fatigued metals.

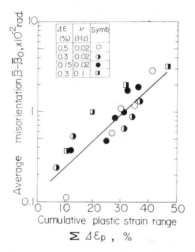

Fig. 7.50. Experimental relation between the average misorientation, β, and cumulative plastic strain range, $\Sigma\varepsilon_p$. Test conditions as for Fig. 7.49. (Reprinted from ref. 24 by permission of Soc. Mater. Sci., Japan)

Figures 7.51 (a) and (b) [24] show the variations of dislocation density, $S_{12}^{\cdot\cdot i}$, and imperfection density, $R_{112}^{\cdot\cdot i}$, with increase in cycles, N, at the unnotched portion. Dislocation and imperfection densities, as well as misorientation, increase with increasing strain cycles in the first stage of rapid cyclic hardening of this material, and then they maintain a nearly constant value as the misorientation increases. Dislocation and imperfection densities at the unnotched portion once again begin to increase rapidly when the crack initiates and propagates at the notched portion. The same increasing behavior of dislocation and imperfection densities is also observed at other strain levels.

In Fig. 7.52 [24], the bottom row shows higher magnification TEM

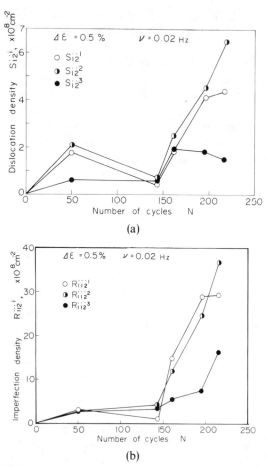

Fig. 7.51. Variation of (a) dislocation density and (b) imperfection density, with increase of the number of strain cycles; $\Delta\varepsilon = 0.5\%$, $v = 0.02$ Hz. Test conditions as for Fig. 7.49. (Reprinted from ref. 24 by permission of Soc. Mater. Sci., Japan)

photographs than the top row. In the coarse-grained specimen tested, at the 50th cycle, dislocations are multiplied but do not form a cell structure. At the 120th cycle, dislocation density does not change from that of the 50th cycle, but cell structure begins to form, and at the 90th cycle dislocation density increases with cell structure. The relation between the misorientation, dislocation/imperfection density and TEM observations is as follows. In the period of the first few cycles, where cyclic hardening is significant, only dislocation multiplication occurs and

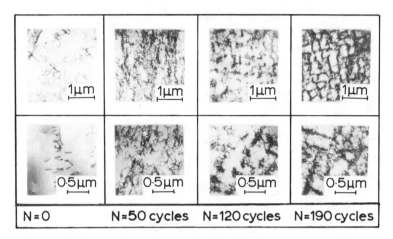

Fig. 7.52. Substructural change during low-cycle fatigue for type 304 stainless steel at 873 K; $\Delta\varepsilon = 0.5\%$, $v = 0.02$ Hz. (Reprinted from ref. 24 by permission of Soc. Mater. Sci., Japan)

the multiplication reflects the increase in dislocation and imperfection densities. In the period where the stress amplitude maintains nearly a constant value, multiplication dislocations begin to form a cell structure so that the dislocation density in Fig. 7.51(a) does not increase and only the misorientation increases due to the formation of cell structure. At the final stage, the dislocation density within the cells and the walls increases, and the fragmentation of the cell and the thickness of the cell wall occur so that the misorientation begins to increase once again at this stage.

Figure 7.53 [24] shows an example of the change in misorientation with the orientation function Γ prior to crack initiation. The orientation function Γ was calculated from Miller's indices. Before loading, and at the 50th cycle of loading, the misorientation does not depend on the diffraction plane, and the misorientations of all the diffraction planes exhibit almost the same value. At the 140th cycle, however, the misorientations of special diffraction planes exhibit the larger value. This indicates that large cumulative plastic deformation occurs on these planes. The planes having a larger misorientation are not the planes of maximum shear stress or the primary slip planes. The notably larger plastic deformation on these planes can be related to the micro mechanism of crack initiation, though the mechanism is not yet clear.

Examining the applicability of the average misorientation, $\bar{\beta} - \bar{\beta}_0$, to the crack opening displacement (COD) and the cycle-dependent J-integral, ΔJ_f, we found that there is a good correlation between the

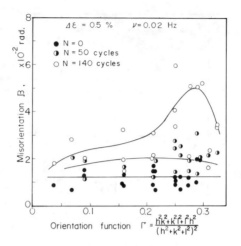

Fig. 7.53. Variation of misorientation, β, with orientation function, Γ. Test conditions as for Fig. 7.49. (Reprinted from ref. 24 by permission of Soc. Mater. Sci., Japan)

COD and $\bar{\beta} - \bar{\beta}_0$, and also between ΔJ_f and $\bar{\beta} - \bar{\beta}_0$. Figure 7.54 [24] shows the experimental relation between the crack propagation rate and the average misorientation. The average misorientation at the crack tip is fairly well correlated with the crack propagation rate at the frequencies v tested. If the correlation between the cumulative plastic strain range and the average misorientation shown in Fig. 7.50 also holds at the crack tip, the local cumulative plastic strain range at the crack tip controls the crack propagation rate in low-cycle fatigue at high temperatures.

Figure 7.55 summarizes the results obtained from the measurement of the average misorientation at the crack tip. In this figure, white arrows indicate the correlation obtained from fracture mechanics and black arrows indicate the correlation obtained from the measurement of misorientation by the micro Laue method in this study. Here the bigger arrows indicate closer relationships than the smaller arrows do. As shown by the black arrows, the crack propagation rate was determined by the cumulative local plastic strain range at the crack tip, and the local plastic strain range could be expressed by COD more accurately than the fatigue J-integral range. This agrees with the previous description wherein the crack propagation rate correlated well with COD rather than with the fatigue J-integral range, ΔJ_f, for low-cycle fatigue of type 304 stainless steel (see Chapter 6). From the above discussion we

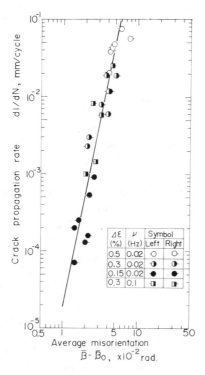

Fig. 7.54. Experimental relation between the crack propagation rate, dl/dN, and the average misorientation, β. Test conditions as for Fig. 7.49. (Reprinted from ref. 24 by permission of Soc. Mater. Sci., Japan)

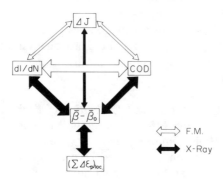

Fig. 7.55. Schematic diagram showing the relationship between misorientation and the parameters in fracture mechanics.

concluded that the average misorientation at the crack tip, measured by the micro Laue X-ray method, is a good parameter for estimating the crack propagation rate in low-cycle fatigue at high temperatures.

In this section, we have concluded the following from both X-ray and fracture mechanical analyses for type 304 stainless steel damaged by total strain-controlled push-pull low-cycle fatigue at 873 K in the atmosphere.

(1) The misorientation at the unnotched part increased monotonously with an increase in stain cycles. The average of misorientations of all the diffraction planes was well correlated with the cumulative plastic strain range at the unnotched part.

(2) Dislocation and imperfection densities obtained from the misorientation at the unnotched part increased in the period of significant cyclic hardening, while in the period of stable stress amplitude those densities maintained constant values in spite of the misorientation increasing. Both densities increased again in the period where the crack initiated and propagated from the center notch hole. These variations of dislocation and imperfection densities correlated well with the TEM observations.

(3) Immediately prior to the crack initiation at the notch root, the misorientation of special diffraction planes increased significantly.

(4) The misorientation at the crack tip correlated well with the crack opening displacement but not with the fatigue J-integral range. From this correlation and the good correlation found between the misorientation and the cumulative plastic strain range, the crack opening displacement was a parameter that expressed the cumulative local plastic strain range much better at the crack tip than for the fatigue J-integral range. Also, the crack propagation rate correlated well with the cumulative local plastic strain range at the crack tip.

(5) X-Ray parameters are useful non-destructive parameters for estimating the crack initiation time and the propagation rate because these parameters can express the cumulative plastic strain range at the crack tip or the notch root.

7.4.2 X-Ray Profile Analysis and Evaluation of Both the Damage and Crack Propagation Life of Austenitic Stainless Steel in Creep–Fatigue at High Temperatures

In this section, as the continuation of the previous section, we describe two series of tests using X-ray profile analysis. First we describe the evaluation of damage for a smooth cylindrical specimen (diameter 7 or

10 mm, gauge length 10 mm) of solution heat treated type 304 stainless steel in three types of test of static tensile creep, total strain range controlled push-pull low-cycle fatigue and that with hold-times at peak tensile strain at 873 K in air. Next we describe the evaluation of crack propagation lives of a round notched bar specimen (60° V-notch, $K_t = 2\cdot6, 4\cdot2, 6\cdot0$) of the same material fractured at 873 K in air under two test conditions. The first is a total strain range controlled push-pull test using triangular strain waves at a frequency of 0·1 Hz, where the nominal strain controlled test for the circumferential notched bar specimen was carried out along a 15 mm gauge length including the notch (see Section 6.5.3 in Chapter 6). The second is the nominal stress range controlled push-pull test using triangular and trapezoidal stress waves at frequencies ranging from 10^{-4} Hz to 5 Hz (see Section 6.5.2 in Chapter 6).

Table 7.3
X-Ray Diffraction Conditions

Characteristic X-ray	Cr-K_α
Diffraction plane	γ-Fe(111)
Filter	V
Tube voltage	30 kV
Tube current	10 mA
Irradiated area	37·7 mm^2
Detector	S.C.
Divergence slit	1°
Receiving slit	0·4 mm
Gonio-scan speed	1°/min
Time constant	2 s

The flat part of the surface of the cylindrical specimen is irradiated with X-rays for the first series of tests, and for the second series of tests only the fracture surface is irradiated where the fatigue crack propagated at the stable rate, by masking the surface of the unstable crack propagation with gum tape. Table 7.3 shows the X-ray diffraction conditions. Particle size, D, and microstrain, $\langle \varepsilon^2 \rangle^{1/2}$, were obtained from the following equation [26]:

$$-\frac{\ln A_L}{L} = \frac{1}{D} + \left(-\frac{1}{2D^2} + \frac{2\pi^2 \langle \varepsilon^2 \rangle h_0^2}{a^2}\right)L \quad (7.24)$$

where A_L is the Fourier cosine coefficient, L is the column length, a is the lattice constant, h_0 is the constant calculated from $h_0^2 = h^2 + k^2 + l^2$, and h, k, l are Miller indices. For calculating particle size and misorientation, a digitizer with a resolving power of 0·1 mm and connected to a microcomputer was used. The resolving power of 0·1 mm corresponds to 0·005° on the profile chart.

Before analyzing the X-ray profile, we pre-checked the influence of the gonio-scanning speed on the X-ray profile shape. From the pre-test results, using three scanning speeds, 0·25°, 0·5° and 1°/min, we arrived at the conclusion that there was no difference in the shape of the profile due to the gonio-scanning speed. So a scanning speed of 1°/min was employed in this study for the efficiency of the experiment. There was also no effect of oxide film on the X-ray parameters, half-value breadth, particle size and microstrain, despite the fact that the oxide film lowered the X-ray diffraction intensity.

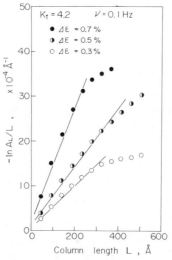

Fig. 7.56. Experimental relation between $-\ln A_L/L$ and column length, L, on the fracture surface of a round notched specimen of type 304 stainless steel ($K_t = 4·2$), in the strain-controlled test at 873 K in air. (Reprinted from ref. 27 by permission of Soc. Mater. Sci., Japan)

Figure 7.56 [27] shows an example of the experimental relation between $-\ln A_L/L$ and column length, L. Since the linearity of the data is good in the range of L less than 200 Å, particle size and microstrain can be calculated from eqn (7.24) with sufficient accuracy in that range.

Figures 7.57 (a) and (b) [25] show the experimental relation between two parameters, particle size D and microstrain $\langle \varepsilon^2 \rangle^{1/2}$, and three kinds

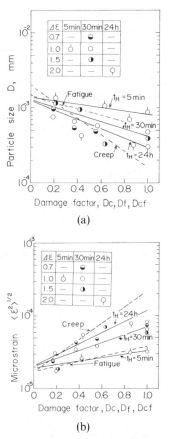

Fig. 7.57. Experimental relation between (a) particle size, D, (b) microstrain $\langle \varepsilon^2 \rangle^{1/2}$, and life ratio, N/N_f, in creep–fatigue tests of type 304 stainless steel at 873 K in air. (Reprinted from ref. 25 by permission of Soc. Mater. Sci., Japan)

of damage factor based on a linear life fraction rule for creep D_c, pure fatigue D_f, and creep–fatigue D_{cf}, for a smooth cylindrical specimen of type 304 stainless steel at 873 K in the atmosphere. In this figure, D_c, D_f and D_{cf} are defined as $D_c = 1 - (t/t_r)$, $D_f = 1 - (N/N_f)$ and $D_{cf} = 1 - (N/N_{cf})$, respectively, where t and N are the arbitrary time elapsed and the number of cycles, and t_r, N_f and N_{cf} are creep rupture time, pure fatigue failure life and failure life under creep–fatigue interaction, respectively. We found from Fig. 7.57(a) that both parameters obtained from the X-ray profile analysis correlate well with the three kinds of damage factor. The relation of particle size, D, to D_{cf} under hold-time t_H of 5 min is similar to that of the test without hold, but the relation under a t_H of 24 h is close to that of static creep. The relation under a t_H of 30 min falls into an intermediate region between the D_c and D_f mentioned above. Since D increases linearly with D_{cf}, we can estimate the creep–fatigue damage D_{cf} of the material if we know the load history, and we can also estimate the inverse relationship. We also found that microstrain, $\langle \varepsilon^2 \rangle^{1/2}$, has the same relation as that of D, as shown in Fig. 7.57(b). We will describe the prediction of remaining life of the structural metal under creep–fatigue interaction, using a non-destructive method of X-ray profile analysis, in the next chapter.

Next, in order to find a parameter which can estimate the crack propagation rate or period at elevated temperatures in transgranular and intergranular fracture cases, we describe fractography using X-ray diffraction, i.e. *X-ray fractography* of SUS 304 stainless steel in high-temperature low-cycle fatigue with and without hold-time.

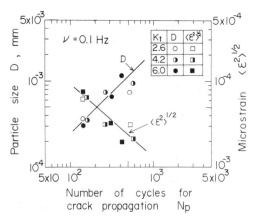

Fig. 7.58. Experimental relation between particle size, D, microstrain, $\langle \varepsilon^2 \rangle^{1/2}$, and the number of cycles for crack propagation, N_p, in strain-controlled tests for type 304 stainless steel at 873 K in air. (Reprinted from ref. 27 by permission of Soc. Mater. Sci., Japan)

Figure 7.58 [27] shows the relation between particle size D, microstrain $\langle \varepsilon^2 \rangle^{1/2}$ and the number of cycles needed for crack propagation N_p of the circumferential notched bar specimen in total strain range controlled tests. Particle size is well correlated with the number of cycles for crack propagation, N_p, without the effect of the stress concentration factor. Since particle size has been reported as a parameter expressing subgrain size [28], the fracture surface with a smaller N_p has a smaller subgrain size. Microstrain, as well as the particle size, is well correlated with N_p without the effect of the stress concentration factor. Microstrain physically expresses the lattice strain of the crystal, so the fracture surface with a smaller N_p has a larger lattice strain than that with a larger N_p. Therefore particle size and microstrain, as well as the half-value breadth, can be good parameters for estimating the number of cycles for crack propagation and the number of cycles to failure for the case of a larger elastic stress concentration factor.

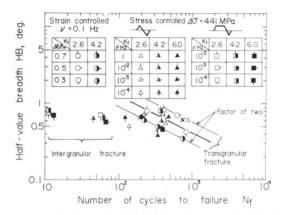

Fig. 7.59. Experimental relation between half-value breadth, HB, and the number of cycles to failure, N_f. Test conditions as for Fig. 7.58. (Reprinted from ref. 27 by permission of Soc. Mater. Sci., Japan)

Figure 7.59 [27] summarizes the relation between the half-value breadth, HB, and the number of cycles to failure for the fracture surface of a circumferential notched bar specimen of type 304 stainless steel, in both the symmetric strain wave tests and symmetric/antisymmetric stress wave tests at 873 K in air. HB is well correlated with N_f, in the transgranular fracture region in strain and stress controlled tests, in a scatter band with a factor of 2. In the intergranular fracture region, however, HB does not correlate well with N_f. In that region the increasing rate of HB, by introducing hold-times, is smaller compared with the reduction rate of N_f. Therefore the half-value breadth can be a good parameter for estimating the number of cycles to failure in the transgranular fracture region, but cannot be used for the intergranular fracture region. Another X-ray parameter is required to estimate the fatigue life in the intergranular fracture region.

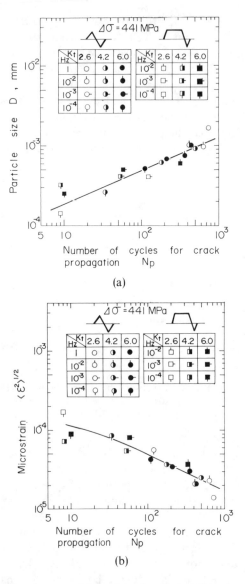

Fig. 7.60. Correlation between (a) particle size, D, and (b) microstrain, $\langle \varepsilon^2 \rangle^{1/2}$, and the number of cycles for crack propagation, N_p, in creep–fatigue tests of type 304 stainless steel at 873 K in air, irrespective of fracture mode of transgranular and intergranular fracture modes. (Reprinted from ref. 27 by permission of Soc. Mater. Sci., Japan)

Figures 7.60 (a) and (b) [27] show the relation between particle size D, microstrain $\langle \varepsilon^2 \rangle^{1/2}$ and the number of cycles for crack propagation N_p in both transgranular and intergranular fracture modes. Here D and $\langle \varepsilon^2 \rangle^{1/2}$ are well correlated with N_p in both transgranular and intergranular fracture modes. In the case of transgranular fracture, since the deformation intensity remaining on the fracture surface correlates to the crack propagation rate and the X-ray parameters, i.e. D and $\langle \varepsilon^2 \rangle^{1/2}$, measured the deformation intensity, the parameters are well correlated with the crack propagation period. In the case of intergranular fracture, on the other hand, deformation intensity near the grain boundary reflects the amount of grain boundary sliding, which plays an important role in the case of intergranular fracture, since the grain boundary could not slide without the deformation along the grain boundary. The X-ray parameters measured the deformation intensity along the grain boundary, and are also well correlated with the crack propagation period. The half-value breadth, on the other hand, is not as sensitive as D and $\langle \varepsilon^2 \rangle^{1/2}$ for the reduction in N_f due to hold-times. Arrangement of the data in Fig. 7.60 with the integral breadth is nearly the same as that with HB, where the presentation of the figure is omitted.

A summary of the above discussion follows. The half-value breadth and the integral breadth were sensitive to a total strain range, or a stress range of the test, but were not sensitive to the reduction in N_f due to hold-times. HB was not a parameter with which to follow the transition of the failure mode from transgranular to intergranular fracture. The transition of the fracture modes did not reflect HB or the integral breadth but reflected the shape of the X-ray profile caused by the microdeformation change remaining on the fracture surface.

In this section, the following conclusions were obtained from the X-ray profile analysis for creep, low-cycle fatigue and creep–fatigue of a type 304 stainless steel at 873 K in air under two test conditions. The first was a total strain range controlled test with and without hold-times up to 24 h, and the second was the stress controlled test using triangular and trapezoidal stress waves at frequencies ranging from 10^{-4} Hz to 5 Hz.

(1) The half-value breadth in X-ray diffraction is an appropriate parameter for evaluating creep damage but is not proper for evaluating fatigue damage. Particle size and microstrain in X-ray profile analysis are good parameters for evaluating both creep and fatigue damages.

(2) In X-ray fractography, the half-value breadth and integral breadth are sensitive to a total range, or a stress range, but are not sensitive to the transition from the transgranular fracture mode to the intergranular

fracture mode. The half-value breadth can be a good parameter for estimating the number of cycles to failure only in the transgranular fracture mode.

(3) Since particle size and microstrain are sensitive to total strain range, stress range and frequency, they can be good parameters for estimating the number of cycles for crack propagation in the intergranular fracture mode.

REFERENCES

1. J. G. Kaufman and H. Y. Hunsicker, Fracture toughness testing at Alcoa Research Laboratories, *ASTM Special Technical Publication*, **381** (1965), 290, Amer. Soc. Test. Mater.
2. M. Ohnami, Influence of hydrostatic pressure on fracture toughness of polycrystalline metallic materials, *Journal of the Japan Society for Technology of Plasticity*, **15** (1974), 769 (in Japanese), CORONA-Sha.
3. M. Ohnami. K. Motoie and T. Yamakage, The effect of hydrostatic pressure on plastic deformation and creep of polycrystalline metals at elevated temperatures, *Journal of the Society of Materials Science, Japan*, **20** (1971), 395 (in Japanese).
4. T. Yokobori, *An Interdisciplinary Approach to Fracture and Strength of Solids*, 2nd Edn, IWANAMI-Shoten, 1974, pp. 141–9 (in Japanese).
5. M. Ohnami and R. Imamura, Study of vacuum environment on high-temperature creep rupture strength of a commercial pure nickel, *Transactions of the Japan Society of Mechanical Engineers*, **46** (1980), 1074 (in Japanese).
6. R. Imamura, Doctoral Thesis, Ritsumeikan University, 1980 (in Japanese).
7. H. Riedel, A Dugdale model for crack opening and crack growth under creep conditions, *Materials Science and Engineering*, **30** (1977), 187, Elsevier Sequoia S.A.
8. A. K. Head, Interaction of dislocations and boundaries, *Philosophical Magazine*, **44** (1953), 92, Taylor and Francis Ltd., London.
9. M. J. Marcinkowski and E. S. P. Das, The relationship between cracks, holes and surface dislocations, *International Journal of Fracture*, **10** (1974), 181, Noodhoff International Publishing, Leyden.
10. K. Jagannadham and M. J. Marcinkowski, Comparison of the image and surface dislocation models, *Physica Status solidi* (a), **50** (1978), 293, Academie Verlag.
11. R. H. Cook and R. P. Skelton, Environment-dependence of the mechanical properties of metals at high temperature, *International Metallurgical Review*, **19** (1974), 199, Institute of Metals, London.
12. M. Ohnami and R. Imamura, Effect of vacuum levels on creep-rupture properties of polycrystalline metals at elevated temperatures (based on the relation between oxide film and the substructure of metals), *Proceedings of 22nd Japan Congress on Materials Research* (1979), p. 41, Soc. Mater. Sci., Japan.

13. A. K. Head and N. Lout, The distribution of dislocations in linear arrays, *Australian Journal of Physics*, **8** (1955), 1.
14. N. T. Muskhelishvili, *Singular Integral Equations*, Noodhoff, Groningen, The Netherlands, 1953.
15. E. Smith, Screw dislocation arrays in heterogeneous materials, *Acta Metallurgica*, **15** (1967), 249, Pergamon Press.
16. M. Ohnami and R. Imamura, A study on the effect of vacuum environments on creep rupture properties of commercial pure copper at elevated temperatures (especially based on grain size dependence), *Proceedings of 23rd Japan Congress on Materials Research* (1980), p. 132.
17. M. Ohnami and R. Imamura, Effect of vacuum environment on creep-rupture properties of SUS 316 stainless steel at elevated temperatures, *Proceedings of 20th Japan Congress on Materials Research* (1977), p. 65.
18. M. Ohnami and R. Imamura, Effect of vacuum environment on creep properties of SUS 316 stainless steel at elevated temperatures, Research Report of 123rd Committee on Heat-resisting Metals and Alloys, **19** (1978), p. 11 (in Japanese), Japan Soc. for Promotion of Science.
19. M. Ohnami, M. Sakane and R. Imamura, Effect of vacuum environment of push-pull cyclic creep fracture of metal at elevated temperatures, *Journal of the Society of Materials Science, Japan*, **30** (1981), 1136 (in Japanese).
20. M. Ohnami and R. Imamura, Effect of vacuum environment on creep rupture properties of Inconel 617 at 1000°C (especially based on crack initiation and propagation), *Bulletin of the Japan Society of Mechanical Engineers*, **24** (1981), 1530.
21. M. Ohnami and M. Sakane, Study of creep-fatigue interaction of polycrystalline metallic material at elevated temperatures '(especially crack initiation and propagation), *Transactions of the Japan Society of Mechanical Engineers*, **44** (1978), 397 (in Japanese).
22. T. Kitamura, R. Ohtani and H. Sugihara, Crack propagation under creep-fatigue interacted conditions, *Proceedings of 34th Annual Meetings of the Society of Materials Science, Japan* (1985), p. 49 (in Japanese).
23. B. A. Bilby and J. D. Eshlby, Dislocations and the theory of fracture, in *Fracture* (ed. H. Liebowitz), Vol. 1, Academic Press, 1968, p. 99.
24. S. Nishiino, M. Sakane and M. Ohnami, An X-ray study of the damage evaluation of high-temperature low-cycle fatigue for austenitic stainless steel, *Proceedings of 29th Japan Congress on Materials Research* (1986), p. 53.
25. S. Nishino, M. Sakane and M. Ohnami, Creep, fatigue and creep-fatigue damage evaluation and estimation of the remaining life of SUS 304 austenitic stainless steel at high temperatures (Creep-fatigue damage evaluation by X-ray analysis), *Journal of the Society of Materials Science, Japan*, **35** (1986), 292 (in Japanese).
26. R. I. Garrod and J. H. Auld, X-ray line broadening from cold-worked iron, *Acta Metallurgica*, **3** (1955), 190, Pergamon Press.
27. S. Nishino, M. Sakane and M. Ohnami, Prediction of high-temperature low-cycle fatigue life of austenitic stainless steel by X-ray fractography, *Proceedings of 29th Japan Congress on Materials Research* (1986), p. 61.
28. T. Tanaka, K. Igawa and K. Hoshino, Structural change during the low-cycle fatigue of SUS 316 at elevated temperatures, *Journal of the Iron and Steel Institute of Japan*, **68** (1982), 80 (in Japanese).

Chapter 8

Technology Development of Metal Plasticity under High Hydrostatic Pressure and of Life Assessment/Remaining Life Prediction Methods for Structural Materials with High-temperature Applications

This chapter describes technologies concerning the metal plasticity and high-temperature strength of structural materials. One is the plastic drawing of pure metals under high hydrostatic pressure and the other is the life prediction of structural materials at elevated temperatures. The former is the application of fundamental research described in Chapters 4 and 5, and also contains the plastic compression of pure metals under hydrostatic pressure. The latter is the application of Chapters 6 and 7, and contains the collaborative R and D efforts on the life prediction of structural metals for nuclear steelmaking, and the reliability evaluation of both structural metals for fossil fuel power plants and heat-resisting alloys for very high temperature gas reactors in Japan.

8.1 PLASTIC DRAWING OF PURE METALS UNDER HIGH HYDROSTATIC PRESSURES

In Chapters 4 and 5, the plastic behaviors of polycrystalline metals under hydrostatic pressure were described from the viewpoint of combined micro and macro machanics. Since the 1960s a high hydrostatic pressure effect has been applied to the technology of metal plasticity. The majority of this technology is applicable to hydrostatic extrusion [1,2] whereas it is relatively scarce concerning hydrostatic drawing [3,4]. In this section, we will describe the experimental and analytical results of the hydrostatic drawing of pure metals under high hydrostatic pressure from the viewpoint of the hydrostatic effect on the plasticity of metals and the forced lubrication effect in pressurized oil. The specially designed drawing test apparatus for use under a high hydrostatic pressure of 9·8 kbar will also be described.

8.1.1 Apparatus for Test under Hydrostatic Pressure

Concerning the high hydrostatic pressure test device, much research has been conducted since Bridgman [2] but there have been no great developments in principle. The small size of the test device can be attributed to the development of new high-strength metals; however, the rate of success is low, so a rather larger sized device has been adopted due to the necessity of elevating hydrostatic pressure. Hence we considered the following in designing a drawing test apparatus for use under hydrostatic pressure. (1) Both the pressure intensifier and static tensile test/drawing test assembly under hydrostatic pressure are separate. (2) The intent is to realize a smaller sized pressure vessel. (3) Commercial steel is used. (4) Both the static tensile test and the drawing test can be performed with one test device. (5) Axial load imposed on the test specimen under a high-confining pressure is directly measured inside the pressure container.

Figure 8.1 [5] shows a general view of the drawing test apparatus under a hydrostatic pressure of 980 MPa. Figure 8.2 [5] shows a block diagram of the test apparatus. In this device special considerations were made to construct a compact test device in order to satisfy the following requirements. (1) A shrinking effect was adopted by using an oil pressure of 196 MPa in the 3 mm gap between the inner and outer cylindrical pressure vessels, and by strengthening the pressure vessel itself. (2) Both static tensile and drawing tests under hydrostatic pressure are performed by using a single specimen grip device. (3) Axial load imposed on the test specimen under hydrostatic pressure can be measured precisely.

The principle of (1) mentioned above is the same as that used by Lengyel et al. [6] with the exception that a very simple structure was realized in the present test apparatus. The inner and outer diameters of the pressure intensifier having a capacity of 98 MPa are 30 mm and 90 mm respectively, and those of the pressure vessel used for the drawing tests are 30 mm and 250 mm respectively. The material used in the inner vessel is Ni–Cr–Mo steel (SNCM 26) and that in the outer vessel is low-carbon steel (S55C).

As shown in Fig. 8.2, the die and the load cell are maintained in pressurized oil and the test specimen is set in a static equilibrium of force

Fig. 8.1. General view of the drawing test apparatus under hydrostatic pressure. (Reprinted from ref. 5 by permission of Japan Soc. Tech. Plasticity)

under hydrostatic pressure to the extent that axial load is not imposed externally on the test specimen. In the present device, since the test specimen is set in pressurized oil, a lubrication effect can be expected during the drawing process. Figures 8.3 (a) and (b) [5]'show the shape and dimensions of the specimen and the grip used for the tensile and drawing tests under hydrostatic pressure, respectively.

The direct measurement of axial load was made by using a load cell in which active and dummy gauges are used in the resistant strain gauge, as shown in Fig. 8.4 [5], and the dummy gauge is not subjected to any axial load in the pressurized oil. From experimental examinations performed under pressurized oil at 196 MPa, we confirmed that load cell readings while under axial load are not influenced by confining pressure.

8.1.2 Test Results and Analysis

Figure 8.5 [5] shows the drawing test results of a 99·97% commercial pure zinc specimen (Fig. 8.3) in oil at 293 K under test conditions where the

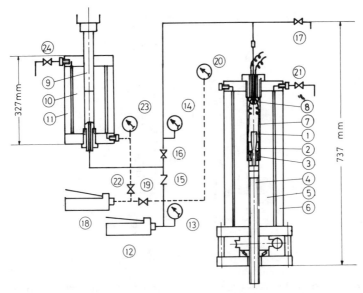

Fig. 8.2. Block diagram of test apparatus. 1, Test specimen; 2, die; 3, load cell; 4, screw plunger; 5, inner vessel for drawing test (250 mm i.d.); 6, outer vessel for drawing test (250 mm o.d.); 7, lead wire; 8, terminal; 9, plunger; 10, inner vessel of intensifier (30 mm i.d.); 11, outer vessel of intensifier (190 mm o.d.); 12, 18, hand pump (200 MPa); 13, 20, 23, pressure meter (200 MPa); 14, pressure meter (1000 MPa); 15, 16, 17, stop valves (1000 MPa); 19, 21, 22, 24, stop valves (220 MPa). (Reprinted from ref. 5 by permission of Japan Soc. Tech. Plasticity)

reduction area (R.A.) is $[1-(A_2/A_1)] \times 100 = 7\%$, the die diameter is 10 mm, the die angle is 10°, the drawing speed is 0·07 mm/min, and A_1 and A_2 are a cross-sectional area of the specimen before and after the drawing respectively. Figure 8.6 [5] also shows the results for a 99·95% commercial pure aluminum specimen under the same conditions as in Fig. 8.5, with the exception of a larger reduction area of 8%. As is seen from Fig. 8.5, the pure zinc is extremely difficult to draw in the atmosphere but is quite plastic under high hydrostatic pressure. It was found that the drawing force for pure zinc increases as pressure increases but, in contrast, the force for pure aluminum decreases with pressure. It is likely that this phenomenon results mainly from the structural changes of the metal under superimposed high hydrostatic pressure, like those discussed in Chapter 5, and the latter is a result of the forced lubrication caused by the pressurized oil. These can be analyzed in the following manner.

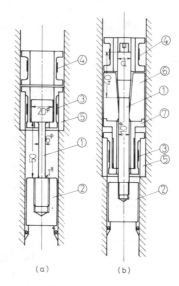

Fig. 8.3. Shape and dimensions (mm) of (a) tensile test specimen and (b) drawing test specimen and grips for tensile test, respectively, under high hydrostatic pressure. 1, Specimen; 2, holder; 3, load cell (active); 4, load cell (dummy); 5, load cell support; 6, die; 7, die support. (Reprinted from ref. 5 by permission of Japan Soc. Tech. Plasticity)

Fig. 8.4. Overview of load cell (bottom) and holder of lead wire terminal (top). (Reprinted from ref. 5 by permission of Japan Soc. Tech. Plasticity)

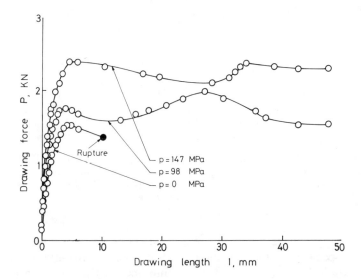

Fig. 8.5. Effect of hydrostatic pressure on the drawing force of 99.97% pure zinc. (Reprinted from ref. 5 by permission of Japan Soc. Tech. Plasticity)

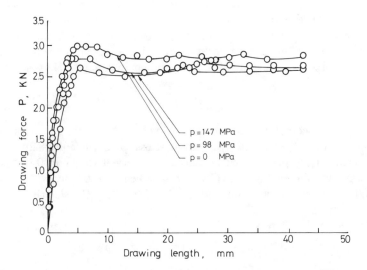

Fig. 8.6. Effect of hydrostatic pressure on the drawing force of 99·95% pure aluminum. (Reprinted from ref. 5 by permission of Japan Soc. Tech. Plasticity)

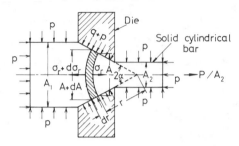

Fig. 8.7. Stress state of cylindrical solid bar in drawing test under hydrostatic pressure (p, hydrostatic pressure; q, contact pressure; P, drawing force; σ_r, radial stress; A, area; α, half vertical angle). (Reprinted from ref. 5 by permission of Japan Soc. Tech. Plasticity)

Assuming a stress state and geometrical configuration of the cylindrical bar specimen in the drawing test while under hydrostatic pressure p as shown in Fig. 8.7 [5], the force equilibrium for the axial direction is given as

$$A d\sigma_r + [\sigma_r + (q + p)(1 + \mu \cot \alpha)] dA = 0 \tag{8.1}$$

where the uniform contact pressure q and the uniform coefficient of friction μ on the die surface are assumed. A perfectly plastic body, in which the yield stress Y in the atmosphere is approximated by the average uniform stress in terms of the area of a true stress/strain diagram charting values from zero to strain of the reduction area, is also assumed, and the following yield condition as shown in Chapter 4 is adopted where

$$J'_2 = k(k + C' J_1) \tag{8.2}$$

and J'_2 is the second invariant of deviatoric stress, J_1 is the first invariant of stress and $3J_1$ denotes hydrostatic stress, p. C' in eqn (8.2) is the coefficient of the pressure effect and has numerical values of 0 to -0.3 for polycrystalline metals (see Chapter 4). When $C' = 0$, eqn (8.2) is a so-called Mises yield condition and $k = Y/\sqrt{3}$ is the yield shearing stress under atmospheric pressure.

The equivalent stress $\bar{\sigma}$ which satisfies eqn (8.2) is written as

$$(1/6)[(\sigma_1 - \sigma_2)^2 + (\sigma_2 - \sigma_3)^2 + (\sigma_3 - \sigma_1)^2]$$
$$= \lambda'^2 \bar{\sigma}^2 + \lambda' C' \bar{\sigma} (\sigma_1 + \sigma_2 + \sigma_3) \tag{8.3}$$

where $\lambda' = (1/2)[-C' + C'^2 + (4/3)]$, and σ_1, σ_2 and σ_3 are principal stresses. Putting $\sigma_1 = \sigma_r$, $\sigma_2 = \sigma_\theta = \sigma_3 = \sigma_\phi = -(q+p)$, eqn (8.3) is written as

$$[\sigma_r + (q+p)]^2 = 3\lambda'^2 \bar{\sigma}^2 + 3\lambda' C' \bar{\sigma} [\sigma_r - 2(q+p)] \quad (8.4)$$

From eqn (8.4), $(q+p)$ is given as

$$q + p \simeq -\left(1 - \frac{3\sqrt{3}}{2} C'\right)\sigma_r - \lambda'(3C' - \sqrt{3})\bar{\sigma} \quad (8.5)$$

where $\sqrt{1 + (3C'\sigma_r/\lambda'\bar{\sigma})} \approx 1 + (3C'\sigma_r/2\lambda'\bar{\sigma})$ and the term C'^2 is neglected. Applying eqn (8.5) to eqn (8.1), the radial stress σ_r is represented as

$$\left. \begin{aligned} \sigma_r &= \frac{1}{a}(CA^a - \bar{\sigma}b) \\ a &= -\frac{3\sqrt{3}}{2} C' \mu \cot \alpha \left(1 - \frac{3\sqrt{3}}{2} C'\right) \\ b &= (1 + \mu \cot \alpha)[\lambda'(3C' - \sqrt{3})] \end{aligned} \right\} \quad (8.6)$$

Adopting a boundary condition of $\sigma_r = -p$ when $A = A_1$, an integral constant C in eqn (8.6) is obtained as $C = (b\bar{\sigma}/A_1^a) - (ap/A_1^a)$. Therefore the radial stress σ_r is written as

$$\sigma_r = \frac{b}{a}\bar{\sigma}\left[\left(\frac{A_2}{A_1}\right)^a - 1\right] - \left(\frac{A_2}{A_1}\right)^a p \quad (8.7)$$

The drawing force P in Fig. 8.7 is given from eqn (8.7) as

$$\frac{P}{A_2 \bar{\sigma}} = \frac{b}{a}\left[\left(\frac{A_2}{A_1}\right)^a - 1\right] - \left(\frac{A_2}{A_1}\right)^a \frac{p}{\bar{\sigma}} + \frac{p}{\bar{\sigma}} = \left[\left(\frac{A_2}{A_1}\right)^a - 1\right]\left(\frac{b}{a} - \frac{p}{\bar{\sigma}}\right) \quad (8.8)$$

Assuming $C' = 0$, that is $\lambda' = 1/\sqrt{3}$, and $p = 0$, eqn (8.8) reduces to

$$\frac{P}{A_2 \bar{\sigma}} = \left(\frac{1}{1 - \mu \cot \alpha}\right)\left[1 - \left(\frac{A_2}{A_1}\right)\mu \cot \alpha\right] \quad (8.9)$$

Equation (8.9) is a so-called Sachs equation.

We found from eqn (8.8) that the drawing force is controlled by a combination of C', μ and p. Figure 8.8 [5] shows an analytical representation of both the influences of nondimensional hydrostatic pressure $p/\bar{\sigma}$ and the coefficient of friction modified by the die angle, $\mu \cot \alpha$, on the nondimensional drawing force $P/(A_2 \bar{\sigma})$. The figure illustrates a case where the reduction area is 10%, and we discovered that when C' takes a comparatively larger value the increase of drawing force with hydrostatic

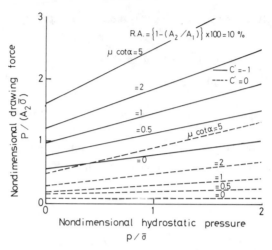

Fig. 8.8. Analytical representation of both influences of nondimensional hydrostatic pressure, $p/\bar{\sigma}$, and coefficient of friction modified by die angle, $\mu \cot \alpha$, on nondimensional drawing force, $P/A_2\bar{\sigma}$. (Reprinted from ref. 5 by permission of Japan Soc. Tech. Plasticity)

pressure is drastically influenced by the coefficient of friction modified by the die angle. We also found that, even if the modified coefficient of friction is zero, the drawing force increases as confining pressure increases when C' is comparatively large. Furthermore, we found that the drawing force increases with pressure when the modified coefficient of friction has a large value even if C' is zero. The decrease in the coefficient of friction modified by the die angle, $\mu \cot \alpha$, means an increase in the lubrication effect due to an increase in oil pressure when the die angle is fixed.

Figure 8.9 [5] shows an analytical representation of the relation between the influence of nondimensional hydrostatic pressure, $p/\bar{\sigma}$, and that of $\mu \cot \alpha$ on the nondimensional drawing force, $P/(A_2\bar{\sigma})$. It was assumed, in the figure, that $\mu \cot \alpha$ assumes values of 5, 2 and 1 when $p/\bar{\sigma} = 0, 1$ and 2, respectively. The figure was drawn by determining $P/(A_2\bar{\sigma})$ satisfying (0, 5) (1, 2) and (2, 1) in terms of the combination $(p/\bar{\sigma}, \mu \cot \alpha)$ in the case of $C' = -1$ and 0, respectively. We found from Fig. 8.9 that $C' = 0$ shows a tendency towards a slight decrease in drawing force due to the forced lubrication with pressurized oil; on the contrary, $C' = -1$ shows an increase in drawing force due to the increase of flow stress which relates to the structural change of the metal with the addition of

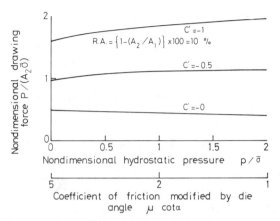

Fig. 8.9. Analytical representation of the relation between the influence of nondimensional hydrostatic pressure, $p/\bar{\sigma}$, and that of coefficient of friction modified by die angle, $\mu \cot \alpha$, on nondimensional drawing force, $P/A_2\bar{\sigma}$. (Reprinted from ref. 5 by permission of Japan Soc. Tech. Plasticity)

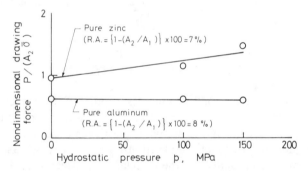

Fig. 8.10. Experimental results of the influence of hydrostatic pressure, p, on nondimensional drawing force, $P/A_2\bar{\sigma}$, for 99·97% pure zinc and 99·95% pure aluminum. (Reprinted from ref. 5 by permission of Japan Soc. Tech. Plasticity)

pressure. In fact, Fig. 8.10 [5] shows the experimental results of the influence of hydrostatic pressure p on the nondimensional drawing force, $P/(A_2\bar{\sigma})$, for pure zinc ($C' = 0.041$–0.178 for the equivalent strain $\varepsilon = 1$–30%) and pure aluminum ($C' = 0$). These experimental results agree well qualitatively with the analytical results of Fig. 8.9.

From these studies of the plastic drawing of commercial pure zinc and commercial pure aluminum under hydrostatic pressure, we can make the following conclusions.

(1) A specially designed drawing test apparatus was successfully constructed to perform under a hydraulic pressure of 980 MPa by adoption of the shrinking effect by using an oil pressure of 196 MPa in the gap between an inner pressure vessel having a 30/40 mm diameter and an outer pressure vessel with a 190/250 mm diameter.

(2) Commercial pure zinc is extremely difficult to draw in the atmosphere but easy under high hydrostatic pressure.

(3) The drawing force of pure zinc under hydrostatic pressure is influenced by the increase of flow stress due to a structural change with pressure rather than forced lubrication, and the drawing force increases as confining pressure increases. On the other hand, the drawing force of pure aluminum decreases slightly with pressure that was mostly due to forced lubrication rather than to structural change with pressure. These phenomena can be easily interpreted from the stress state of a cylindrical bar specimen in a drawing test under hydrostatic pressure in which the yield condition $J'_2 = k(k + C'J_1)$ is adopted, where J'_2 is the second invariant of deviatoric stress, k is yield shearing stress, J_1, is the first invariant of stress and C' is the coefficient of the pressure effect.

8.2 PLASTIC COMPRESSION OF PURE METALS UNDER HIGH HYDROSTATIC PRESSURES

The frictional boundary condition and the rupture of the lubrication film in plastic compression under atmospheric pressure were studied by Pearsall et al. [7,8], but the study of plastic compression under hydrostatic pressure was not discussed. In the present section, special interest is placed on analyzing the correlation between the effect of the structural change of metals under hydrostatic pressure and that of friction in deformation processes under hydrostatic pressure.

8.2.1 Test Apparatus for Plastic Compression under Hydrostatic Pressure

Figure 8.11 [9] shows the test apparatus for plastic compression in pressurized oil of 490 MPa in which the pressure intensifier and the compression test assembly are unified. The high-pressure vessel 1 has a pressure capacity of 490 MPa and is made of Ni–Cr–Mo steel (SNCM 8). Oil pressure can be intensified by a balancing valve 2 with an intensifying ratio of 1:10. The test method is as follows. First, the balancing valve 2 is set at the lowest position and a compression test specimen 3 is set between the upper and lower anvils 4 and 5; at the same time a sleeve 6 is

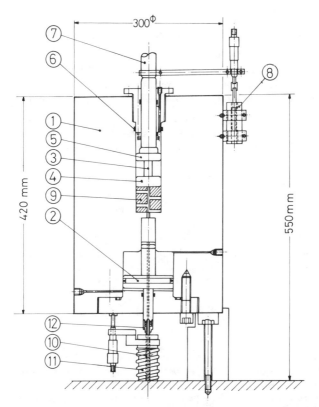

Fig. 8.11. Test apparatus for plastic compression under hydrostatic pressure. 1, Pressure vessel (300 mm o.d.); 2, balancing piston; 3, specimen; 4, lower anvil; 5, upper anvil; 6, O-ring; 7, plunger; 8, DTF; 9, coil spring with rectangular cross-section for loading; 10, core; 11, DTF; 12, rod to transmit displacement of 9. (Reprinted from ref. 9 by permission of Japan Soc. Tech. Plasticity)

set by inserting a plunger 7 into the oil chamber. Oil pressure is adjusted in the lower oil chamber by using a regulator valve. Axial compression is imposed on the test specimen by using a press with the capacity of 100 tons, and the displacement of the plunger 7 is measured by DTF 8.

Oil pressure is measured by a 0·1 mm diameter manganin wire inside the pressure vessel. Axial load on the specimen is also measured precisely by using a specially designed spring coil 9 with a rectangular cross-section and its displacement is measured by using a core 10 and DTF 11 outside the pressure vessel through a connecting rod 12 (the principle is the same as in Fig. 4.46).

8.2.2 Test Results and Analysis

Figure 8.12 [9] shows the experimental results of the plastic compression of 99·95% commercial pure aluminum under the conditions of (a) dry friction and (b) wet friction in the atmosphere, respectively. Figure 8.13 [9] shows the results under hydrostatic pressures of (a) 74 MPa and (b) 149 MPa. As seen from the comparison between both figures, nominal compressive stress σ_n under hydrostatic pressure has a gentle slope to the specimen radius/height ratio a/h compared with that under atmospheric pressure in oil, but there is no great difference between them.

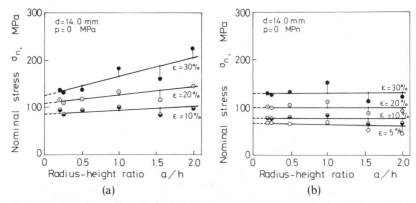

Fig. 8.12. Plastic compression of 99·95% pure aluminum, under conditions of (a) dry friction, 296 K in air, and (b) wet friction, 296 K in oil, in the atmosphere. (Reprinted from ref. 9 by permission of Japan Soc. Tech. Plasticity)

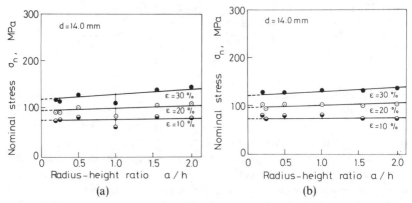

Fig. 8.13. Plastic compression of 99·95% pure aluminum, at 296 K in oil, under hydrostatic pressure of (a) $p = 74$ MPa and (b) $p = 149$ MPa. (Reprinted from ref. 9 by permission of Japan Soc. Tech. Plasticity)

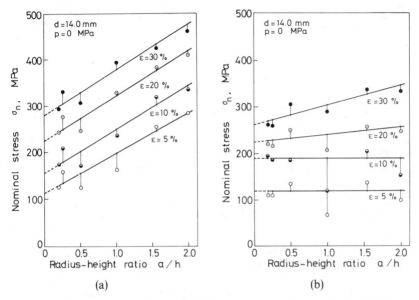

Fig. 8.14. Plastic compression of 99·97% pure zinc, under conditions of (a) dry friction at 296 K in air and (b) wet friction at 296 K in oil, in the atmosphere. (Reprinted from ref. 9 by permission of Japan Soc. Tech. Plasticity)

On the other hand, Fig. 8.14 [9] shows the change of σ_n for 99·97% commercial pure zinc under the conditions of (a) dry friction and (b) wet friction in the atmosphere. In the case of dry friction there is no remarkable change of σ_n with increase of compressive strain ε but in the case of wet friction σ_n is smaller than that in dry friction and the change in σ_n is large with the advance of ε. Figure 8.15 [9] also shows the results under hydrostatic pressures of (a) 74 MPa and (b) 149 MPa. We discovered that Fig. 8.15(b) shows a steep slope compared with Fig. 8.13(b) and σ_n at $a/h = 0$ increases as pressure increases. To discover why σ_n increases as a/h increases, the following two problems should be examined. (1) If the frictional boundary condition is the same for both pure aluminum and pure zinc, the slope for pure aluminum under dry and wet friction in the atmosphere and that of pure zinc are the same or the former must be larger than the latter [10]. (2) A steep slope under hydrostatic pressure means an apparent increase in friction but it is impossible to consider the increase in friction as due to hydrostatic pressure. These are analyzed as follows.

As a point of reference let us examine the stress state for a solid

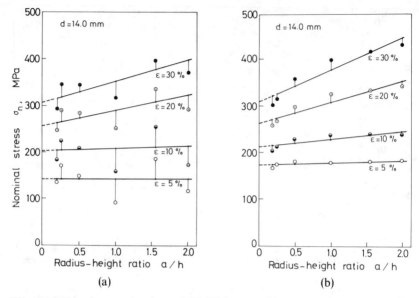

Fig. 8.15. Plastic compression of 99·97% pure zinc, at 296 K in oil, under confining pressure of (a) $p = 74$ MPa and (b) $p = 149$ MPa. (Reprinted from ref. 9 by permission of Japan Soc. Tech. Plasticity)

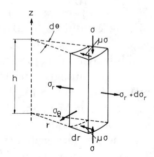

Fig. 8.16. Force equilibrium condition for solid cylindrical specimen.

cylindrical bar specimen subjected to axial compressive stress σ under hydrostatic pressure p, shown in Fig. 8.16; the force equilibrium condition is given as

$$\frac{d\sigma_r}{dr} - \frac{2\mu\sigma}{h} = 0 \qquad (8.10)$$

where μ is the coefficient of friction on the end surfaces. Putting $\sigma_1 = \sigma_r = \sigma_2 = \sigma_\theta$ and $\sigma_3 = \sigma_z = -\sigma - p$ in eqn (8.3), the following equation is obtained:

$$(\sigma_r - \sigma - p)^2 = 3\lambda'\bar{\sigma}^2 + 3\lambda'C'\bar{\sigma}(2\sigma_r - \sigma - p) \tag{8.11}$$

Therefore σ_r is derived from eqn. (8.11) as

$$\sigma_r = -\left(1 + \frac{3\sqrt{3}}{2}C'\right)\bar{\sigma} + \sqrt{3}\lambda'(\sqrt{3}C' + 1)\bar{\sigma} - \left(1 + \frac{3\sqrt{3}}{2}C'\right)p \tag{8.12}$$

where $[1 - (3C'/\lambda')(\sigma + p)/\bar{\sigma}]^{1/2} \simeq 1 - (3C'/2\lambda')(\sigma + p)/\bar{\sigma}$. Thus we obtain the following equations from eqns (8.10) and (8.12):

$$\sigma = \exp\left[\frac{2\mu A}{h}(C_1 - r)\right]$$
$$A = \frac{2}{2 + 3\sqrt{3}C'} \tag{8.13}$$

Putting $\sigma_r = -p$ when $r = a$, that is $\sigma = B\bar{\sigma} - Dp$, the integral constant C_1 in eqn (8.13) is written as $\exp(AC_1) = (B\bar{\sigma} - Dp)\exp[(2\mu A)/h]a$ and eqn (8.13) is given as

$$\sigma = (B\bar{\sigma} - Dp)\exp\left[\frac{2\mu A}{h}(a - r)\right] \tag{8.14}$$

where $B = 3\lambda'(\sqrt{3}C' + 1)A$ and $D = 3\sqrt{3}C'/(2A)$. Putting $A = B = 1$ and $D = 0$ when $p = 0$ and $C' = 0$, that is $\lambda' = 1/\sqrt{3}$, eqn (8.14) reduces to

$$\sigma = \bar{\sigma}\exp\left[\frac{2\mu}{h}(a - r)\right] \tag{8.15}$$

Equation (8.15) is the so-called Siebel equation. The average compressive stress σ_m is obtained from eqn (8.14) as

$$\sigma_m = \int_0^a 2\pi r\sigma \, dr/(\pi a^2) = (B\bar{\sigma} - Dp)\left(1 + \frac{2\mu A}{3}\cdot\frac{a}{h}\right) \tag{8.16a}$$

or

$$\frac{\sigma_m}{\bar{\sigma}} = \left(B - D\frac{p}{\bar{\sigma}}\right)\left(1 + \frac{2\mu A}{3}\cdot\frac{a}{h}\right) \tag{8.16b}$$

As is seen from eqn (8.16b), a relation between $\sigma_m/\bar{\sigma}$ and a/h is

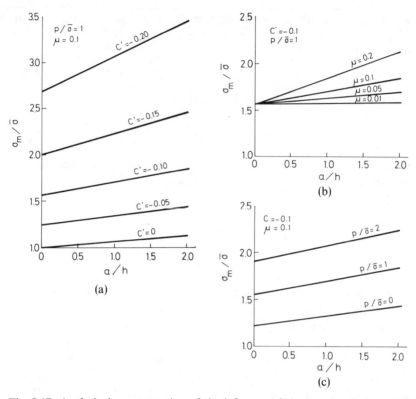

Fig. 8.17. Analytical representation of the influence of the structural change of material, subjected to combined hydrostatic pressure, on the $\sigma_m/\bar{\sigma}$ versus a/h relation. (a) Influence of C'; (b) influence of μ; (c) influence of p. (Reprinted from ref. 9 by permission of Japan Soc. Tech. Plasticity)

influenced by a combination of C', μ and p. Figure 8.17 [9] shows an analytical representation of the influence of the structural change of material subjected to combined hydrostatic pressure in the $\sigma_m/\bar{\sigma}$ versus a/h relation. Figures (a), (b) and (c) demonstrate the influences of the coefficient of the pressure effect C', and the coefficient of friction μ and pressure p, respectively. Commercial pure aluminum has a remarkably small value of C' and a gentle slope in the $\sigma_m/\bar{\sigma}$ versus a/h relation appears. On the other hand, commercial pure zinc has a comparatively larger value of C' and the value increases as plastic strain increases. Therefore we found that the experimental results of Figs 8.14 and 8.15 are in agreement with the analytical results of Fig. 8.17 qualitatively. We also found that, though Fig. 8.17(b) has a steep slope in the analytical relation of $\sigma_m/\bar{\sigma}$ to a/h as the coefficient of friction increases, it has a steep slope in the experimental relation under high hydrostatic pressure. This can be interpreted as follows.

The magnitude of the slope in the analytical relation of $\sigma_m/\bar{\sigma}$ versus a/h is controlled mainly by the coefficient of the pressure effect of C' which relates to the structural change of material under pressure. This can be interpreted as follows: the steep slope results from both the increase in the coefficient of friction, μ, and the coefficient, C', but μ in pressurized oil lessens the slope due to forced lubrication. In Fig. 8.17(c) we can see that superimposed loading of hydrostatic pressure p contributes to the increase in σ_n at $a/h = 0$ rather than a change in slope. Thus the above mentioned is recognized by a comparison of Fig. 8.15(b) and Fig. 8.17(a), and we can see that the change in the σ_n versus a/h relation is mainly due to the structural change of the material under hydrostatic pressure.

From these studies of the plastic compression of commercial pure aluminum and commercial pure zinc under hydrostatic pressure, we can make the following conclusions.

(1) The high hydrostatic pressure of turbine oil No. 90 does not contribute to an increase in forced lubrication.

(2) The nominal stress σ_n versus the specimen radius/height ratio (a/h) relation of commercial pure aluminum is not influenced by pressure. On the contrary, commercial pure zinc is markedly influenced by pressure and the cause is interpreted by the structural change of the metal under hydrostatic pressure.

8.3 RESEARCH AND DEVELOPMENT OF NUCLEAR STEELMAKING IN JAPAN AND THE LIFE PREDICTION METHOD OF STRUCTURAL MATERIALS FOR VERY HIGH TEMPERATURE GAS REACTOR (VHTGR) APPLICATIONS

'Save energy.' 'Diversify energy resources.' 'Develop new energy technology.' These are among the most pressing technical imperatives facing the world today.

A study of so-called 'nuclear steelmaking' was performed as one of Japanese national projects under the heading of 'research and development of direct steelmaking technology by utilizing high-temperature reducing gas' under the sponsorship of the Japanese Ministry of

International Trade and Industries. Its cost during 1973–1980 was ¥13·7 billion and the research was carried out in cooperation with governmental laboratories, private industries and universities. The collaborative efforts resulted in truly integrated research which produced a 'total system' design.

The aim of this research was to find a method of eliminating air pollution in steelmaking as we know it today, freedom from dependence on coal as a raw material, increasing energy efficiency and improving the utilization of energy by such means as the multi-utilization of energy sources. In order to develop the technical utility of *high-temperature gas-cooled reactors* (HTGR) for steelmaking processes and realize a nuclear steelmaking plant base having a 'closed system' in the future, research efforts focused on establishing a technical base to realize a pilot nuclear steelmaking plant which will be connected to the 50 MW (thermal) HTGR. In developing the HTGR, multi-purpose applications are being sought for nuclear energy and it is now a national project promoted by the Japan Atomic Energy Research Institute (JAERI, governmental laboratory). The author had participated in the nuclear steelmaking project as chairman of the Expert Committee of Material Strength, on the Special Committee on the Structural Strength of High Temperature Heat Exchangers (chairman Dr. T. Udoguchi, a former professor at the University of Tokyo) sponsored by the Engineering Research Association of Nuclear Steelmaking (ERANS). The author is also a member of the Committee on HTGR for Multi-Purpose Use (chairman Dr. Y. Mishima, a former professor at the University of Tokyo) in JAERI. In this section we will describe the R & D of a high-temperature heat exchanger from the viewpoint of the life prediction of structural materials at elevated temperatures.

8.3.1 Material Strength of the High-temperature Gas-cooled Reactor (HTGR) Helium-to-Helium Heat Exchanger

Figure 8.18 [11] illustrates the concept of a nuclear steelmaking system where very high temperature helium gas is used as an energy source. In this system, reduced gas (H_2/CO) is heated to 1123 K by using helium gas, and reduced iron is produced 'directly' from iron oxide pellets by this hot gas. The reduced gas is produced as follows. Asphalt is dissolved by high-temperature steam, and naphtha, pitch and fuel oil are produced. Thus the reduced gas is produced by both the steam reformation of naphtha and pitch gasification. Very high temperature helium gas

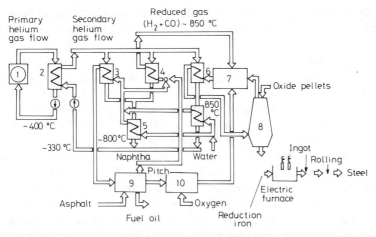

Fig. 8.18. System concept of the nuclear steelmaking process. 1, Very high temperature gas-cooled reactor; 2, He–He intermediate heat exchanger; 3, steam heater; 4, steam reformer; 5, steam generator; 6, reducing gas heater; 7, reducing gas production system; 8, shaft furnace; 9, steam dissolving equipment; 10, pitch gasification equipment. (Reprinted from ref. 11 by courtesy of Ishikawajima-Harima Industry Co. Ltd.)

(1198 K, 4·1 MPa) is supplied to the steam reformer. Thirty-three patents were obtained for this nuclear steelmaking system and there are 30 working models throughout the world, which had great impact on the European Nuclear Club (ENC) and Prototype-anlage Nukleare Prozesswarme (PNP), West Germany, and others. The system has the following characteristics:

1. Output conditions of helium gas are 1273 K and 4·1 MPa in pressure. These are the most severe conditions in the world. In order to prevent the mixture of radioactive matter into the products and maintain safety, a *helium-to-helium intermediate heat exchange system* is employed.
2. Asphalt is used as the raw material for the reduced gas.
3. A shaft furnace is used as the means of direct reduction of iron pellets.

The results of these studies were reported in an integrated report (1981) [12]. We will describe the R & D of the high-temperature intermediate heat exchanger (abbreviated IHX) in connection with the study of the life prediction method of structural material as follows.

Since the IHX used for the 1·5 MW (thermal) plant was exposed to very high temperature helium gas, a nickel-based superalloy was used for the structural material. It was necessary to develop a design guide by modifying the ASME Boiler and Pressure Vessel Code Case N-47 (1979) [13]. The maximum temperature allowed under the new guidelines was 1273 K and the material used for temperatures ranging from 1073 K to 1273 K was a nickel-based superalloy Inconel 617. Figure 8.19 [14] summarizes the problems encountered for each step of the design procedure for the high-temperature components.

The first step of the design was to determine the system structure and to select the materials for each component based on the integrity requirements. At this stage decisions were made on the basis of material data, but data concerning the *effect on the environment were scarce and insubstantial*. The second stage of the design is the stress and strain analysis of each component. At this stage it was necessary to know the deformation properties of the material. A *constitutive equation for detailed inelastic analysis* was important for determining the average behavior of the material. The third stage of the design was to evaluate the strength of the components by comparing the calculated stress or strain values with design limits. At this stage the *high-temperature failure criteria*, the material data for each failure mode and a procedure to use the material data converged in the evaluation of the strength of components. We will describe several problems concerning the high-temperature strength of structural materials for the IHX.

A Problem Associated with the Large Temperature Dependence of Yield Stress

Figure 8.20 [14] shows an example of the temperature dependence of the yield strength of some high-temperature superalloys. We can see that the temperature dependence of the yield strength is large at temperatures above 1023 K. The yield strength of austenitic stainless steel shows relatively small temperature dependence at the temperature allowed in the Code Case N-47. The *thermal ratchet* design procedure should be revised because of the large temperature dependence of the yield strength [15]. The yield strength depends also on the deformation rate, as shown in Fig. 8.21 [14]. Due to the large degree of strain rate dependence of the yield strength, especially at high temperatures, the strain rate dependent yield strength should be incorporated into the thermal ratchet design procedure [14].

TECHNOLOGY DEVELOPMENT OF METAL PLASTICITY

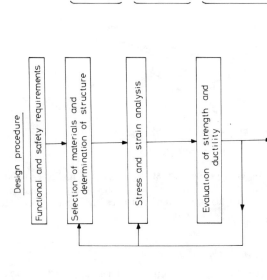

Fig. 8.19. Problems in the development of a high-temperature design code. (Reprinted from ref. 14 by permission of Amer. Soc. Mech. Engrs)

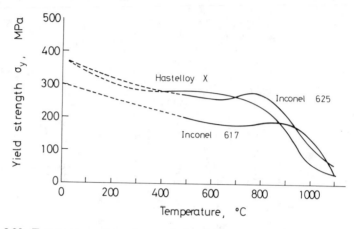

Fig. 8.20. Temperature dependence of the yield strength of three superalloys. (Reprinted from ref. 14 by permission of Amer. Soc. Mech. Engrs)

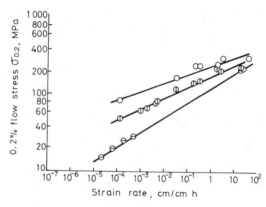

Fig. 8.21. Strain rate dependence of the yield stress of Inconel 617 at (○) 800°C, (⌽) 900°C and (⊖) 1000°C in air. (Reprinted from ref. 14 by permission of Japanese Ministry of International Trade and Industries and Amer. Soc. Mech. Engrs)

A Problem Associated with the Tertiary Creep Curves

Figure 8.22 [16] illustrates typical creep curves of Inconel 617 at 1273 K in helium gas. (The material tested was supplied by INCO.) Here we can see that, unlike the creep curves at the elevated temperatures, the creep curves of the superalloy at very high temperature are mainly repre-

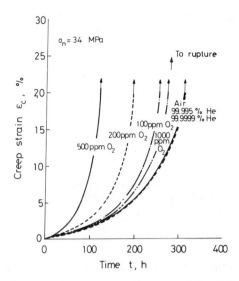

Fig. 8.22. Effect of oxygen content on creep curves of Inconel 617 at 1273 K in helium. (Reprinted from ref. 16 by permission of Amer. Soc. Metals)

sentations of tertiary creep. (The environmental effect of the oxygen content on creep as shown in Fig. 8.22 will be discussed in the next section.)

On the other hand, the point of onset of tertiary creep is used as the allowable stress such as stress limit S_t in Code Case N-47. If the purpose of determining S_t using the tertiary creep strength is to prevent the *creep instability* of structural components in high-temperature applications, then the tertiary creep strength can be replaced by some other creep properties to the extent that the new creep properties can prevent creep instability. If we choose a ratio of 1/10 or 1/5 as the fraction, then residual strain is 80% or 90% of the total elongation. Figure 8.23 [14] shows the comparison between the above-mentioned creep strength (one-tenth of total elongation), σ_δ, the 1% creep strength, $\sigma_{1\%}$, and the creep rupture strength, σ_r. The test was performed by using Inconel 617 supplied by INCO at 1173 K in air. Notice that σ_δ is between $\sigma_{1\%}$ and σ_r. Similar results were also observed for other nickel-based superalloys.

If we replace the tertiary creep strength with the elongation-based creep strength, σ_δ, then the new S_t may be determined as the lesser of the 1% creep strength, 67% (2/3) of the creep rupture strength and 80% (4/5) of the elongation-based creep strength. The advantages of using the

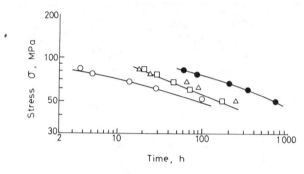

Fig. 8.23. Creep and creep rupture strength of Inconel 617 at 1173 K in air. (●) σ_r–t_r; (□) σ_{3rd}–t_{3rd}; (△) σ_o–t_o; (○) $\sigma_{1\%}$–$t_{1\%}$. (Reprinted from ref. 14 by permission of Japanese Ministry of International Trade and Industries and Amer. Soc. Mech. Engrs)

elongation-based creep strength in determining the S_t are: (1) less chance of error in reading the values from creep curves, (2) the effectiveness of utilizing the metal strength, especially of some nickel-based superalloys, and (3) a choice of the S_t limit for the metals which have a fracture elongation of less than 10%. The new S_t results in more severe limits for brittle alloys and prohibits the use of non-ductile materials.

A Problem Associated with the Environmental Effect
The consequences of the environmental effect on the creep and fatigue of metals at elevated temperatures is an important problem as discussed in Chapter 6. Our understanding of the *effect of helium gas* on the high-temperature strength of structural metals is very limited.

Figure 8.24 [16] shows the experimental results of the effect of impure oxygen in helium and a vacuum on the high-temperature strength of Inconel 617 at 1273 K. The test was successfully performed by Hosoi and Abe using a smooth round bar specimen of Inconel 617. We found that the creep rupture life is greatly reduced with an oxygen content (ppm) in helium gas or oxygen partial pressure (μPa) in a vacuum. We can also see that the creep rate increases considerably with the same oxygen content in Fig. 8.24. It should be noted that only a small change in the impurity level decreases the creep strength drastically. This reduction is closely related to the *decarburization* of the alloy as follows.

If the decarburization rate is assumed to be controlled mainly by carbon diffusion, then the following experimental results may be used

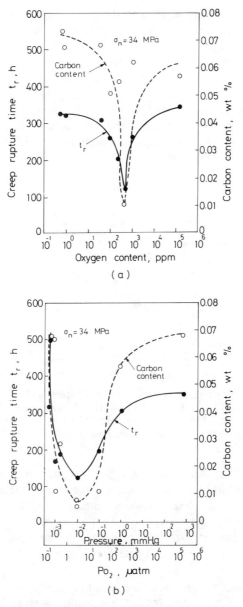

Fig. 8.24. Effect of oxygen content in (a) helium gas and (b) a vacuum, on creep rupture time and decarburization of Inconel 617 at 1273 K. (Reprinted from ref. 16 by permission of Amer. Soc. Metals)

quantitatively to predict weakening due to helium gas or a vacuum.†

The relation between the carbon content of Inconel 617 and the exposure time to high-temperature helium was obtained, and the relation between the carbon content and the *environmental strength reduction factor* (the ratio of creep strength in helium to that in air) was obtained as shown in Fig. 8.25[14]. This factor may be obtained from the following figure. Figure 8.26[14] shows the 1% creep strength in various environments. The carbon content was calculated as a function of time and the plate thickness by a diffusion equation using a fictitious diffusion coefficient obtained experimentally.

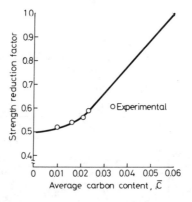

Fig. 8.25. Relation between average carbon content and strength reduction factor of Inconel 617 in 4-nine helium at 1273 K. (Reprinted from ref. 14 by permission of Japanese Ministry of International Trade and Industries and Amer. Soc. Mech. Engrs)

Figure 8.27[17a] shows the *notch effect* on the creep crack initiation time t_i of Inconel 617 both in air and in a vacuum of 40 Pa. The test was carried out by using a double-sided notched plate specimen with elastic stress concentration factors K_t of 1·9, 3·8 and 6·4. A vacuum of 40 Pa was employed as the decarburizing environment. In the figure the analytical results are also represented. These were determined from both the

---

†Inconel 617 contains both solute carbon and carbide $M_{23}C_6$ or M_6C at the grain boundaries. For example, the former accounts for about 20% of the total carbon content and the latter for about 80%. If the decarburization due to diffusion of the solute carbon occurs, the carbide $M_{23}C_6$ is resolved as $M_{23}C_6 \to (23M + 6C)$ in order to maintain the chemical equilibrium.

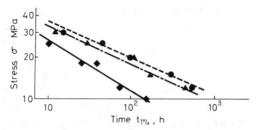

Fig. 8.26. Effect of helium gas atmosphere on the 1% creep strength of Inconel 617 at 1273 K; 1% strain in (●) air, (▲) pure helium, (♦) 4-nine helium. (Reprinted from ref. 14 by permission of Japanese Ministry of International Trade and Industries and Amer. Soc. Mech. Engrs)

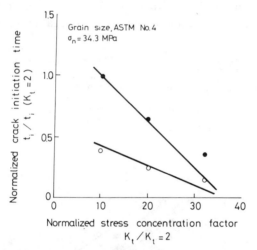

Fig. 8.27. Effect of the elastic stress concentration factor, K_t, on the creep crack initiation time, t_c, of Inconel 617, in (●) the atmosphere and (○) a vacuum of 40 Pa at 1273 K. – denotes analytical result. (Reprinted from ref. 17a by permission of Japan Soc. Mech. Engrs)

diffusion equation and the relationship between the carbon content and the creep rate of the alloy tested. In the analysis one of the criteria of crack initiation is used where the crack occurs when the local strain at the notch root reaches a critical magnitude. We can see that the analytical result was in agreement with the experimental result. It is suggested that the decrease in the crack initiation life of the alloy due to the increase in local creep rate at the notch root in the vacuum is mainly a result of the decarburization in a vacuum environment lacking oxygen.

A similar effect was also observed in the creep crack propagation. As shown in the experimental notch opening displacement, L_{NOD}, versus time curves of the notched specimens in both the atmosphere and a vacuum of 40 Pa in Fig. 7.40 in the previous chapter, we found that the reason the crack initiates earlier in 40 Pa than in air is due to a higher rate of L_{NOD}, of which critical L_{NOD} is nearly independent of the environment but dependent on K_t. On the other hand, we found from Fig. 7.40 that the creep crack propagation rate is well represented by a single curve on the basis of a creep J-integral J_c independent of the environment. In Fig. 7.41, J_c increases as the local creep rate increases, while in 40 Pa the rate of increase is greater than that exhibited in air. Thus we concluded that the decrease in the creep rupture life of Inconel 617 at 1273 K in the decarburizing vacuum environment of 40 Pa is a result of both the early initiation and fast propagation of the crack.

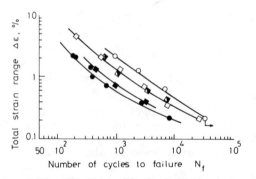

Fig. 8.28. Push-pull low-cycle fatigue life of solution heat treated and carburized Inconel 617 at 1273 K in a vacuum, 4-nine helium and air. (Reprinted from ref. 18 by permission of Japanese Ministry of International Trade and Industries)

On the other hand, Fig. 8.28 [18] shows the results of a push-pull strain-controlled low-cycle fatigue test conducted under several environments. The tests were carried out by using a smooth round bar specimen of Inconel 617 supplied by INCO at 1273 K. We can see from this figure that the failure lives are largest in a vacuum of 6.7×10^{-3} Pa, and that the life in helium gas with a CH_4 level of about 1000 ppm is nearly equal to that in 99·995% helium gas and falls between the lives in air and in a vacuum. The environment of the HTGR is most likely to be helium gas with carburizing elements. It is known that corrosion of the alloy changes from decarburizing to carburizing if the CH_4 content in the helium gas increases. We found that the effect of a vacuum or helium gas on the high-temperature low-cycle fatigue is quite the contrary in the case of creep rupture (see Chapter 7). In air, fatigue cracks were seldom observed, but in other environments cracks were frequently observed. The majority of the cracks were of the intergranular type, and there was a decarburizing layer on the surface of the specimen in helium gas which was not present on the specimen in the gas containing CH_4. Therefore we can see that it was advantageous for the prevention of decarburization of the superalloy.

8.3.2 Lifetime Test of a Partial Model of the HTGR Helium-to-Helium Heat Exchanger under Creep, Fatigue Interaction Conditions and Various Environments and the Life Prediction Method

The above-mentioned design guide was intended to cover the conceivable fracture modes of the HTGR high-temperature components and to include a safety margin that would assure the total safety of the designed components.

However, since no prior data or experience in developing IHX components exists, it is conceivable that some unexpected failure mode may occur at the design stage. Furthermore, since the life prediction of the high-temperature components is usually difficult, especially under interaction conditions between creep, fatigue and the environment of helium gas flow, predicting the necessary safety margin of the proposed design guide is extremely difficult. To increase the reliability of the guide, a *lifetime test of the to-scale model* of the IHX components was needed.

We will describe some lifetime tests of the partial heat exchanger center pipe and tube models briefly. The center pipe and tube models were selected for the lifetime tests because the preliminary strength analysis revealed that the highest temperature portion of the center pipes and the heat exchanger tube assembly are subjected to severe damage during the life test.

Mock-up Model and the Lifetime Test
Figure 8.29 [15, 19] is a schematic of the *mock-up model* tested. The

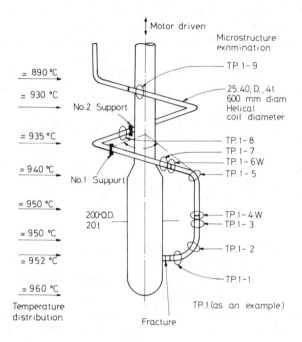

Loading

Tested tube number	Differential pressure p, MPa (kgf/cm²)	Support location θ, deg
1	0·98(10)	117·25
2	0·98(10)	117·25
3	2·94(30)	163·25
4	5·88(60)	140·25
5	8·83(90)	140·25
6	7·85(80)	207·25
7	0·98–10·78(10–110)	297·25
8	0·98–10·78(10–110)	320·25

Fig. 8.29. Loading condition of a mock-up test model. (Reprinted from refs 15 and 19 by permission of Japanese Ministry of International Trade and Industries and Amer. Nuclear Soc., respectively)

model consists of eight 600 mm diameter helical coils (25·4 mm outer diameter and 4 mm thick tubes) and 200 mm diameter and 20 mm thick center pipes made of Inconel 617. The eight tubes were welded to the stubs† machined on the lower part of the center pipe, and the outer ends were fixed to the tube supports connected to the vessel wall. The center pipe is guided and is capable of being electrically driven along the vessel axis (150 mm maximum stroke δ, $\dot{\delta}=0$–25 mm/h). This model was a scaled construction of the partial heat exchanger center pipe and tubes, and the lifetime test was performed under accelerated loading conditions in the HTL-40 helium test loop located at Isjhikawajima-Harima Heavy Industries.

To apply hoop stress to the tubes, 99·995% helium gas was charged to the predetermined pressure. The pressure for each tube was varied to produce a variety of damage conditions. The load conditions of each tube are summarized in the table in Fig. 8.29. All of the employed load conditions were set to be more damaging than load conditions of actual components in order to accelerate the failure of the mock-up model. The deflection stress of tubes 1 and 2 was higher than the actual stress. In tubes 3 and 6, both the deflection stress and the internal pressure stress were set higher. The load conditions for tubes 7 and 8 represent the faulted conditions.

The load history model can be seen in Fig. 8.30 [19]. First, several cycles under a 50 mm stroke were applied to confirm that the loading equipment was functioning properly. The helium gas temperature was controlled by an electric heater to maintain the metal temperature of the lowest portion of the helical coil at 1223 K.

The fractured model is shown in Fig. 8.31(a) [19]. The fracture occurred at the stub portion of the center pipe as shown in Fig. 8.31(b) [19] (not on the welded joint between the stub and the tube). Tubes 1 to 5 failed and leaked during the test at the times shown in Fig. 8.30. The times shown are the accumulated time at the maximum operating temperature of 1223 K. For tube 6 the cracks were found on the inside surface of the tube. Tubes 7 and 8 did not fail even after prolonged exposure to 14·7 MPa.

The damage to other parts of the tested model, in general, was small. A slight amount of decarburization was observed on the outside and inside

---

†The stub and the attachment nozzle were machined on the outer surface of the center pipe. Each stub was blanked off to maintain different internal pressures from one tube to another.

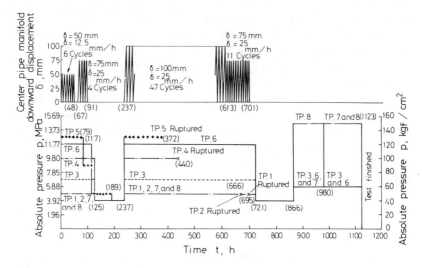

Fig. 8.30. Loading histories of tested heat exchanger tubes; the numbers in parentheses denote elapsed time. (Reprinted from ref. 19 by permission of Japanese Ministry of International Trade and Industries and Amer. Nuclear Soc.)

surfaces of the tubes. An internally oxidized layer was observed near the inside surface of the tubes. This may have been due to the fact that the helium inside the tubes was not flowing, and as a result the partial oxygen pressure was low enough to cause the internal oxidation. This internal oxidation does reduce the material ductility and therefore may be related to the crack initiation from the inside surface of the tubes.

All of the welded joints between the tubes and stub/tubes were found to be sound after the test. The oxidation on the surface of the tubes under the tube support was thinner than at the outer part for the same temperature, and a slight amount of sliding wear was observed. Both of these were considered to be insignificant.

Life Prediction

In order to evaluate the accuracy of the life prediction method, the lives of the five fractured tubes were calculated using the following life prediction criteria and structural analysis method.

The amount of damage due to diametral expansion of the tube is so large that the linear creep–fatigue damage summation rule of the Code

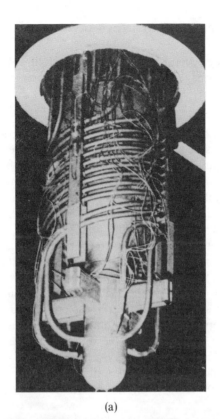

(a)

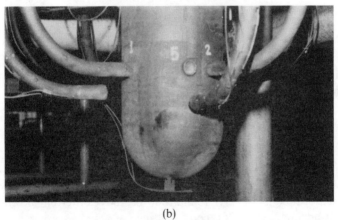

(b)

Fig. 8.31. General appearance of the fractured model: (a) overview; (b) central pipe manifold. (Reprinted from ref. 19 by permission of Japanese Ministry of International Trade and Industries and Amer. Nuclear Soc.)

Case N-47 cannot be used here. Therefore the following failure criterion, which includes damage due to strain accumulation, was employed:

$$\phi_f + \phi_c + \phi_D = 1 \tag{8.17}$$

where ϕ_f and ϕ_c are fatigue and creep damage, respectively, as in the Code Case N-47, and ϕ_D is the damage due to unidirectionally accumulated strain. Usually the creep damage is calculated by a time ratio on the basis of the static creep rupture time. However, under cyclic strain conditions the cyclic creep rupture time is more adequate than the static creep rupture time. Therefore the creep damage should be calculated on the basis of the cyclic creep rupture time.

On the other hand, the cyclic creep rupture time is obtained by repeating the creep test in a fixed strain range, and therefore the cyclic creep rupture property includes some fatigue damage. In addition, the number of strain cycles in the lifetime test was small. Under these conditions, the first term of eqn (8.17) was considered to be included in ϕ_c. This is a creep–fatigue interaction and the ϕ_f was neglected.

The third term may be called the ductility exhaustion damage and unidirectional creep rupture ductility, ε_r, may be used as a base. To cover the unidirectional creep failure as well as the cyclic creep failure, the ductility exhaustion damage ϕ_D can be written as

$$\phi_D = \frac{t_{rc} - t_{rs}}{t_{rc}} \cdot \frac{\Delta \varepsilon}{\varepsilon_r} \tag{8.18}$$

where t_{rc} and t_{rs} are respectively the cyclic and static creep rupture times under a given stress. For Inconel 617 the cyclic rupture time is approximately ten times as large as the static creep rupture time; the damage criteria may be written as [15]

$$\int \frac{dt}{t_{rc}} + 0.9 \int \frac{d\varepsilon}{\varepsilon_r} = 1 \tag{8.19}$$

Figure 8.32 [19] shows the results of the strain-controlled low-cycle fatigue test with an increasing mean strain for a small round bar specimen of Inconel 617. We found that the above criterion shown in the figure is adequate for this type of loading.

To formulate an accurate method of predicting the life of the mock-up model, the finite element method with three-dimensional solid elements may be employed to analyze details of the structural behavior including

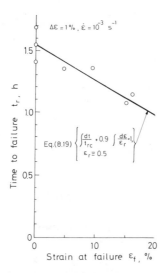

Fig. 8.32. Comparison of experimental (o) and predicted (—) lives under cyclic loading with increasing mean strain for Inconel 617 at 1273 K in air. (Reprinted from refs 15 and 19 by permission of Japanese Ministry of International Trade and Industries and Amer. Nuclear Soc., respectively)

the analysis of the crack propagation. However, this kind of approach is usually impractical because it takes too much computer time. Here a simplified analysis method was proposed and utilized under a justified assumption. The outline of the method is described as follows [20].

First, the creep behavior of the model was analyzed under unidirectional center pipe motion. The tested tubes (25·4 mm outer diameter, 4 mm thickness) behave as a beam because the ratio of thickness to diameter is high enough. However, for the creep deformation analysis, the effort of radial and tangential stresses caused by internal pressure on the creep behavior of the tube should be calculated. For this purpose, an incremental stress/strain relationship of the three-dimensional solid element was obtained using the stress distribution of the thick-walled cylinder under internal pressure. This relation was modified to fit the strain field of the beam theory, and was included in the beam elements of a general-purpose three-dimensional thermoelastic-plastic creep analysis program, *ISTRAN/3D* [21].

Using this program, the creep behavior of the helical coil structure under loading conditions was analyzed. The results of the calculation revealed that (1) the maximum strain occurs at the stub portion of the center pipe, and (2) the main strain is axial strain caused by bending deformation of the tube. The maximum strain rate of the stub after the steady-state creep was found to be proportional to the center pipe displacement rate and to be independent of internal and external pressures. This finding was used to estimate the axial strain rate of the stub under cyclic straining of the center pipe. On the other hand, the distribution of the radial and tangential stresses can be approximated by the stress distribution for the thick-walled cylinder. Furthermore, the stress and strain resulting from these stress conditions can be easily calculated using the von Mises flow rule; consequently the creep damage may be determined from the stress history calculated in this manner.

To calculate the ductility exhaustion damage, ϕ_D, an extensive strain analysis may be necessary if one wants to be exact. However, a simplified method was employed. First, the deformation was assumed to be axially symmetric, and then the tangential expansion of the tubes was calculated using the tangential strain calculated for triaxial stresses and the incompatibility assumption. Finally, the damage ϕ_D was calculated by considering the changes in the outer and inner diameters.

The predicted lifetime obtained using this analysis method under the fracture criteria and the material constants are summarized in Table 8.1 [15]. The results clearly show that in tubes 1 and 2, where the thermal expansion stress is high, the creep damage ϕ_c is large. In tube 5, where internal pressure is high, the ductility exhaustion damage ϕ_D is the life-controlling factor. The life was calculated for crack initiation at the outer and inner surfaces of the tube. The total life calculated for the leakage of

Table 8.1
Results of Simulated Calculations (Reprinted from refs 15 and 19 by permission of Japanese Ministry of International Trade and Industries and Amer. Nuclear Soc., respectively)

T.P. number	Simulated calculation						Experimental rupture time, $(t_f)_{exp}(h)$
	Outer surface			Inner surface			
	ϕ_c	ϕ_D	Failure time, $(t_f)_o$ (h)	ϕ_c	ϕ_D	Failure time, $(t_f)_i$ (h)	
1 and 2	0·620	0·380	793	0·652	0·348	1 033	695 and 666
3	0·359	0·641	787	0·404	0·596	1 117	>700
4	0·209	0·891	421	0·245	0·755	557	>400
5	0·153	0·847	341	0·184	0·816	445	372

the tube is expected to be between the calculated lives of the inner and outer surfaces because there is only a small damage gradient throughout the thickness of the tube wall.

We can see that the predicted lifetime is in good agreement with the experimental failure time t_f shown in Table 8.1. Although the calculation shows that the outer surface fails earlier than the inside surface, the actual failure occurred from the inside surface. As noted above, this may have been caused by internal oxidation on the inside surface.

Because this lifetime test was performed under enhanced loading, it is difficult to meet the design requirements specified in the guide. However, the safety margin of the guide may be estimated by calculating the creep life or creep–fatigue life on the bases of the guide method and comparing them with the experimental life.

The creep damage limit $\Sigma t_c/t_{mi} \leqslant 1$ for primary (hoop) stress was evaluated for tubes 4 and 5 in Table 8.1, because these tubes failed mainly under primary stress. The results are summarized in Table 8.2 [19] and the safety margins are 8·3 and 90, respectively. In addition to the safety margin of these limits, other proposed rules were checked here. From the results of these calculations it was found that the most likely fracture mode was the creep–fatigue fracture and that the minimum safety margin of the proposed guide is more than 200 in each lifetime as shown in Table 8.2. Generally speaking, life prediction requires more scientific precision than is found in the design guide. The present mock-up model tests confirmed the appropriateness of the life prediction method.

Table 8.2
Safety Margin of Proposed Code (Reprinted from refs 15 and 19 by permission of Japanese Ministry of International Trade and Industries and Amer. Nuclear Soc., respectively)

Limit	Tube number	Allowable time (h)	Time to failure (h)	Safety margin
$\Sigma(t_c/t_{mi}) \leqslant 1$	4	40	330	8·3
For primary stress	5	2·4	216	90
$\phi_f + \phi_c \leqslant 1$	1	1·0	584	584
	4	1·0	330	330
	5	1·0	215	215

8.3.3 Construction of a 1·5 MW (thermal) He–He Intermediate Heat Exchanger

A 1·5 MW (thermal) *helical-coil, one-through counter flow type helium-to-helium heat exchanger* was selected by the Expert Committee on Heat Exchangers (chairman Dr. Y. Mori, a professor at the Technical University of Tokyo), the Expert Committee on Heat Resistant Alloys (chairman Dr. R. Tanaka, a professor at the Technical University of Tokyo) and the Special Committee on the Strength of Heat Exchangers (chairman as mentioned previously). A nickel-based superalloy of Inconel 617 was employed as the high-temperature heat transfer tube material for temperatures above 1073 K and Incoloy 800 alloy and 2·25%–Cr–1% Mo steel were used in the intermediate or lower temperature materials, respectively. Figure 8.33 [15] illustrates the general arrangement and specification of the heat exchanger and a general view of its tube bundle.

Here 1273 K/3·9 MPa primary helium gas flows upward from the primary helium inlet on the bottom of the heat exchanger and proceeds downward to the bottom outlet along the outside of the tubes, and 600 K/4·1 MPa secondary helium gas flows in the opposite direction to the primary flow. The structural arrangement is a type of *manifold* consisting of an assembly of tubes. Stress or thermal expansion of the components is designed to be uniformly relaxed due to the relatively small difference in the thermal expansion of the tubes and the axisymmetrical arrangement of the high-temperature components. Both the stress and strain of each structural component were analyzed under each operating condition of thermal load, self weight, internal and external pressures, and earthquake in the states of normal, upset and emergency, respectively. Cooperative and integrated research efforts have resulted in a successful load test of the system in helium gas at 1273 K which received much attention from other countries. The design guide of the high-temperature heat exchanger is especially indebted to advanced studies of high-temperature material strength and structural analysis.

8.3.4 New Developments in Heat-resisting Superalloys

As specified by the Japanese government, the He–He IHX tube material must have a 100 000 h creep rupture strength above 9·8 MPa in helium gas at 1273 K and must be plastically worked into a tube having a 25 mm outer diameter, 5 mm thickness, and length of 7 m. The majority of heat-resisting superalloys have been developed for use in jet engine parts, and hence the period following the 1940s was marked by rapid development.

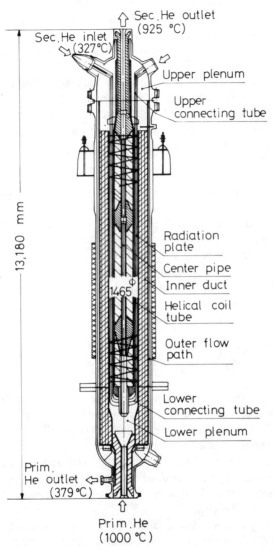

Fig. 8.33. General arrangement of 1·5 MW (thermal) He–He intermediate heat exchanger. (Reprinted from ref. 15 by permission of Japanese Ministry of International Trade and Industries)

In the 1970s, when the present project started, several alloys with a 100 000 h creep rupture strength above 9·8 MPa at 1273 K had already been developed. However, the creep rupture strength of Inconel alloy 617 used in the tube material was estimated as 6·9 MPa at most, and Hastelloy X was also established to have an even lower creep rupture strength.

When the development of a nuclear steelmaking system was discussed in 1977, three types of high melting point alloys, iron-based wrought alloy, and nickel-based wrought alloy were selected. These candidate alloys were selected in order of priority with the following development goals, in addition to the target of 100 000 h creep strength above 9·8 MPa at 1273 K:

1. Creep rupture property in air, helium and reduced gas
2. Hot corrosion
3. Hydrogen permeation
4. Welding
5. High-temperature short-term tensile properties
6. High-temperature fatigue properties
7. Creep–fatigue interaction
8. Plastic working
9. Resistance to carburization
10. Heat conductivity
11. Surface treatment
12. High-temperature impact properties

At the beginning of 1973, Japanese steelmakers developed several alloys which they proposed for use and submitted to the Science and Technology Agency, Japan, for examination. These alloys were examined using the above-mentioned criteria at the National Institute of Research of Materials (NIRM) and by the steelmakers. The steelmakers also tested the tube production and examined the welding, bending-working, and surface treatment. Five alloys were accepted by the agency. The chemical compositions are shown in Table 8.3 [15]. In 1975 the five alloys were once again evaluated and the public subscription was reopened with the exception of R4286 alloy previously selected. As a result, six new alloys were selected as shown in Table 8.4 [15] and the development of the seven alloys was continued. Through the second (1977) and the third (1980) evaluations both *SSS113MA* and *KSN* were the final selections as the candidate alloys.

Table 8.3
Chemical Compositions (wt %) of Alloys Selected in 1973 (Reprinted from ref. 15 by permission of Japanese Ministry of International Trade and Industries)

	C	Ni	Cr	Mo	W	Nb	Fe
NSC-1	0·06	Bal.	18·0	0·50	15·0		
SZ	0·03	Bal.	27·0	5·5	5·5		
KSN	0·02	Bal.	16·0		26·0	0·4	
SSS 113 MA	0·02	Bal.	23·0		18·0		
RS-513	0·20	30·0	20·0	3·0	1·0	1·0	Bal.
MA-X7	0·08	Bal.	20·0		20·0		

Table 8.4
Chemical Compositions (wt %) of Alloys Selected in 1975 (Reprinted from ref. 15 by permission of Japanese Ministry of International Trade and Industries)

	C	Ni	Cr	Mo	W	Co	Fe	Al	Ti
SA	0·25	25·0	25·0				Bal.	0·30	0·35
SB	0·25	31·0	25·0		1·7		Bal.	0·10	0·10
CNF	0·38	35·0	26·0		7·0	14·5	15·0		
SSS 410	0·05	Bal.	16·0		20·0	30·0			0·50
R 4286	0·05	Bal.	18·0	4·0	6·0	10·0		2·00	2·50

Table 8.5
1273 K/50 000 h Creep Rupture Strength (MPa) of Alloys (Reprinted from ref. 15 by permission of Japanese Ministry of International Trade and Industries)

		KSN	113MA	R4286	Inconel 617
In air	Larson–Miller method	13·4	13·0	12·9	10·0
	Extrapolation method	12·5	10·9	12·4	8·6
In helium	Extrapolation method	11·1	11·8	9·5	—

Table 8.5 [15] demonstrates the 50 000 h creep rupture strength of two alloys in helium gas at 1273 K compared with that of Inconel 617. The chemical content (ppm) of helium gas was as follows: H_2 300, H_2O 3, CO 100, CO_2 1, CH_4 44, N_2 5. The flow rate was 300 ml/min. We can see

that the long-term creep rupture strength of the new alloys was more than 9·8 MPa in air, which exceeds that of Inconel 617. On the other hand, in helium the SSS113MA alloy demonstrated an equivalent strength to that in air and it is estimated that the strength in helium gas flow was close to the target of this project as well as to that of KSN alloy.

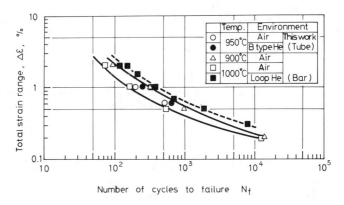

Fig. 8.34. Comparison between high-temperature low-cycle fatigue lives of solid cylindrical bar specimens of SSS113MA alloy and the heat exchanger tube. (Reprinted from ref. 21)

Figure 8.34 [21] shows an example of the symmetrical fast-fast strain-controlled low-cycle fatigue data for SSS113MA at high temperatures in air and in B-type helium gas. The comparison between the data of the solid cylindrical bar specimen and those of the heat exchanger tube is also shown in this figure. The strength of the tube at 1223 K is between the strengths of the solid bar specimen at 1173 K and 1273 K respectively. We found that the failure life of SSS113MA under the fast-fast strain wave form in helium gas is larger than that in air as well as the other nickel-based superalloys SZ, NSC-1, Hastelloy XR and Inconel 617. These interpretations were made based on the examination of the specimen surface oxidation behavior of the alloys. The failure lives under an unsymmetrical strain waveform of slow-fast were markedly smaller than those under the symmetrical strain waveform of fast-fast but there was no difference in the failure lives in both air and helium. Since creep damage of these alloys controls mainly slip accumulation on the surface

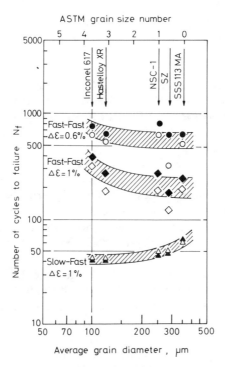

Fig. 8.35. Data from high-temperature low-cycle fatigue test, in (open symbols) air and (closed symbols) B type He, of heat exchanger tubes for HTGR, on the common comparison basis of the grain size. (Reprinted from ref. 21)

and in the interior of the alloys, we can estimate that it will hasten crack initiation.

Figure 8.35 [21] shows the low-cycle fatigue lives of Inconel 617, Hastelloy XR, NSC-1, SZ and SSS113MA nickel-base superalloys on the common comparison basis of the effect of the average grain size. We find that when alloys with a fine grain are subjected to a fast-fast symmetrical strain wave they exhibit a longer failure life compared with those alloys with a coarser grain. In contrast, the coarse-grained alloys subjected to a slow-fast unsymmetrical strain wave exhibited a longer life than those with a fine grain.

These new alloys are derived from a Ni–Cr–W base. Nowadays there is no harmful precipitation such as σ-phase even if compositions of Cr and W are increased; on the contrary, an effective W-based precipitation (body centered cubic crystal) is observed. Effective $M_{23}C_6$ and harmful M_6C have also been studied.

8.4 COLLABORATIVE R AND D EFFORTS ON THE RELIABILITY OF STRUCTURAL MATERIALS FOR HIGH-TEMPERATURE APPLICATIONS IN JAPAN

8.4.1 The Total Plan

Industrial plants such as fossil fuel power plants and chemical plants, and public facilities such as bridges, which are larger in size and higher in effectiveness, have been designed for long-term use. The structural materials used in these applications are exposed to severe environments and must maintain long-term integrity. Some fossil fuel power plants are getting older, and strong interest has recently been placed on the *prediction of the residual life of the structural materials.*

In 1982 a national feasibility study to determine the research program was made by a collaborative effort of the industries involved, the universities and the government, under the sponsorship of the Science and Technology Agency, Japan [22]. In this feasibility study the following matters were investigated: (1) the service environment and damage/deterioration of structural metals in fossil fuel power plants, chemical plants and public facilities such as bridges, (2) updating measurement technology used to assess material damage, and (3) the life/remaining life prediction method. This national collaborative research was started in 1983 and the Expert Group on the High Temperature Strength of Structural Materials (chairman the author) aimed at solving the following problems:

1. Clarification of the damage/fracture mechanism of structural metallic materials in high-temperature and corrosive environments, and of material damage/crack growth laws and life prediction.
2. Development of measurement technology used to assess the damage and cracks in structural metallic materials.
3. Development of analytical evaluation methods for the damage/life of structural metallic materials and establishment of a data base concerning life/remaining life prediction.
4. Reliability evaluations of structural metallic materials by using the technologies developed in problems 1–3.

The reliability study of structural metals at elevated temperatures in this project aims at performing the following tasks:

1. Applying the laws of damage/crack growth in structural metals used in steam boiler tube/pipe and steam turbine rotors and newly developed measurement technology; a practical prediction method of life/remaining life is proposed.
2. Metallurgical and micromechanical factors of material damage, crack behavior of mechanical and thermal fatigue in the creep range, and mechanical factors controlling the life, such as multi-axiality of stresses, size effects and others, are to be studied from a viewpoint of the precision and expansion of the application of the life prediction method.
3. The development of new methods for assessing material damage, and of a special apparatus for testing structural metals used in high-temperature applications.

In the following pages the essence of this collaborative effort, which began in 1983, is reported [23].

8.4.2 Research Concerning the Damage of Structural Metals used in Very Long Term Applications

The material properties, metallurgy and corrosion resistance of boiler tubes used for 100 h and 10 000 h were investigated for specimen alloys: 9Cr–1 Mo steel (STBA26), 2·25 Cr–1 Mo steel (STBA24), 18 Cr–8 Ni steel (SUS 304HTB) and 18 Cr–8 Ni–Ti steel (SUS 321HTB). The results are as follows.

(1) Characteristic material change during a very long term was caused only by metallurgical changes and no disparity between portions of the structure was observed.

(2) High-precision measurements of density and metallographical observations of the metal surface by means of replicas were made and no remarkable damage was observed.

(3) The long-term service environment was estimated from the high-temperature corrosive behavior on the outer surface and from oxidation on internal surfaces.

For categories (1) to (3) a service temperature of 833 K to 873 K was estimated, and it was possible to predict no remarkable variation in the service environment. The creep rupture life of the tube used for a long term was shorter than that of the virgin tube and the life fraction to rupture life was 70–80%. The longer the service, the sooner creep damage occurs due to creep voids.

8.4.3 Fundamental Research Performed from a Viewpoint of the Precision and Expansion of the Application of the Remaining Life Prediction Method

Quantitative Evaluation Method of Estimating Creep Damage/Rupture Life

(1) Creep damage tests of Cr–Mo–V steel, 2·25 Cr–1 Mo steel, and types 304 and 316 stainless steel used for fossil fuel power plants were performed and a creep fracture mechanism diagram of these metals was made. Using a density measurement apparatus with a precision of 10^{-5} in this study, the change in density, $-\Delta\rho/\rho$, correlated well with the parameter $\varepsilon_c t \sigma^3 \exp(-234\,000/RT)$, where ε_c is creep strain, t is creep time, σ is creep stress and T is absolute temperature, as shown in Fig. 8.36 [23–25]. The half-value breadth, β, in an X-ray reflection was also

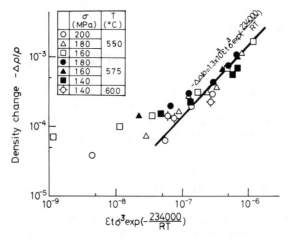

Fig. 8.36. Representation of density change, $\Delta\rho/\rho$, in Cr–Mo–V crept steel, by use of the parameter proposed. (Reprinted from ref. 25)

correlated with the above parameter, as shown in Fig. 8.37 [23–25]. Figure 8.38 [23–25] shows an example of a creep damage diagram of Cr–Mo–V steel at 823 K in air, in which the half-value breadth, β, and the density change, $-\Delta\rho/\rho$, were specified as the creep damage.

(2) As mentioned previously in Fig. 7.57 in Chapter 7, microstrain, $\langle \varepsilon^2 \rangle^{1/2}$, and particle size, D, of type 304 stainless steel analyzed from an X-ray profile analysis correlated well with the damage factors D_c, D_f and

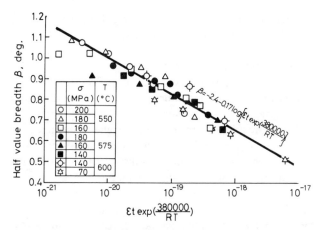

Fig. 8.37. Representation of half-value breadth, β, in Cr–Mo–V crept steel, by use of the parameter proposed. (Reprinted from ref. 25)

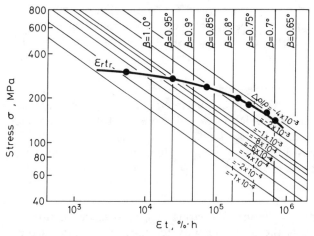

Fig. 8.38. Creep damage diagram for Cr–Mo–V steel at 823 K in air. (Reprinted from ref. 25)

D_{cf} under static creep, pure fatigue and creep–fatigue interaction conditions respectively, where D_c, D_f and D_{cf} are defined, based on a linear life fraction rule, as $D_c = 1 - (t/t_r)$, $D_f = 1 - (N/N_f)$ and $D_{cf} = 1 - (N/N_{cf})$.

(3) In laboratory tests, a decrease of acoustic velocity in an ultrasonic measurement correlated well with that in density due to creep voids, but

in 2·25 Cr–1 Mo steel and 18 Cr–8 Ni–Ti steel used long-term, almost no change was observed compared with virgin materials. However, in the crept metals the noise energy value was effective for the creep damage evaluation of the metals.

(4) From the creep tests of 1 Cr–1 Mo–0·25 V steel (used for steam turbine rotors) and type 304 stainless steel which were aged or stress-aged in the long term, a marked decrease in the creep resistance of 1 Cr–1 Mo–0·25 V steel resulted from the inhomogeneous structural change in subgrains near pre-austenitic boundaries, and that of type 304 steel was estimated from the formation of an inhomogeneous domain without carbide precipitation near the grain boundaries.

Quantitative Evaluation Method of Creep–Fatigue Damage/Life

(1) Alternation tests of creep (stress-controlled) and fatigue (strain-controlled) for type 304 and 316 stainless steels and 1 Cr–1 Mo–0·25 V steel were performed and the damage mode of these metals correlated well with those under creep conditions. Therefore we clarified that the relation of creep damage, $\phi_c = \Sigma(t/t_r)$, to fatigue damage, $\phi_f = \Sigma(N/N_f)$, was influenced by the static creep fracture mode of the metals. A good correlation between the decrease of failure life due to the hold-time at peak tensile strain and the fraction of intergranular facets was also observed.

(2) It was clarified that the damage mode of metals in the alternation tests under the creep–fatigue interaction condition was estimated and the ϕ_c versus ϕ_f relation was also predicted if the creep fracture mode was known. Thus two kinds of life prediction method were proposed. One is based on the following creep ductility, δ_c, modified creep–fatigue life equation [26]:

$$(\Delta\varepsilon_{in}/\delta_c)N_f^n = C \tag{8.20}$$

where $\Delta\varepsilon_{in}$ is the inelastic strain range, N_f is the number of cycles to failure, δ_c is the creep ductility for $t_r = t_H N_f$, and t_H is the hold-time. Figure 8.39 [26] shows the experimental relation between the inelastic strain range normalized by δ_c and the failure life of austenitic stainless steels and NCF 800 alloy at 873 K in air. Therefore, if ϕ_c for rupture time t_r is known, we can predict the long-term creep–fatigue life. The other type of prediction method is a procedure based on the creep fracture mode as shown schematically in Fig. 8.40 [27]. Considering the fracture mode of the metals during creep deformation, we can predict the low-cycle fatigue failure life of the metals by using the following ϕ_c

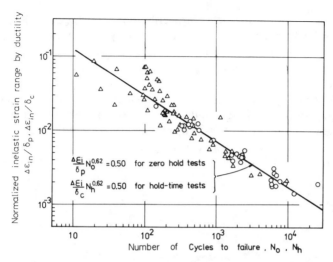

Fig. 8.39. Experimental relation between rupture ductility modified inelastic strain range, $\Delta\varepsilon_i/\delta_p$, $\Delta\varepsilon_i/\delta_c$, and failure life, N_f, of austenitic stainless steel and NCF 800 alloy at 873 K in air: ○, zero hold tests; △, hold-time tests. (Reprinted from ref. 26 by permission of Iron and Steel Inst. Japan)

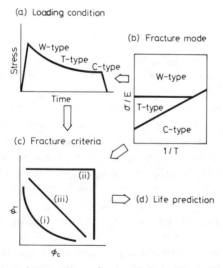

Fig. 8.40. A proposed procedure of creep–fatigue life prediction in which the creep fracture mechanism is considered. (Reprinted from ref. 27 by permission of Soc. Mater. Sci., Japan).

versus ϕ_f relations: (i) the failure criterion $\phi_c + A(\phi_c \phi_f)^{1/2} + \phi_f = 1$ or $\phi_c + \phi_f < 1$ which corresponds to the wedge-type fracture mode, (ii) the criterion $\phi_c = C$ or $\phi_f = C'$ which agrees with the transgranular fracture mode, and (iii) the criterion $\phi_c + \phi_f = 1$ which relates to the cavity-type fracture mode.

(3) For SUS 304 stainless steel, both remaining creep rupture life and remaining low-cycle fatigue failure life diagrams were drawn by using stepwise variational load tests. The remaining creep–fatigue life diagram was also drawn from both remaining creep and fatigue diagrams by using criterion (i) or (iii) mentioned above. Figure 8.41 [28, 29] shows the

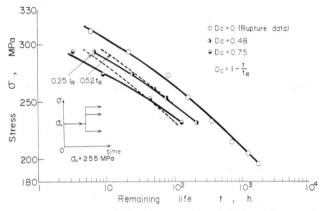

Fig. 8.41. Creep versus remaining life curves of type 304 stainless steel at 873 K in air. (Reprinted from ref. 28 by permission of Soc. Mater. Sci., Japan)

remaining creep rupture life diagram at 873 K in air, where the creep damage factor, ϕ_c, is calculated from $1 - (t/t_r)$. Figure 8.42 [28, 29] also shows the remaining low-cycle fatigue life diagram at the same temperature, where the fatigue damage factor, ϕ_f, is calculated from $1 - (N/N_f)$. Figure 8.43 [28, 29] shows the remaining fatigue life diagram in the tests with a hold-time of $t_H = 30$ min at peak tensile strain for a creep–fatigue damage factor of $D_{cf} = 0$ and 0·5. In the figure the solid curves illustrate calculations determined from both Figs 8.41 and 8.42 by using the failure criterion of (i), $\phi_c + \phi_f = 0.6$. The calculated residual life was in good agreement with the experimental life. If a small specimen cannot be extracted from the structural materials, the remaining life diagram cannot be used for life prediction and non-destructive examinations such as those in Fig. 7.57 must be adopted.

(4) Regarding macroscopic through-cracks which are larger than

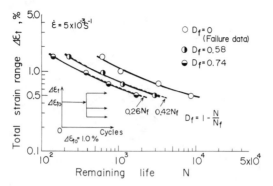

Fig. 8.42. Low-cycle fatigue versus remaining life curves of type 304 stainless steel at 873 K in air. (Reprinted from ref. 28 by permission of Soc. Mater. Sci., Japan)

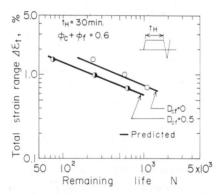

Fig. 8.43. Estimation of creep–fatigue versus remaining life curves of type 304 stainless steel, from both the creep versus remaining life curves (Fig. 8.41) and fatigue versus remaining life curves (Fig. 8.42), by use of linear damage summation rule. (Reprinted from ref. 28 by permission of Soc. Mater. Sci., Japan)

about 1 mm, two kinds of propagation behavior were observed as shown schematically in Fig. 8.44 [30]. These can be written by

$$dl/dN = C_f (\Delta J_f)^{m_f} \qquad (8.21)$$

$$dl/dt = C_c (J_c)^{m_c} \qquad (8.22)$$

$$dl/dN = C_c \Delta J_c \quad (m_c \simeq 1) \qquad (8.23)$$

where ΔJ_f is the fatigue J-integral range, J_c is the creep J-integral (C^* parameter), $\Delta J_c = \int_0^t J_c \, dt$ is the creep J-integral range, and C_f, m_f, C_c, m_c

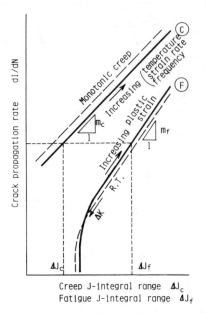

Fig. 8.44. Schematic representation of correlation between crack propagation rate, dl/dN, time-dependent J-integral range, ΔJ_c, and cycle-dependent J-integral range, ΔJ_f. (Reprinted from ref. 30 by permission of Iron and Steel Inst. Japan)

are material constants. Equation (8.21) includes a dl/dN versus ΔK relation for high-cycle fatigue and usually $1 < m_f < 2$. On the other hand, eqn (8.22) is equivalent to the static (monotonic) tensile creep with a usual value of $m_c = 1$. Figures 8.45 (a) and (b) [31] show the effectiveness of the ΔJ_f and ΔJ_c parameters for determining the macroscopic propagation rate of several metals and alloys under a creep–fatigue interaction condition. These two kinds of parameter were used to evaluate the failure life under the so-called P type and C type tests of thermal and isothermal low-cycle fatigue of 1 Cr–1 Mo–0·25 V steel and Mar M247CC nickel-based superalloy [31].

(5) Characteristics of the surface cracks on a smooth bar specimen were as follows: (i) crack initiation occurs at 10% of the failure life, (ii) the scattering of the crack initiation life is very large, (iii) the propagation rates of short cracks were characteristically distributed; in particular, no deterministic law was observed in short cracks smaller than 3–5 grain size, but cracks longer than 200 μm have a correlation with the macroscopic through-crack mentioned above.

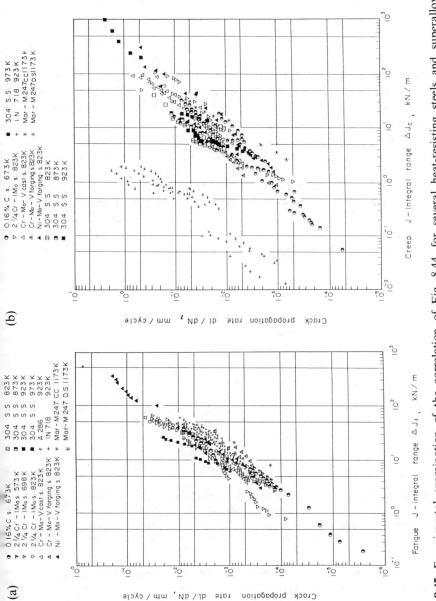

Fig. 8.45. Experimental examination of the correlation of Fig. 8.44, for several heat-resisting steels and superalloys. (Reprinted from ref. 31 by permission of Japan Soc. Mech. Engrs)

8.4.4 Development of a New Creep–Fatigue Damage Test Apparatus

Cam-and-Lever Type Long-term Creep–Fatigue Testing Machine
In order to obtain long-term data for the life/residual life prediction method under creep–fatigue interaction conditions at elevated temperatures, the following cam-and-lever type creep–fatigue testing machine with long-term stability and maintenance-free operation was developed. The specifications are: (i) axial strain-controlled push–pull loading, (ii) use of a cylindrical solid bar specimen, (iii) a temperature of 1073 K in air, (iv) hold-test of up to 100 h at peak tensile strain, (v) continuous and unattended test performance up to 10 000 h, (vi) uselessness of water/air cooling, (vii) small size and low cost, and (viii) safety features in case of power failure. Figure 8.46 [23] shows the concept of the test apparatus. From a comparison between the data, by using the present apparatus and an electro-hydraulic servo-type fatigue testing machine, the operational soundness of the present apparatus developed was confirmed. The present testing machine was developed by the Tokyo Koki Co. Ltd.

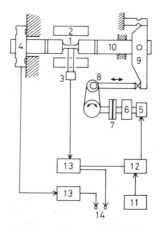

Fig. 8.46. Cam-and-lever type long-term creep–fatigue test apparatus. 1, Specimen; 2, furnace; 3, extensometer; 4, load cell; 5, servo-motor; 6, reduction gear; 7, clutch; 8, cam; 9, lever; 10, rod; 11, generator; 12, controller; 13, amplifier; 14, recorder. (From ref. 23 by courtesy of Science and Technology Agency, Japan)

Heat-actuator Type Long-term Creep–Fatigue Damage Test Apparatus
A heat-actuator type creep–fatigue test apparatus was developed in which the specimen is subjected to combined thermal and strain cycling. The mechanical strain cycling is imposed on the specimen by the heat-

actuator of the metal rod which is directly connected to the specimen and is heated and cooled cyclically, and thermal strain cycling is also performed by temperature cycling the specimen. Thus so-called thermo-mechanical low-cycle fatigue tests under combined strain cycling, with and without a phase, can be made. Since there is no electro-hydraulic servo mechanism or cam-and-lever type mechanism in this apparatus, the structure is very simple and the cost is low. Absence of trouble in the long-term operation of the present test apparatus was confirmed by using type 321 austenitic stainless steel at 873 K in air. This test apparatus was developed by NRIM.

Weld-structure Type Creep Damage Testing Apparatus
A new strain-controlled type creep damage test apparatus was recently developed. A weld-structural specimen of which the gauge length was composed of 18 Cr–8 Ni–Ti steel and the sides of the grips were composed of 2·25 Cr–1 Mo steel was welded together with SUS 316 stainless steel plates, as shown in Fig. 8.47 [23]. The specimen was then subjected to thermal strain cycling due to the difference in the thermal expansion of 2·25 Cr–1 Mo steel and 18 Cr–8 Ni steel. Creep damage of the gauge material was observed by using a transmission electron microscope. No creep voids were observed after 120 thermal cycles in the 18 Cr–8 Ni–Ti steel (SUS 321HTB, commonly used in fossil fuel power plants) during 20 000 h, but occurred in material used for 100 h and 10 000 h after 120 cycles. It was confirmed that this simple test apparatus was useful for observing the creep void of materials under strain-controlled conditions, and a comparatively large sized specimen was tested. The apparatus was constructed at Mitsubishi Heavy Industries.

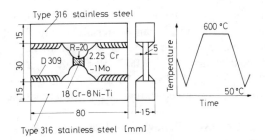

Fig. 8.47. Shape of specimen and temperature cycle, in a thermal cycling type long-term creep–fatigue damage test apparatus. (From ref. 23 by courtesy of Science and Technology Agency, Japan)

High-temperature Biaxial Creep–Fatigue Testing Machine for Cruciform Specimens

In the previous combined push-pull and reversed torsional biaxial low-cycle fatigue tests at elevated temperatures, the test condition of principal strain ratio ϕ ($=\varepsilon_3/\varepsilon_1$) is limited in the region $-1 \leqslant \phi \leqslant -0.5$, therefore the tests under $-1 \leqslant \phi \leqslant 1$ cover a wide range of biaxial stressing. In order to perform the tests under $-1 \leqslant \phi \leqslant 1$ it is necessary to perform a biaxial push-pull test with precision and stability by using a cruciform test specimen at elevated temperatures. As previously mentioned in Figs 6.17–6.23 in Chapter 6, the high-temperature biaxial low-cycle fatigue testing machine for use with cruciform specimens was successfully developed in the author's laboratory. We also described the data by using the present testing machine in Fig 6.24 and 6.25.

8.5 COLLABORATIVE R AND D EFFORTS EXAMINING THE CORROSIVE AND MECHANICAL BEHAVIOR OF HEAT-RESISTING ALLOYS FOR 50 MW (THERMAL) VHTGR APPLICATIONS IN JAPAN

JAERI has performed research and development of multi-purpose applications for HTGR since 1969. The main goal of JAERI's HTGR is to perform (i) the total testing of fundamental technology for HTGR, (ii) irradiation tests of fuel/material for HTGR in the future, (iii) examination of high-temperature nuclear thermal sources for multi-purpose applications and the development of systems for use. In this government enterprise the following tests have been performed: the test of nuclear characteristics by Very High Temperature Reactor Critical Assembly, the testing of facilities in high-temperature applications by use of the Helium Engineering Demonstration Loop (HENDEL), and the testing of the Oarai Gas Loop-1 in the Japan Material Testing Reactor (JMTR). Corrosive and mechanical behaviors of structural alloys, particularly Hastelloy XR (C 0·07, Mn 0·83, Si 0·32, P <0·005, S <0·006, Cr 21·84, Co 0·19, Mo 9·06, W 0·53, Fe 18·20, Ni Bal, Al <0·05, Ti <0·05 wt%), have been studied and the creep–fatigue behavior was studied as follows.

8.5.1 Long-term Corrosive Behavior of Hastelloy XR in Hot Helium Gas

Hastelloy XR was recently produced in order to improve the corrosion resistance of Hastelloy X in environments with low oxidation potential.

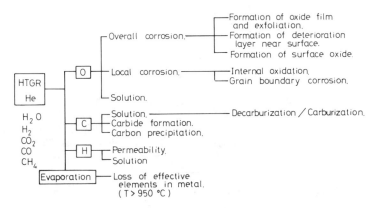

Fig. 8.48. High-temperature reactions between HTGR helium gas and super-alloy. (Reprinted from ref. 32)

Long-term corrosion tests under temperature cycling were performed for Hastelloy XR in B type helium gas (H_2 210, H_2O 0·8–1·2, CO 100–110, CO_2 2–3 μatm). The HTGR helium gas generally has a high-temperature reaction with the alloy, as shown in the schematic illustration in Fig. 8.48 [32]. In order to clarify the stability of the protective oxide film from the viewpoint of thermodynamics, exfoliation resistance and self-repairability, time-dependence in weight increase, depth in a chromium deficiency, the depth of intergranular corrosion and carburizing/decarburizing were investigated, and a long-term corrosion test of 30 000 h was performed under temperature cycling between room temperature and 1273 K in helium gas. We found that the corrosion resistance of the alloy was controlled mainly by the protection of the surface oxide film. The following conclusions were made [32].

(1) The oxidation rate of Hastelloy XR was considerably smaller than that of Hastelloy X.

(2) Remarkable exfoliation of the surface oxide film and good corrosion resistance were observed in Hastelloy X at 1273 K, whereas wholly corrosive behavior was observed in Hastelloy XR after 10 000 h at 1273 K.

(3) The difference in the corrosion resistance between both alloys is dependent on whether a thick-spinel oxide surface layer of $MnCr_2O_4$ is formed or not.

(4) The free oxide increase that Hastelloy XR exhibited was about 1000 times that of Hastelloy X.

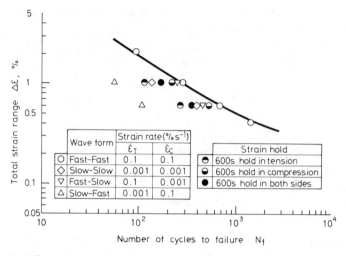

Fig. 8.49. Effect of strain waveform on low-cycle fatigue life of Hastelloy XR at 1223 K in B type helium gas. (Reprinted from ref. 33 by permission of Soc. Mater. Sci., Japan)

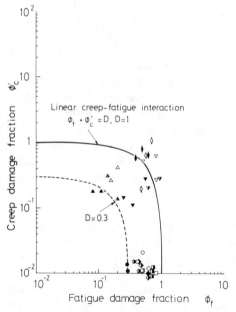

Fig. 8.50. Creep–fatigue/environment interaction for Hastelloy XR at 1223 K in (open symbols) air, (closed symbols) a vacuum (10^{-2} Pa) and (half-closed symbols) B type helium gas. Waveform: ◇, slow–slow; ▽, fast–fast; △, slow–fast; ○, hold in tension; □, hold in compression; ◊, hold in both sides. (Reprinted from ref. 33 by permission of Soc. Mater. Sci., Japan)

8.5.2 Failure Life Evaluation of Hastelloy XR under Creep–Fatigue Interaction Conditions

Figure 8.49 [33] shows the total strain versus the number of cycles to failure relation in the strain-controlled low-cycle test with and without hold-time in a strain cycle for Hastelloy XR at 1223 K in B type helium. The figure clearly shows that the smallest failure life results under a slow-fast strain wave. The test was performed at Ishikawajima-Harima Heavy Industries. Figure 8.50 [33] shows the evaluation result based on the linear cumulative damage rule, $\phi_c + \phi_f = D$, where ϕ_c is based on the cyclic creep rupture data and absence of damage under hold-time at peak compressive strain is assumed. The figure includes the data in air and in a vacuum of 10^{-2} Pa. We can see from Fig. 8.50 that the linear creep–fatigue interaction rule is applicable for Hastelloy XR in helium gas provided that the cyclic creep rupture data are adopted. This was also confirmed in the other test at 1173 K performed in JAERI. In the test, with hold-time at peak tensile strain, internal intergranular cracks or cavities were observed, and in the tests with hold-time at peak compressive strain or both peak tensile and compressive strain a marked spreading of cracks into branches was observed.

REFERENCES

1. B. I. Beresner, L. F. Versehagin, Yu. N. Ryabinin and L. D. Livshits, *Some Problems of Large Plastic Deformation of Metals at High Pressure*, Pergamon Press, Oxford, 1963.
2. P. W. Bridgman, *Studies in Large Plastic Flow and Fracture*, Harvard University Press, 1964.
3. H. L. D. Pugh and K. A. Schcroft, NEL Report No. 32 (1962), No. 142 (1964), Her Majesty's Stationery Office.
4. M. Ishi and M. Yajima, The effect of hydrostatic pressure on the wire drawing of aluminum, *Journal of Japan Society for Technology of Plasticity*, **8** (1986), 74 (in Japanese), CORONA-Sha, Tokyo.
5. Y. Awaya, A. Takada and M. Ohnami, Test devices and the test results of plastic drawing of pure zinc and pure aluminum—a study on plastic working under hydrostatic pressure, *Journal of Japan Society for Technology of Plasticity*, **15** (1974), 751 (in Japanese).
6. B. Lengyel et al., *Proceedings of 7th International MTDR Conference* (1966), p. 319.
7. G. W. Pearsall and W. A. Backofen, Frictional boundary conditions in plastic compression, *Transactions of the American Society of Mechanical Engineers*, **B85** (1963), 68.
8. G. W. Pearsall and W. A. Backofen, Size and effects from lubricant breakdown in plastic compression, *Transactions of the American Society of Mechanical Engineers*, **B85** (1963), 329.

9. M. Ohmura, M. Ohnami and Y. Awaya, Plastic compression of metallic materials under hydrostatic pressure, *Journal of Japan Society for Technology of Plasticity*, **13** (1972), 668 (in Japanese).
10. F. P. Broon and D. Tabor, *The Friction and Lubrication of Solids*, 1st Edn, Clarendon Press, Oxford, 1954.
11. Nuclear Power Development Office, Research and development of a high-temperature heat exchanger, *Ishikawajima-Harima Review*, **19**–1 (1979), 18 (in Japanese), Ishikawajima-Harima Heavy Industries.
12. Agency of Industrial Science and Technology, The Japanese Ministry of International Trade and Industries and ERANS, Integrated Report of Research and Development of Direct Steel Making Technology by Utilizing High-temperature Reduced Gas (Nuclear Steelmaking, 1982), pp. 1–209 (in Japanese).
13. American Society of Mechanical Engineers, ASME Boiler and Pressure Vessel Code Case N-47, Components in Elevated Temperature Service, Section III, Division 1 (1979).
14. K. Kitagawa, K. Mino, H. Hattori, A. Ohtomo, M. Fukagawa and Y. Saiga, Some problems in developing the high-temperature design code for 1·5 MWt helium heat exchanger, *Elevated Temperature Design Symposium* (1976, Mexico City), p. 33, Amer. Soc. Mech. Engrs.
15. Ref. 12, pp. 45–77.
16. Y. Hosoi and S. Abe, The effect of helium environment on creep rupture properties of Inconel 617 at 1000°C, *Metallurgical Transactions*, **6A** (1975), 1171, Amer. Soc. Metals.
17. (a) M. Ohnami and R. Imamura, Effect of vacuum environment on creep rupture properties at 1000°C (especially based on crack initiation and propagation), *Transactions of the Japan Society of Mechanical Engineers*, **A47** (1981), 45 (in Japanese). (b) *Bulletin of the Japan Society of Mechanical Engineers*, **24** (1981), 1530.
18. M. Kitagawa, J. Hamanaka, T. Umeda, T. Goto, Y. Saiga, M. Ohnami and T. Udoguchi, A new design code for 1·5 MWt helium heat exchanger, *5th International Conference on Structural Mechanics in Reactor Technology* (SMIRT), **F9**/1 (1979), p. 1, Berlin.
19. M. Kitagawa, H. Hattori, A. Ohtomo, T. Terayama and J. Hamanaka, Lifetime test of a partial model of high-temperature gas-cooled reactor helium–helium heat exchanger, *Nuclear Technology*, **66** (1984), 665, Amer. Nuclear Soc.
20. T. Terayama and H. Ohya, Analysis and design of structures and machinery employed at high temperature: 1st report, inelastic program of three dimensional structures and creep buckling, *Ishikawajima-Harima Review*, **14**–2 (1981), 69, Ishikawajima–Harima Heavy Industries (in Japanese).
21. H. Hattori, M. Kitagawa and A. Ohtomo, High temperature low cycle fatigue properties of Ni-based superalloys for heat exchanger tubes of HTGR, *Research Report of 123rd Committee on Heat-resisting Metals and Alloys*, **25** (1984), p. 41 (in Japanese), Japan Soc. for Promotion of Science.
22. Science and Technology Agency, Japan, Report of Research of Severe Environments and Damage of Structural Materials in Fossil Fuel Power Plants, Chemical Plants and Public Structures (Feasibility Study, 1982) (in Japanese).

23. Science and Technology Agency, Japan, Report of the Development of Evaluation Techniques for the Reliability of Structural Materials by the Fund of Promotion of Science and Technology (1986) (in Japanese).
24. N. Shinya, S. Yokoi and J. Kyono, Prediction of residual rupture life and quantitative measurement of creep cavities in Cr-Mo-V steel, *Research Report of 123rd Committee on Heat-resisting Metals and Alloys*, **22** (1981), p. 69 (in Japanese).
25. S. Shinya, T. Nonaka, J. Kyono and T. Takahashi, Evaluation of creep damage of structural metals at elevated temperatures, *Proceedings of the Committee on the High Temperature Strength of Materials*, No. 85-4-1 (1986), p. 7, Soc. Mater. Sci., Japan.
26. K. Yamaguchi, N. Suzuki, K. Ijima and K. Kanazawa, Prediction of creep-fatigue life by use of creep rupture ductility, *Journal of the Iron and Steel Institute of Japan*, **71** (1985), 1526 (in Japanese).
27. K. Yagi, O. Kanemaru, K. Kubo and C. Tanaksa, Life prediction under creep-fatigue loading condition for SUS 316 stainless steel, *Journal of the Society of Materials Science, Japan*, **35** (1986), 176 (in Japanese).
28. S. Nishino, M. Sakane and M. Ohnami, Creep-fatigue damage evaluation and estimation of remaining life for a type 304 stainless steel, *Journal of the Society of Materials Science, Japan*, **35** (1986), 292 (in Japanese).
29. M. Ohnami, M. Sakane and S. Nishino, A proposal of predictive methods of crack propagation life and remaining life of structural metal under creep-fatigue interacted conditions by use of X-ray profile analysis, in *Proceedings of International Conference on Role of Fracture Mechanics in Modern Technology* (Fukuoka, Japan, 1986) (ed. G. C. Sih and H. Nishitani), North-Holland, Amsterdam, 1987.
30. R. Ohtani, T. Kitamura and T. Kinami, A concept of life and remaining life prediction of high temperature structural materials based on the characteristics of creep-fatigue crack propagation (Review), *Journal of the Iron and Steel Institute of Japan*, **72** (1986), 711 (in Japanese).
31. R. Ohtani, Characteristic of high temperature strength of materials from viewpoint of creep-fatigue crack propagation, *Transactions of the Japan Society of Mechanical Engineers*, **52A** (1986), 1461 (in Japanese).
32. T. Tsukada, M. Shindo, T. Suzuki and T. Kondo, Long-term corrosion of Ni-base heat-resistant alloy under thermal cycling in helium gas, *Research Report of 123rd Committee on Heat-resisting Metals and Alloys*, **26** (1985), p. 79 (in Japanese).
33. H. Hattori, M. Kitagawa and A. Ohtomo, Evaluation of creep-fatigue/environment interaction in Ni-based wrought alloys for HTGR applications, *Journal of the Society of Materials Science, Japan*, **35** (1986), 305 (in Japanese).

Appendix I

Table A
Fluctuation of Strength of Materials and the Distribution Function

Phenomenon	c.o.v. (%) (N, sample size)	Distribution function	Characteristics of parameters in distribution function and features of fluctuation	Researcher(s) (Year)
Tensile strength, σ, of SM50A structural steel at room temperature	4.3 [1] ($N = 472$)	$f(\sigma)$ is normal distribution ($\bar{\sigma} = 535$ MPa; $s = 23$ PMa)	(1) Scattering is comparatively small [1] Fig. A [1]	Nishimura [1] (1969)
Low-temperature brittle fracture strength, σ_c, of 0.1% carbon steel	7·6 [2] (N 158)	(1) $f(\sigma)$ of the material with n cracks is Weibull distribution with two parameters [2]: $f = \beta m \sigma^{m-1} \exp(-\beta \sigma^m)$	(1) σ_c is written from mode $f(\sigma)$ as $$\sigma_c = \frac{1}{(\alpha n)^{1/m}} \left(\frac{m-1}{m}\right)^{1/m} \quad (3)$$ where $n = \rho V$, ρ is crack density, V is specimen volume, and $\sigma_c \propto V^{-1/m} \quad (4)$ Thus σ_c shows *size effect* [3]	Davidenkov et al. [2] (1947) Weibull [3] (1939) Yokobori [4] (1956)

Tensile strength, σ_B, of cast iron at room temperature

$F = 1 - \exp(-\beta\sigma^m)$ (1)

where $\beta = n\alpha > 0$, $\alpha > 0$, $m > 1$

(2) That above is also obtained from stochastic process of two states and one step [4]:

$$f = \frac{L}{\dot\sigma}\sigma^\delta \exp\left[-\frac{L}{(\delta+1)\dot\sigma}\sigma^{\delta+1}\right]$$

$$F = 1 - \exp\left[-\frac{L}{(\delta+1)\dot\sigma}\sigma^{\delta+1}\right]$$ (2)

where $m(t) = L(\dot\sigma t)^\delta$

(2) Scattering increases as V and m decrease [2–4]

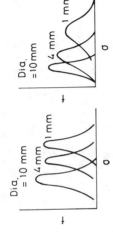

Fig. B [2]

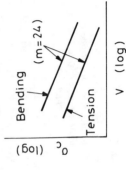

Fig. C [2]

Table A—contd.

Phenomenon	c.o.v. (%) (N, sample size)	Distribution function	Characteristics of parameters in distribution function and features of fluctuation	Researcher(s) (Year)
Rotating bending fatigue limit, σ_w, of low-carbon steels [5] S25C: N, 885 S35C: N, 865 Q, 865 T, 650 S35C: N, 865 Q, 865 T, 600 S35C: N, 865 Q, 865 T, 550	 5·7 (N 11) 5·3 (N 12) 4·4 (N 12) 5·0 (N 12)	$f(\sigma_w)$ is normal distribution for smooth specimen, and Weibull distribution for notched specimen	(1) The mean $\bar{\sigma}_w$ is proportional to the carbon content, and is correlated with tensile strength, hardness (H_v), ferrite grain size and inclusions [6]. (2) Standard deviation, s, increases with larger σ_w. s of quenched and tempered materials is larger than that of normalized ones. s is dependent on heat treatment and specimen size [6]. (3) c.o.v. increases with larger H_v, and c.o.v. of rolled and normalized steels is larger than that of annealed ones. c.o.v. is influenced by heat treatment and specimen size. c.o.v. of the notched specimen is 1·5–2 times as large as that of the smooth specimen [6]. (4) Distribution of σ_w and σ_a (see Fig. D) is approximated to Weibull distribution, but one will be unable to extrapolate the properties determined from σ_a and σ_w.	National Research Institute for Metals, Japan (NRIM) [5] (1975–) Tanaka et al. [6] (1980)

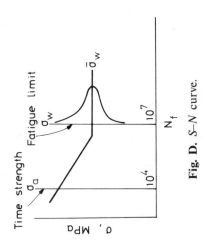

Fig. D. S–N curve.

S45C:		
N, 845	3·7	
Q, 845		(N 11)
T, 650		
S45C:		
N, 845	5·0	
Q, 845		(N 11)
T, 600		
S45C:		
N, 845	5·5	
Q, 845		(N 11)
T, 550		
S55C:		
N, 825	2·1	
Q, 825		(N 11)
T, 650		
S55C:		
N, 825	2·6	
Q, 825		(N 11)
T, 600		
S55C:		
N, 825	3·7	
Q, 825		(N 11)
T, 550		

Rotating bending fatigue limit of alloy steels [5]

Table A—contd.

Phenomenon	c.o.v. (%) (N, sample size)	Distribution function	Characteristics of parameters in distribution function and features of fluctuation	Researcher(s) (Year)
SCM435: N, 870 Q, 855 T, 650	4·1 (N 14)			
SCM435: N, 870 Q, 855 T, 600	4·7 (N 14)			
SCM435: N, 870 Q, 855 T, 550	4·8 (N 14)			
SCM440: N, 870 Q, 855 T, 650	2·3 (N 15)			
SCM440: N, 870 Q, 855 T, 600	3·2 (N 15)			
SCM440: N, 870 Q, 855 T, 550	4·3 (N 15)			

Item	Value	Distribution	Notes	Reference
SCR440: N, 870 Q, 855 T, 650	2·6 (N 8)			
SCR440: N, 870 Q, 855 T, 600	1·5 (N 8)			
SCR440: N, 870 Q, 855 T, 550	3·8 (N 8)			
Rotating bending fatigue limit, σ_w, of cast steel [7] (specimen was sampled from a large-sized crank arm)	13 [7] (N 30–35; smooth specimen) 11·6 [7] (N 30–35; notched specimen)	$f(\sigma_w)$ is normal distribution	(1) Scattering of cast steel is 3·2–3·5 times as large as that of forged steel [7]. (2) Adopting confidence limit as $\sigma_w - 2s$, it takes 161 MPa for cast steel and 255 MPa for forged steel. In the case of $\sigma_w - 3s$, it takes 133 MPa for cast steel and 243 MPa for forged steel, and the difference between σ_w of large-sized cast steel and that of forged steel is unexpectedly large [7]. (3) Scattering as above will be due to microcavities in cast steel and comparatively large inclusions in forged steel [7].	Tokuda et al. [7] (1970) Fatigue Committee, Soc. Mater. Sci. Japan [8] (1959) Nakazawa [9] (1971)
Rotating bending fatigue limit, σ_w, of forged steel [7]	4·1 [7] (N 30–35; smooth specimen) 3·1 [7]	$f(\sigma_w)$ is normal distribution	(4) Scattering of carbon steel is considerably smaller than that in both cast and forged steels. Factor of safety, S, under reliability of 99·9% is 1·05, and it is concluded that factor of safety for	

Table A—contd.

Phenomenon	c.o.v. (%) (N, sample size)	Distribution function	Characteristics of parameters in distribution function and features of fluctuation	Researcher(s) (Year)
	(N 30–25; notched specimen)		low-carbon steel is remarkably small [9].	
Rotating bending fatigue limit of 0·39% C steel [8]	14	$f(\sigma_w)$ is normal distribution		
Machine structural steels [10]	13 [9] (N 161)	$f(\sigma_w)$ is normal distribution		

(1) N, Q and T denote normalizing, quenching and tempering, respectively, and the numbers show the temperature in °C.
(2) When the *cumulative distribution function*, $F(x)$, is continuous and differentiable regarding x, $f(x) \equiv dF(x)/dx$ is called the *probability density function*. *Probability of survival*, $p(x)$, or *reliability*, $R(x)$, is written as $p(x) \equiv R(x) = 1 - F(x)$. For continuous probability variable, t, *transitional probability of fracture*, $\mu(t)$, or *failure rate*, $\lambda(t)$, is written as $\mu(t) \equiv f(t)/p(t)$ [eqn (1)]. Combining eqn (1) with $f(t) = -dp/dt = -dR(t)/dt$ [eqn (2)], the following relations are obtained: $\mu(t) \equiv (t) = -d \ln p(t)/dt = -d \ln R(t)/dt$ [eqn (3)] and $p(t) \equiv R(t) = \exp[-\int_0^t \mu(t)dt] = \exp[-\int_0^t \lambda(t)d\tau]$ [eqn (4)]. Thus, respective $F(t), f(t), p(t), \mu(t)$ are determined from each other and eqns (1)–(4) are fundamental equations in reliability engineering.

Table B
Fluctuation and Distribution of the Lifetime of Materials

Phenomenon	c.o.v. (%) (N, sample size)	Distribution function	Characteristics of parameters in distribution and features of fluctuation	Researcher(s) (Year)
Delayed fracture time of glass [[11]	100 [11] (N 400)	$f(t)$ is exponential function [11] $p = \exp(-mt)$ $f = m\exp(-mt)$ $\mu = m = A\exp(\alpha\sigma)$ $\bar{t} = 1/m$ (1)	Fig. E [11]	Hirata [11] (1948) Kubota [12] (1948)
Bending strength of glass [13]	24	Under loading of $f = kt$ (k is load rate) $\mu = A\exp(\alpha t)$ and $p = \exp\left[\dfrac{A}{\alpha k}(1 - e^{\alpha k t})\right]$ $f = Ae^{\alpha k t} \cdot \exp\left[\dfrac{A}{\alpha k}(1 - e^{\alpha k t})\right]$ (2)	Fig. F [13]	Hirata and Terao [13] (1952)

Table B—contd.

Phenomenon	c.o.v. (%) (N, sample size)	Distribution function	Characteristics of parameters in distribution and features of fluctuation	Researcher(s) (Year)
Stress corrosion cracking of 7075 aluminum alloy (test by J. H. Harshbarger (1971))	45 (A alloy) 26 (B alloy) 36 (C alloy) [14] (N 20 for each stress level)	$f(t)$ is Weibull distribution [11] $$f = \frac{\alpha}{\beta}\left(\frac{t}{\beta}\right)^{\alpha-1} \exp\left[-\left(\frac{t}{\beta}\right)^{\alpha}\right]$$ $$\mu = \frac{\alpha}{\beta}\left(\frac{t}{\beta}\right)^{\alpha-1}$$ where $\alpha > 0$, $\beta > 0$	Constant load [graph showing distributions A, B, C vs t, day; and curve vs t, day] **Fig. G** [14]	Shibata [14] (1978)
Tensile creep rupture of copper wire at room temperature [15,16]	70 [15] (N 8–100 for each stress level)	$f(\ln t)$ is normal distribution	(1) Creep rupture time is understood as a two-step succession probability process, and each step is a rate process with thermal activation [16]. (2) Scattering of rupture time is interpreted from logarithmic normal distribution, and the scattering increases as both stress and temperature increase [16].	Yokobori [15] (1951) Yokobori and Ohara [16] (1958)
Tensile creep rupture of	4·3–164 (N 13–20	$f(\ln t)$ is normal distribution	(3) For low-carbon steel, variance is nearly independent of stress level and there is a	Japan Soc. for Promot. of

Material/Test	c.o.v. (%) [ref] (N)	Distribution	Description	Reference
0·17% C steel at 450°C [17]	(for each stress level)		significant relation between the minimum creep rate and rupture time [17].	Sci. [17] (1966)
Tensile creep rupture at high temperatures [18]			(4) c.o.v. of the same steel (the same heat) is nearly 10%, and there is no large difference between data [19].	NRIM [18] (1970)
0·15% C steel (450°C)	6·5 [17] (N 28)	$f(t)$ for carbon steel is normal distribution but not for 2·25%Cr–1%Mo steel and 18%Cr–12%Ni–2·5%Mo steel. $f(\ln t)$ and $f(t)$ are normal distributions	(5) Scattering of rupture time of the same steel with different heat is considerably large (nearly 20% for rupture strength). For carbon steel and alloy steel the scattering is influenced by inclusions in the steels [20,21].	
2·25%Cr–1% Mo steel (550°C)	12 [18] (N 57)			
18%Cr–12% Ni–2·5% Mo steel (700°C)	8·2 [18] (N 54)		(6) Characteristics of the distribution obtained by a comparatively short-term test are not necessarily extrapolated to that of long-term. However, the extrapolation with a standard deviation smaller than 0·1 in terms of logarithmic rupture time is possible by means of standard TTP, provided that the data are classified by individual heat [22].	NRIM [19] (1980)
Tensile creep rupture of 18%Cr–8% Ni stainless steel [19]	12·8 [19] (650°C; N 63) 11·1 [19] (732°C; N 406)			

Fig. H [15]

| Rotating bending fatigue life | ~100 [23] (N 70 for each stress | $F(N)$ for definite life is normal. The transitional | (1) Variance V of life N is written as $V = 1/\mu$ for an exponential distribution. For hard steel, the scattering increases | Yokobori [23] (1951, 1953) |

Table B—contd.

Phenomenon	c.o.v. (%) (N, sample size)	Distribution function	Characteristics of parameters in distribution and features of fluctuation	Researcher(s) (Year)
of 0·41% C steel [23] Push-pull fatigue life of 0·19% C steel [23]	level) 100 [23] (N 80–102 for each stress level)	probability of fracture μ is written as $$\mu = \beta \exp(B\sigma_a) \quad (4)$$ $$\bar{N} = 1/\mu$$	as stress amplitude, σ_a, decreases, and the rougher the specimen is finished the more scattering decreases [23]. (2) For soft steel, μ increases gradually, but for hard steel it increases rapidly and comes to a constant value. The work-hardening is similar to the change of μ, and the saturation point agrees with that of work-hardening [23].	Kawamo and Nakagawa

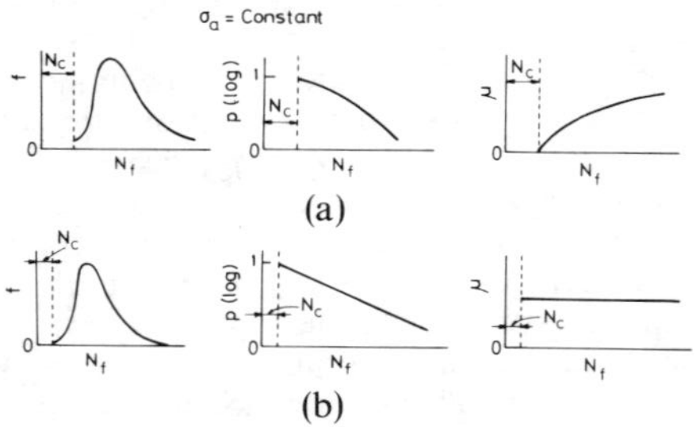

Fig. 1 [23] (a) Soft steel; (b) hard steel.

Aluminum Alloy (Al2024 BE-T4)		For definite life N, $f(\ln N)$ is normal distribution	(1) Standard deviation for steels is related to grain size, the ferrite path and non-metallic inclusion [6]. (2) c.o.v. of definite N is five times as large	Nakagawa

Material	Range	Description	Reference
Carbon steels (SS41, S25C, S35C)	2–10 for $\ln N$	of press-fitting on c.o.v. are small. In general, c.o.v. increases as stress amplitude σ_a decreases and comes to nearly 10% in the vicinity of fatigue limit σ_w.	Nakamura and Tanaka Tanaka Tokuda [7] Shimokawa
Cast steel			
Aluminum alloy (Al2024-4)			Tanaka and Fujii
Carbon steels (S25C, S35C, S45C, S50C)		(3) Shape population parameter in Weibull distribution takes nearly constant value, and both scale and position parameters are dependent on stress level [6]. (4) The following method is most favorable to estimate the parameters in Weibull distribution. Fixing the position parameter so as to maximize the correlation coefficient in the Weibull probability diagram, and then estimating both the shape and scale parameters from linear regression. In order to clarify the distribution with three parameters, at least 30–50 test pieces are necessary for a stress level [24].	Soc. Mater. Sci., Japan
Maraging steel (CH10R)			Sakai [24] (1980)
Carbon steel (S25C)		(5) Mathematical representation of life distribution with probability of survival in the region of long life is proposed, and the three parameters estimation method is also proposed for time cut-off sampling from the population [24].	
Low alloy steel (SCM 4)			
Maraging steel			

REFERENCES

1. A. Nishimura, Fluctuation of mechanical property of steel, *Journal of Structural Steel and Construction*, **5** (1969), 68 (in Japanese), Soc. Struct. Steel Constr., Japan.
2. N. Davidenkov, E. S. Schevandin and F. Witmann, The influence of size on the brittle strength of steel, *Journal of Applied Mechanics*, **14** (1947), A63, Amer. Soc. Mech. Engrs.
3. W. Weibull, The phenomenon of rupture in solids, *Ingeniors Vertenskaps Akademien Handlinger*, **153** (1939), 1–55, Generalsstabens Litografiska Anstalts Forlag, Stockholm.
4. T. Yokobori, Statistical interpretation of the results of testing of materials (Review), *Journal of the Japan Society for Testing and Materials*, **5** (1956), 266 (in Japanese).
5. S. Nishijima, Strength reliability of steels for machine structural use, *Journal of the Materials Science Society of Japan*, **17** (1980), 12 (in Japanese).
6. S. Tanaka and G. Miyashita, Fatigue strength and reliability, Text in 500th Lecture of Japan Society of Mechanical Engineers (1980), pp. 27–38 (in Japanese).
7. A. Tokuda, H. Tsukada and K. Sakabe, On the endurance limits and reliability of large casings and forgings, *Journal of the Japan Society of Mechanical Engineers*, **73** (1970), 1500 (in Japanese).
8. Committee of Materials Fatigue, Research of fatigue strength of metallic materials under two-level two-step wave, *Journal of the Japan Society for Testing and Materials*, **8** (1959), 684 (in Japanese).
9. H. Nakazawa, Statistical safety factor of fatigue limit, *Journal of the Japan Society of Mechanical Engineers*, **74** (1971), 1264 (in Japanese).
10. Japan Society of Mechanical Engineers (Ed.), Metallic Materials: Design Data of Fatigue Strength (I) (1961) (in Japanese).
11. M. Hirata, Statistical phenomena in science and engineering, *Science of Machine*, **1** (1949), 231 (in Japanese), YOKENDO.
12. H. Kubota, Times of rupture of glass, *Journal of Physics*, **17** (1948), 286 (in Japanese), Phys. Soc. Japan.
13. M. Hirata and S. Terao, On the breaking strength of glass, *Journal of Applied Physics, Japan*, **20** (1952), 234 (in Japanese), Soc. Appl. Physics, Japan.
14. T. Shibata, Probabilistic Evaluation of Localized Corrosion, BOSHOKU GIJITSU, **27** (1978), 23 (in Japanese).
15. T. Yokobori, Delayed fracture in creep of copper, *Journal of the Physical Society of Japan*, **6** (1951), 78.
16. T. Yokobori and H. Ohara, Statistical aspect in accelerating creep and fracture of OFHC copper, *Journal of the Physical Society of Japan*, **13** (1958), 305.
17. S. Taira, Collaborative test results on scattering of creep rupture of low carbon steel at elevated temperature, *Journal of the Iron and Steel Institute of Japan*, **52** (1966), 1791 (in Japanese).
18. S. Yokoi, N. Shinya, A. Miyazaki and Y. Kitagawa, Studies of creep rupture data by using multiple-type tester, *Journal of the Iron and Steel Institute of Japan*, **56** (1970), S217 (in Japanese).

19. S. Yokoi and Y. Monma, Scattering of creep-rupture data for high-temperature steels and alloys, *Journal of the Materials Science Society of Japan*, **17** (1980), 3 (in Japanese).
20. S. Yokoi, N. Shinya, M. Kohri and H. Tanaka, Some causes of scattering in rupture strength of silicon-killed carbon steel for boiler tubes, *Journal of the Society of Materials Science, Japan*, **25** (1976), 249 (in Japanese).
21. S. Yokoi, N. Shinya and M. Kohri, Long term creep-rupture properties of 12 Cr–Mo–W–V steel for turbine blades, *Journal of the Society of Materials Science, Japan*, **26** (1977), 241 (in Japanese).
22. S. Yokoi and Y. Monma, Prediction of long-time creep-rupture strength for high-temperature materials, *Journal of the Iron and Steel Institute of Japan*, **65** (1979), 831 (in Japanese).
23. T. Yokobori, Fatigue fracture of steel, *Journal of the Physical Society of Japan*, **6** (1951), 81.
24. T. Sakai, Reliability engineering study on fatigue crack propagation behavior and distribution characteristics of fatigue life of metallic materials, Doctoral Thesis, Ritsumeikan University, Kyoto, Japan, 1980 (in Japanese).

Appendix II

Cartesian Localized Coordinates

Consider that a cube having three edges which are represented by unit vectors, $\overset{\circ}{\mathbf{e}}_1$, $\overset{\circ}{\mathbf{e}}_2$ and $\overset{\circ}{\mathbf{e}}_3$, deforms to a rhombus with edge vectors, $\mathbf{e}_1$, $\mathbf{e}_2$ and $\mathbf{e}_3$, as shown in Fig. A. Here, radial circumferential and axial directions of this thin-walled cylindrical specimen are taken as Cartesian

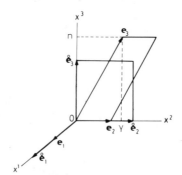

Fig. A. Coordinate system employed in the analysis of deformation of a thin-walled tubular specimen.

localized coordinates $\mathbf{x}^1$, $\mathbf{x}^2$ and $\mathbf{x}^3$ respectively. In the figure, $\overset{\circ}{\mathbf{e}}_3$ changes to $\mathbf{e}_3$ and the change in the direction $\mathbf{x}^3$ is n-times $\overset{\circ}{\mathbf{e}}_3$ and that in the direction $\mathbf{x}^2$ is γ, or shearing strain in engineering notation. Therefore shortening in both $\mathbf{x}^1$ and $\mathbf{x}^2$ directions is given by $1/\sqrt{n}$ from the assumption of the incompressibility in plastic strain. From this relation, strain and strain rate tensors in the localized coordinate system are connected with n and γ as

$$\varepsilon_{23} = (1/4)\int (\gamma/n^2)\mathrm{d}n + (1/2)\int (\mathrm{d}\gamma/n)$$

$$\dot{\varepsilon}_{23} = (\gamma/4n)\dot{\varepsilon}_{33} + (\dot{\gamma}/2n) \qquad (\text{a})$$

The anisotropic parameter A is determined as follows. From eqns (4.2), (4.3) and $\dot{\varepsilon}_{ij} = \lambda(\partial f/\partial S_{ij})$, the following flow-rule is obtained:

$$\dot{\varepsilon}_{ij} = \lambda[\{G_{ijkl} + A\{L_{ijkl} - (2/3)\delta_{ij}\varepsilon_{kl}\} + A^2\{M_{ijkl} - (1/3)M_{rskl}\delta^{rs}\delta_{ij}\}\}S^{kl} - B\varepsilon_{ij}] \quad \text{(b)}$$

We can derive the following equation from eqn (b) and the volume constant law:

$$\begin{aligned} 2\dot{\varepsilon}_{23}/\dot{\varepsilon}_{33} = [&6\tau\{1 + (1/2)A\varepsilon_{33} - (1/2)A^2\varepsilon_{33}^2 + A^2\varepsilon_{23}^2\} \\ &+ \sigma\{2A\varepsilon_{23} + 3A^2\varepsilon_{33}\varepsilon_{23}\} - 2B\varepsilon_{23}]/[2\sigma\{1 + A\varepsilon_{33} \\ &+ (3/4)A^2\varepsilon_{33}^2 - (2/3)A^2\varepsilon_{23}^2\} \\ &+ \tau(2A\varepsilon_{23} + 5A^2\varepsilon_{33}\varepsilon_{23}) - B\varepsilon_{33}] \quad \text{(c)} \end{aligned}$$

The parameter of the Bauschinger effect B is given as

$$B = \{(P\sigma^2 + Q\sigma\tau + 3R\tau^2) - \bar{\sigma}^2\}/\{3(\varepsilon_{33}\sigma + 2\varepsilon_{23}\tau)\} \quad \text{(d)}$$

where P, Q and R are given in eqn (4.5). Eliminating B in eqn (c) by using eqn (d), eqn (c) is given by a quadric equation in terms of $\dot{\varepsilon}_{33}$, $\dot{\varepsilon}_{23}$, ε_{33} and ε_{23}. Therefore A is determined from the quadric equation for an arbitrary strain history. In the same way, B is determined from eqn (d) by the use of the numerical value of A.

Index

A286 steel
  fatigue J-integral range versus crack propagation rate, 479
  notch effect in, 318–20
Abe, S., 450
Absolute reaction rates, theory of, 29
Acknowledgements and thanks, 3–4
Action
  at a distance, 86, 89
  through a medium, 86, 89
Active slip systems
  aluminum under tension, 122–3
  nonuniqueness in choice of, 64, 72–3
Aircraft engine designs, 109–10
Aircraft safety, 40
Alcoa Research Laboratory, tear test specimen, 359
Allowable stress, 38
Alloy steels
  fluctuation of
    fatigue life of, 499
    rotating/bending fatigue limit of, 491–3
Aluminum
  alloys
    fluctuation of fatigue life of, 498, 499
    fluctuation of stress corrosion cracking of, 496
    compression under hydrostatic pressure, 438
    creep affected by hydrostatic stress, 170, 175, 176, 178, 179

Aluminum—*contd.*
  dislocation and imperfection densities affected by hydrostatic pressure, 230–2, 235
  drawing under hydrostatic pressure, 429, 431, 435, 436
  flow stress affected by hydrostatic pressure, 228–9
  fracture toughness affected by hydrostatic pressure, 365–6, 368
  microstructural change during low-cycle fatigue, 217
  misorientation affected by hydrostatic pressure, 229–30, 234
  necking strain affected by hydrostatic pressure, 367
  plastic deformation of, 121–3
  pressure soaking of, 220–3, 225–7
  principal strain directions in, 125–6
  X-ray study of flow stress in, 199–210
Amari, S., 82, 217
American Society of Mechanical Engineers (ASME)
  boiler and pressure vessel code, 40, 41, 43
  Boiler and Pressure Vessel Code
    high-temperature design code developed from, 446, 447
    life prediction method, 327–31, 333, 338, 339, 340

American Society for Testing and Materials (ASTM), publications on fatigue, 6
Amerlinckz, S., 194
Analogs, 44
Anisotropic hardening, 139, 140, 141, 144
Austenitic stainless steels
  creep–fatigue life prediction for, 474, 475
  fatigue fracture surface of, 9, 10
  see also Stainless steel, type 304; ... type 316

Babcock–Hitachi, finite element analysis program produced by, 333
Bach, C. von, 38
Back-stress, pile-up dislocations, 233, 237
Bairstow, L., 105, 106–7, 109
Barba, M. J., 47
Bauschinger effect, 139, 140, 144, 162, 165, 166, 180
  change with repeated bending, 210, 211
  meaning of term, 210, 211
  observed in unloading path, 218, 219
  stages of, 212
Beran, B. J., 82
Biaxial creep–fatigue testing machine, 482
  see also Cruciform specimens
Biaxial fatigue
  alternating push-pull/reversed torsion (APT) loading in, 153
  apparatus used, 147
  companion specimen test method used, 145, 146
  crack direction in, 274–80
  cyclic combined axial and torsional (CPT) loadings used, 158–9
  incremental/decremental (ID) test method used, 146, 148–9
  polycrystalline metals, 145–62

Biaxial fatigue—*contd.*
  polycrystalline metals—*contd.*
    cyclic stress–strain relations, 145–52
    microstructural considerations, 152–5
    strain history effect on stress response, 156–7
    stress amplitude ellipses resulting, 158–61, 162
  principal stress axes changed, effect of, 296–303
  proportional/non-proportional loading used, 147
  push–pull (P) loading in, 153
  results summarized for type 304 stainless steel, 265
  reversed torsion (T) loading in, 153
  strain hold-times, effect of, 293–6
  strain paths used, 147
  strain-controlled mode, 259
    BMK strain range used, 260
    COD strain and strain ranges used, 260, 263
    LE strain range used, 260
    maximum principal strain and stress ranges used, 261, 264
    maximum shear strain and shear ranges used, 261, 263
    von Mises strain and stress ranges used, 261, 264
Biaxial stress
  fatigue affected by, 250–80
  high-temperature low-cycle fatigue criterion, 254–66
Bilby, B. A., 21, 75, 76, 185, 232, 404
Bilby–Cottrell–Swindeman (BCS) model, 77
Bishop, J. F. W., 64, 67, 68, 69
BMK strain, fatigue data plotted against, 261
Bošković, R. J., 86–7
Bošković's theory (of material particles), 86
Boundaries, types of, 21–3
Brandon, D. G., 23

Bridgman, P. W., 427
Britain, creep research in, 100
Brittle fracture, 8, 9, 10
  fluctuation of low-temperature
    strength in carbon steel, 488
  Griffith's theory for, 26, 31, 92
  weakest-link model for, 36, 37, 51, 489
Brown, M. W., 262, 274
Buckley, S. N., 210
Budianski, B., 69, 70
Bullet train, fatigue tests for, 50
Burchgrenze (fatigue limit), 11
Burgers, J. M., 90
Burgers circuit, 188
Burgers vectors, 77, 188, 190

Cam-and-lever creep–fatigue testing
    machine, 480
Carbon steels
  creep J-integral range versus crack
    propagation rate, 479
  creep testing of, 163–8
  crystal rotation in, 123, 124
  cyclic stressing of, 130, 133, 135, 136
  excess dislocation density
    variation, 127–8, 129
  fatigue J-integral range versus
    crack propagation rate, 479
  fluctuation
    fatigue life of, 497–8, 499
    low-temperature brittle fracture
      strength of, 488
    rotating/bending fatigue limit of,
      490–1, 494
    tensile creep rupture of, 496–7
  fracture surface of, 9, 10
  misorientation
    principal strain direction and,
      126
    variation within grains of, 127,
      128
  necking strain affected by
    hydrostatic pressure, 367
  strain concentration factor in, 123,
    124, 125

Carbon steels—*contd.*
  stress cycling of, 138, 140, 141,
    142, 143
  thermal fatigue of, 346–53
Cartan, E., 76, 185
Cartesian localized coordinates, 140,
    502–3
Cast steel
  fluctuation
    fatigue life of, 499
    rotating/bending fatigue limit of,
      493
  strength of, 489
Chemical environment, effects of,
    379
Cherepanov, G. P., 26, 93
Chromium–molybdenum steels
  creep J-integral range versus crack
    propagation rate, 479
  decarburization of, 380
  fatigue J-integral range versus
    crack propagation rate, 479
  fluctuation of tensile creep rupture
    of, 497
Chromium–molybdenum–vanadium
    steels
  biaxial low-cycle fatigue of,
    267–74, 277, 279
  creep damage tests for, 472–4
  creep J-integral range versus crack
    propagation rate, 479
  creep rupture affected by notch,
    304–10
  fatigue cracks in, 274–5
  fatigue J-integral range versus
    crack propagation rate,
    479
  material properties of, 267
Chromium–nickel–molybdenum
    steels
  fluctuation of tensile creep rupture
    of, 497
Chromium–nickel steels
  fluctuation of tensile creep rupture
    of, 497
  long-term damage studies, 471
  see also Stainless steel, type
    304HTB

Chromium–nickel–titanium steels
  long-term damage studies, 471
Circular disk specimens
  heat transfer calculations for,
    340–3
  thermal crack initiation and
    propagation on, 343–5
  thermal fatigue of, 340–53
    experimental apparatus used,
      345–7
    hollow cylindrical specimens
      compared with, 351–3
    specimen configuration used, 346
Circular disks
  maximum free energy required to
    create crack with shape of,
    30
  thermal fatigue of, 340–53
Climbing dislocations, 190, 191, 369
CNF heat-resisting steel, 467
Cobalt-based superalloys,
    creep–fatigue interaction of,
    281–5
Coefficient of variation (c.o.v.),
    meaning of term, 34
Coffin, L. F., 103, 283, 300
Cohesion forces, 25, 33
Coincidence boundary model, 23
Collaborative research
  notch effect in creep–fatigue
    interaction, 333–9
  reliability of high-temperature
    materials, 470–82
  VHTGR materials, 482–5
Combined micro and macro fracture
    mechanics (CMMFM),
    368–9
  compared with fracture mechanics,
    97
  stratificational cognition in, 96
Complex body, meaning of term,
    19–20
Complexity of configuration, 20, 81
Compressive strain hold, fatigue
    affected by, 282–3
Computer programs
  creep analysis, 461
  finite element analysis, 333, 334

Conceptual material space
  analysis by non-Riemannian
    geometry, 184–90
  changes with pressure soaking
    effect, 223–4
Configuration, complexity of, 20, 81
Connection, coefficient of, 186
Constant deformation model, 66
Constitutive equations, 286, 323
Continuously dislocated continuum
    (CDC)
  geometrical aspects of, 184–98
  hydrostatic pressure effects studied
    by means of, 228–36
  microstructural study of
    cyclic constitutive relation by
      means of, 210–20
    flow stress by means of, 199–210
    pressure soaking effects by
      means of, 220–8
Continuum
  approximation approach, 18–21
  history of, 78–90
  meaning of term, 80
  meaning of term, 15
  mechanical approach
    classification of, 83
    fracture laws in, 240–353
    historical development of, 82–3
    stratificational cognition in, 93–5
Coordinate systems
  Cartesian localized, 140, 502–3
  holonomic, 185
Copper
  fluctuation of tensile creep rupture
    of, 496
  fracture surfaces of, 9, 10, 363–4
  fracture toughness affected by
    hydrostatic pressure, 362–5,
    368
  hydrostatic stress, effect on
    creep–fatigue, 241–50
  load wave shape, effect on fatigue
    of, 401–8
  oxide layers in
    creep rate testing
      thickness effects in vacuum,
        374–5

Copper—*contd.*
  oxide layers in—*contd.*
    creep rupture specimens
      atmospheric testing, 384–5
      vacuum conditions, 382, 383
    vacuum effects on
      creep rate, 374–5
      creep rupture, 379–85, 387–91
        electron probe microanalysis of, 382, 383, 384
        fracture mode effects, 387–9, 390, 391
        grain size effects, 387–9
        optical micrographs of, 382, 383, 384
        scanning electron micrographs of, 382
        void effects, 389–90
Correlation functions, 21, 82, 85
Cosserat continuum mechanics, 88–9
  see also Continuously dislocated continuum (CDC)
Cosserat, E. M. P., 87–9
Cosserat, F., 87–9
Cottrell, A. H., 33, 109
Couple stress vector, 81
Covariance, 35
Crack
  analogy to electron, 31–3
  direction, biaxial fatigue, 274–80
  extension force, hydrostatic pressure, effect on, 248
  growth, X-40 superalloy, strain waveform effects on, 283–5
  initiation
    energy required (UIE), 360, 362, 363
    hydrostatic pressure, effect on, 243, 244
  initiation damage, 406, 407, 408
  mode I cracking, 274
  mode II cracking, 274
    transition to mode I, 274–5
  nucleation theory, 29–31
  propagation
    damage, 406, 407, 408
    energy required (UPE), 360, 362, 363

Crack—*contd.*
  propagation—*contd.*
    hold-time effects on, 288–9
    hydrostatic pressure, effect on, 244–5
    incubation period for, 12
    J-integral used, 245–8
    load-sequence effects on, 288, 289
    mechanical model of, creep–fatigue interaction conditions, 286–92
    rate concept, 13–14
    rate, relation to misorientation, 414, 415–6
    thermal fatigue, 344–5
    types of in biaxial low-cycle fatigue, 275, 280
Crack opening displacement (COD)
  biaxial stress state equation for, 258
  correlation with crack propagation rate, biaxial low-cycle fatigue, 255, 257
  experimental details for investigation, 255
  relation to crack length, 362, 364
  strain range, fatigue data plotted, 262, 266, 273
  stress range, fatigue data plotted against, 264, 266, 295, 301
Creep
  crack growth in, 13–14
  damage
    calculation for HTGR heat exchanger, 460, 462
    diagrams, 472, 473
    relation to fatigue damage under hydrostatic pressure, 248–9
  fracture
    equation for, 28
    mechanism in creep–fatigue life prediction, 474–6
    hydrostatic pressure effects on, 234–7, 241–50
    instability at high temperatures, 449
  J-integral, 13, 14

Creep—*contd.*
  J-integral—*contd.*
    crack propagation represented by, 393, 399
    vacuum effect on creep rupture, 393–4
  notch strengthening/weakening in creep rupture, mechanical factors determining, 304–10
  polycrystalline metals
    hydrostatic stress, effects of, 168–79, 180
    plasticity laws for, 162–80
    strain history, effects of, 163–8
  tertiary creep at high temperatures, 448–9
Creep–fatigue damage test apparatus, 480–2
  cam-and-lever type, 480
  cruciform specimens used, 482
  heat-actuator type, 480–1
  weld-structure type, 481
Creep–fatigue damage/life, evaluation/prediction of, 474–9
Creep–fatigue interaction
  crack propagation, 402, 408
  frequency–elastic stress concentration factor modified equation used in life prediction, 327–33
  life prediction for notched specimens, 327–40
  collaborative research on, 333–9
  frequency effect on, 328
  hold-time effect on, 328–9
  methods used, 327–33
  material constants, effects of, 288–91
  mechanical model of crack propagation, 286–92
  notch effect on low-cycle fatigue, 304–27
  notch weakening-to-strengthening transition in, 310–20
  strain wave shapes, effect of, 281–5
Critical crack length, 48–9

Critical *J*-integral, 26, 96, 369
Critical nucleus size, 29, 30
Critical shear stress law, 63
Cruciform specimens
  biaxial low-cycle fatigue testing, 266–74
  hysteresis loop illustrated, 271
  inelastic analysis of, 267–9
  shape and dimensions of, 269
  test rig used, 269–71
  creep–fatigue damage testing by, 482
Crystal
  disclinations, 78–9
  grain interaction, 73–4
  plasticity
    cyclic stressing conditions, 130–7
    inhomogeneity in, 121–6
    interaction between neighboring grains, 135, 136
    tension conditions, 120–30
  rotation, 123, 124
Cycle-dependent *J*-integral range, 407–8, 478
Cyclic constitutive relation
  microstructural study by means of continuously dislocated continuum, 210–20
  microstructure and, 217–20
  misorientation and, 210–16
Cyclic hardening, 107
  Bauschinger effect, relation to, 210
  type 304 stainless steel, 145–62, 298, 303–4, 333
  types of, 333, 334
Cyclic softening, 107
Cyclic stress–strain relation, proportional/non-proportional loading, 145–52
Cyclic stressing
  crystal plasticity under, 130–7
  materials structural changes during, 131–3
  polycrystalline metals affected by strain history under, 137–45
  yield affected by microscopic stress and strain, 133–5

INDEX

D.C. potential measurement, crack initiation observed by, 326–7
Damage
  evaluation of, 41, 42, 409–24
  measures, 41, 42
  research into, 471
  tolerance designs, 40, 92
Davidenkov, N., 488
Debris accumulation, first noted, 105
Decarburization, 380
  Inconel 617 alloy, 395–8, 451
Dederichs, P. H., 82
Dekeyser, W., 194
Design code, high-temperature materials, 446
  problems in development of, 447
Diffusional creep, 178
Dimple fracture surface, 9, 363
Disclination, 189–90
  density tensor, 190
  types of, 190
Disclination density tensor, 80
Discontinuity (in space and time)
  continuum approximation for, 18–21
  stochastic process involved, 6, 26–8
  stratificational cognition of, 6, 14–18
  structural boundary as, 21–3
  surface energy considerations, 23–6
Discrete distribution of dislocations, models for, 77, 376
Dislocation
  climbing motion, 190, 191, 369
  density
    change during bending of aluminum, 215, 219
    distributions in nickel under vacuum conditions, 377
    hydrostatic pressure, effects on, 230–2, 235
    meaning of term, 201
    pressure soaking, effect on, 220–3
    relation to
      Schmid factor, 202, 204
      shear strain, 206–8

Dislocation—contd.
  density—contd.
    tensor, 21, 76, 80, 188
    variation with number of cycles, 411–12
  dipole, 85, 188
  lines, crystallographic model of, 188
  network, 85
  pile-ups, 94, 95, 131, 132, 218
    theory of, 77
  structure, effect on, of
    biaxial stress cycling, 155
    loading mode change, 156–8
  velocity tensor, 78
Distant parallelism space, 198, 224, 232
Distributed dislocations, theory of
  creep crack behavior in vacuum, 370–5
  history of, 74–8
  transitional phenomenon at reloading in stress-hold tests described by, 403–6
Distribution functions, 34–7, 488–99
  extreme, 36–7
  normal, 36–7, 488, 493, 494, 496, 497, 498
  Weibull, 37, 488, 489, 490, 496
Ductility exhaustion damage, calculation for HTGR heat exchanger, 460, 462

Einstein's relativity theory, 88
Elastic constant effect, 374
Elastic stress concentration factor
  creep crack initiation of Inconel 617 affected by, 396, 453
  fatigue affected by, 321–4, 325–7
  life prediction using, 327–33, 339, 340
Elasticity, stratificational concepts of, 15
Elber, E., 97
Electron probe microanalysis (EPMA)
  type 316 stainless steel, 392

Electron probe microanalysis (EPMA)—*contd*.
  vacuum effect on creep rupture of copper, 382, 383, 384
Energy momentum tensor, 25, 93, 111
Entwistle, K. M., 210
Environmental effects
  creep rupture affected by, 379, 380
  high-temperature gas-cooled reactor heat exchanger, 446, 450–5
  interactions listed, 380
  see also Chemical; Vacuum
Eshelby, J. D., 25, 69, 75, 93, 110–11
Euclidean space, properties lost by imperfect crystal, 20
Ewald's reflection sphere, 192
Ewing, J. A., 62, 105–6, 107–8
Exponential functions, 495
Eyring, H., 29

Factor of safety, 38
Fail-safe designs, 40, 92
Failure rate, meaning of term, 494
Fatigue
  advances in study of, 96–105
  crack opening and closure, 97
  damage
    HTGR heat exchanger, calculation for, 460
    relation to creep damage under hydrostatic pressure, 248–9
  fracture toughness, 48
  frequency effect on notched specimens, 320–4
  frequency–elastic stress concentration factor modified equation, 321–4, 327–33, 339, 340
  hold-time, effects of, 242–3, 253, 281–2
  hydrostatic stress, effect on, 241–50
  life prediction methods for, 299–303, 304

Fatigue—*contd*.
  life prediction methods for—*contd*.
    frequency separation method, 299, 300, 301, 302
    Ostergren's method, 299–300, 301, 302, 304
    Tomkin's method, 299, 300, 302, 303
  limit, 11
  low-cycle fatigue, stainless steel, 9, 10
  publications on, 6
  research methodology for, 6
  research studies made, 98
  strain wave shapes, effect of, 281–3
  strength reduction factor, meaning of term, 49
  threshold stress for, 11
  world-wide research programmes, 100–1
Fictive length of structure, 52
Finite element analysis
  computer programs used, 333, 334
  cruciform (biaxial fatigue) specimens, 267–9
  life prediction based on, 337–40
  notched specimens under creep–fatigue interaction conditions, 333–6
Flow stress
  hydrostatic pressure effects on, 228–37
  microstructural study by means of continuously dislocated continuum, 199–210
Forged steel, fluctuation of rotating/ bending fatigue limit of, 493
Form factor, meaning of term, 48
Format (of this book), 2–3
Fractography, X-ray diffraction, 420–3
Fractology, 60
  advances in, 91–105
  formation of, 90–1
  historical account of, 90–111
  men associated with, 105–11

Fracture
  brittle fracture, 8, 9, 10
    see also main entry: Brittle fracture
  chiselpoint fracture, 9
  classification of characteristic
      modes, 8–9
  cup-and-cone fracture, 8–9
  ductile fracture, 8–9, 10
  fatigue fracture, 8, 9, 10
  fibrous fracture surfaces in, 9
  interaction between fracture
      processes, 6, 11–14, 91
  intergranular fracture, 8
  laws, continuum mechanical
      approach to, 240–353
  mechanics (FM)
    advances in, 91–6
    comparison with combined
        micro and macro fracture
        mechanics, 97
    meaning of term, 91
    relation to J-integral in
        continuum mechanics,
        358–424
  modes
    listed, 8–11
  notch effect transition, 316–17
  nucleation theory for, 29–31
  serpentine glide, 8, 9, 364
  slip-off fracture, 9
  stratificational concepts of, 17
  toughness, 48, 92
    hydrostatic pressure, effect of,
        359–68
  transgranular fracture, 8–9, 10
  variety of, 7–11
France, creep research in, 100
Frank's definition of misorientation,
    194–5
Frequency–elastic stress
      concentration factor
      modified equation, 321–4,
      327
  life prediction by, 327–33, 339, 340
Frequency-modified fatigue life
      prediction equation, 300
Friction stress, pressure soaking,
      effect of, 225–7

Fujii (researcher), 499

Gas turbine
  disks, thermal fatigue of, 340–53
  rotors, creep–fatigue interaction of
      metal in, 281–5
Gibson, A. H., 109
Gilman, J. J., 77
Glass, fluctuation of
  bending strength, 495
  delayed fracture time, 495
Goldhoff, R. M., 99, 103
Grain
  boundary properties, 21–6
  interaction between grains, 73–4
  variation of X-ray parameters
      within and between, 127–30
Griffith, A. A., 26, 92, 108–10
Griffith's brittle fracture theory, 26,
    31, 92
Gumbel, E. J., 37

Halford, G. R., 287
Half-value breadth, 421, 423–4
  Cr–Mo–V steel, 472, 473
Hall–Petch relation, 77
  see also Petch's relation
Hall relationship, creep of nickel in
    vacuum, 372, 373
Hanafee, J. E., 232
Harshbarger, J. H., 496
Hashin, Z., 69
Hastelloy X
  creep rupture properties of, 466
  temperature dependence of yield
      strength, 448
Hastelloy XR
  chemical composition of, 482
  fatigue life data, 468, 469
  life evaluation under creep–fatigue
      interaction conditions, 484,
      485
  long-term corrosive behavior in
      hot helium gas, 482–3
Head, A. K., 77, 376

Heat-actuator creep–fatigue test
    apparatus, 480–1
Heat-resisting steels
  chemical composition of, 467
  load wave shape, effect on fatigue
    of, 281–92
  new developments in, 464, 466–70
Helium
  creep properties of Inconel 617
    affected by, 450–2, 453
  Engineering Demonstration Loop
    (HENDEL), 482
  Hastelloy XR affected by, 482–3
  methane content of, 455
  nuclear reactor use of, 444, 445
Hershey, A. V., 69
Heterogeneous materials, meaning of
    term, 15
Hierarchy cognition, 14–18, 80
  see also Stratificational concepts
High-temperature failure criteria, 446
High-temperature gas-cooled reactor
    (HTGR)
  heat exchanger
    construction of, 464, 465
  environmental effects in, 450–5
  high-temperature design code
    developed, 446, 447
  life prediction for, 458, 460–3
  lifetime test for partial model,
    455–8, 459
  material strength of, 444–55
  mock-up model made, 455–8
  temperature dependence of yield
    stress in, 446, 448
  tertiary creep curves for, 448–50
  steelmaking use of, 444, 445
High-temperature low-cycle fatigue
    criterion
  biaxial stress states, 254–66
  see also Crack opening
    displacement (COD)
High-temperature strength,
    interaction between
    properties, 99
Hill, R., 64, 67, 68, 69
HINAPS finite element analysis
    computer program, 333, 334

Hirata, M., 27, 109, 495
Historical account
  continuum approximation
    approach, 78–90
  fractology, 90–111
  plasticity studies, 61–74
  SFM studies, 59–61
  theory of distributed dislocations,
    74–8
Hitachi, finite element analysis
    computer program produced
    by, 333
Hold-times (in fatigue testing),
    242–3, 253, 281–2
  see also Stress-hold tests
Holonomic coordinate system, 185
Honda, K., 108
Hosoi, Y., 450
Hutchinson, J. W., 68, 69
Hydrostatic pressure
  fracture toughness affected
    aluminum, 365–6
    copper, 365
    polycrystalline metals, 359–68
  maximum tensile load affected by,
    363
  necking strain affected by, 365–6,
    367
  plastic compression of pure metals
    under high pressures,
    436–43
    force equilibrium condition for
      solid cylindrical specimen,
      440
    material structural effects, 442–3
    test apparatus for, 436–7
  plastic drawing of pure metals
    under high pressures, 428–9,
    431–6
    factors affecting, 433–4
    stress state of bar in, 432
    test apparatus for, 427–8, 429,
    430
  soaking. See Pressure soaking
Hydrostatic stress
  creep affected by, 168–79, 180
  creep rupture life affected by,
    241–50

INDEX 515

Hydrostatic stress—contd.
  experimental apparatus for, 170, 171, 173
  low-cycle fatigue affected by, 241–50
  notch effect on, 306–7
Hysteresis
  magnetization, 108
  stress cycling, 106–7
Hysteresis loops
  biaxial fatigue testing of Cr–Mo–V steels, 271
  notched SUS 304 specimens, 333, 335–6
  superalloy X-40 fatigue, 284

Ideal fracture strength, 18
Image dislocation model, 77, 376
Impact fracture, 9, 10
Imperfect crystals
  analysis of, 20–1
  plasticity laws of, 184–237
Imperfection density
  change during bending of aluminum, 216, 219
  hydrostatic pressure effects on, 230–2, 235
  meaning of term, 202
  pressure soaking, effect on, 220–3
  relation to
    Schmid factor, 202, 204
    shear strain, 206–8
  variation with number of cycles, 411–12
Incoloy 800, HTGR heat exchanger constructed from, 464
Incompatibility tensor, 76
Inconel 617
  creep rupture properties of, 450, 466, 467
  decarburization of, 380, 451
  fatigue life data, 469
  helium gas, effects on creep rupture, 451
  HTGR heat exchanger constructed from, 464

Inconel 617—contd.
  life prediction for, 460, 461
  methane effects on, 455
  notch effect on creep crack initiation, 452–3
  notch-opening displacement curves for, 398, 454
  oxygen effect on creep, 449
  push-pull fatigue tests, 454–5
  strain rate dependence of yield stress, 448
  temperature dependence of yield strength, 448
  vacuum effects on creep rupture of, 395–9, 400, 451
Inconel 625
  temperature dependence of yield strength, 448
Inconel 718
  creep J-integral range versus crack propagation rate, 479
  fatigue J-integral range versus crack propagation rate, 479
Inelastic finite element analysis
  notched specimens under creep–fatigue interaction conditions, 333–6
  see also Finite element analysis
Inhomogeneity
  crystal plasticity, 121–6
  plastic strain, 21, 76
Inoue, T., 86
In-service inspection requirements, 41
Integrity evaluation method, 41, 43
Interaction (of fracture processes), 6, 11–14, 91
Interdisciplinary approaches
  safety, 40
  strength and fracture of materials, 5, 59–61
International Civil Aviation Organization (ICAO), accident study, 40
International Conference on Fracture (ICF), 61
Interstitial atom, crystallographic model of, 189

Iron, hydrostatic stress–strain curves for, 169–70
Iron and Steel Institute of Japan (ISIJ)
   Notch Effects Test Subcommittee, 333
   on Oliver's law, 47
Irreversible processes, 29–33, 95–6
   crack nucleation and, 29–31
Irwin, G. R., 46, 92
Isjhikawajima-Harima Heavy Industries, testing by, 457, 485
Isotropic hardening rule, 161, 162
ISTRAN/3D creep analysis program, 461
Ito, M., 69, 70
Iwasaki, C., 44

J-integral, 25–6, 93–5
   crack propagation in creep, 245–8
   cycle-dependent, 407–8, 478
   fracture toughness and, 361–2, 368
   modified (J')
      crack propagation in creep, 246–8
      definition of, 246
      resultant forces summed in, 361–2
      time-dependent, 407, 408, 478
      usefulness of, 362
   see also Creep; J-integral
J_c criterion, 26, 96, 369–70
J_c-integral. See Creep . . .; Critical J-integral
Japan
   Atomic Energy Research Institute (JAERI), research by, 444, 482
   creep research in, 99, 101, 102
   Material Testing Reactor (JMTR), 482
   Ministry of International Trade and Industries
      sponsorship of nuclear steelmaking project, 443–4

Japan—contd.
   National Symposium on Strength, Fracture and Fatigue of Materials (JNSSFM), 61
   Science and Technology Agency, collaborative research, 470–82
   Society of Materials Science, 493, 499
   Society for Promotion of Science, 496–7
Jimma, T., 74, 121
Johnston, W. G., 77
Johnston–Gilman equation, 77

Kawamo (researcher), 498
Kennedy, A. J., 61
Kikukawa, M., 97
Kinematic hardening, carbon steel, 166, 167, 168
Kinematic hardening rule, notched specimens, 333–6, 340
Kinetic theory of fracture of solids, 32
Kitamura, T., 402, 407
Kochendorfer (plasticity) model, 70
Kondo, K., 20, 75–6, 185
Kondo's natural state, 185
Kröner, E., 21, 69, 70, 75, 76, 82, 185
Kröner–Budianski–Wu (KBW) model, 70, 72
KSN superalloy, 466, 467, 468
Kubota, H., 495
Kunin, I. A., 20

Ladder type dislocation structures, 154, 155, 162, 299
Langyel, B., 427
Lattice
   defects
      Riemann–Christoffel curvature tensor and, 84
      torsion tensor and, 84
   rotation, 73

Lattice—contd.
vacancy, crystallographic model of, 189
Laue X-ray diffraction method
asterism changing during bending, 212, 213–14
discoverer of, 33
micro method used to evaluate damage, 409–16
conditions used, 409
misorientation measured by, 123, 127, 129, 190–2, 199–200, 221, 409–16
schematic representation of, 200
Layout (of this book), 2–3
LE strain, fatigue data plotted against, 260, 273
Lead, creep resistance of, 166
Leibnitz's law of continuity of matter, 86
Leonardo da Vinci, 45–6
Levi-Civita's parallelism, 217
Li, J. C. M., 206
Life prediction, HTGR heat exchanger, 458, 460–3
Lifetime test, HTGR heat exchanger, 455–8, 459
Lifetimes, fluctuation of, 495–9
Lin, T. H., 68, 69, 70
Linear damage rule, 407, 408
Linear fracture mechanics (LFM), stress intensity factor used, 27
Lippman, H., 80
Load factor, meaning of term, 38
Load wave shape, effect on fatigue of
heat-resisting metals, 281–92
pure metals, 400–9
superalloys, 281–92
Local equilibrium, principle of, 32, 96
Locking force of dislocation, pressure soaking, effect of, 225–7
Lode's parameters, correlation during creep of carbon steel, 165–6

Lord, B. C., 283
Louat, N., 77

MA-X7 superalloy, 467
McEvily, A. J., 99, 103
Machine structural steel, fluctuation of strength of, 494
Manson, S. S., 287
Mar M247 superalloy
creep J-integral range versus crack propagation rate, 479
fatigue J-integral range versus crack propagation rate, 479
Maraging steels, fluctuation of fatigue life of, 499
MARC type combined hardening rule, notched specimens, 333–6, 340
MARC-J2 finite element analysis computer program, 333, 334
Marcinkowski, M. J., 77, 376
Markovic, Z., 86, 87
Marsh, D. J., 345
Material point, meaning of term, 15
Materially homogeneous body, meaning of term, 20
Materially isomorphic body, meaning of term, 20
Materially uniform body, meaning of term, 20
Maximum principal strain and stress, fatigue data plotted against, 260, 261, 264, 266
Maximum shear strain, fatigue data plotted against, 260, 261, 266
Maze type dislocation structures, 154, 155, 162, 299
Measures, 16, 80
Mekal, A. G., 109
Methane, corrosion affected by, 455
Micro-support effect, 51
Microbeam X-ray diffraction patterns, biaxial stress state, 155
Microcrack
propagation mechanisms, 12

Microcrack—*contd.*
  threshold, 33
  type A and type B, 11
Microstrain
  relation to
    damage factors, 419
    number of cycles for crack propagation, 420–1, 422
Microstructure, and cyclic constitutive relation, 217–20
Mild steel
  decarburization of, 380
  phenomenological yielding of, 53
Miller, K. J., 262, 274
Milton, J., 107
Minagawa, H., 78, 82, 88
Minagawa, S., 21
Mises, *See* Von Mises . . .
Misorientation, 21, 76, 84
  changes under cyclic stress, 132–3
  cyclic constitutive relation and, 210–16
  definition of, 190–5
  experimental method analyzed, 195–8
  Frank's definition of, 194–5
  hydrostatic pressure effects on, 229–30
  measurement by Laue X-ray diffraction method, 123, 127, 190–2, 199–200, 221, 409–16
  relation to
    crack propagation rate, 414, 415
    orientation function, 413, 414
    principal strain direction, 126–7
  Schmid factors related to, 135, 136
  variation with number of cycles, 410–11
Mitsubishi Heavy Industries, test apparatus developed by, 481
Miyahara, S., 44
Miyamoto, H., 73, 74
Mock-up model
  HTGR heat exchanger, 455–8, 459
  fracture illustrated, 459
  loading condition of, 456
  loading histories of, 458

Model
  HTGR heat exchanger, 455–8, 459
  meaning of term, 43, 44
Modeling, 6, 43–4
Modified J-integral, 246–8
  definition of, 246
Mott, N. F., 92
Multi-point correlation function, 21, 82, 85
  tensor, 21
Mura, T., 78
Muskhelishvili's inversion theorem, 386

Nabarro, F. R. N., 75
Nabarro–Herring type creep, 178
Nagaoka, H., 108
Nakagawa (researcher), 498
Nakamura (researcher), 499
Nakanishi, F., 53
Nakazawa, H., 493
National Research Institute of Metals (NRIM, Japan), testing by, 99, 102, 466, 481, 490, 497
Natural state, meaning of term, 185–7
NCF 800 alloy, creep–fatigue life prediction for, 474, 475
Neuber, H., 51–2
Neuber's rule, 323, 324
  life prediction using, 327–33
  modified rule, 324–5
Nichrome V, vacuum effect on creep rupture, 382
Nickel
  creep crack behavior in vacuum, 370–4
  oxide films, 377
  vacuum effect on creep rupture of, 370–4, 378–9, 382
Nickel–molybdenum–vanadium steels
  creep J-integral range versus crack propagation rate, 479
  fatigue J-integral range versus crack propagation rate, 479
Nickel-based superalloys
  chemical composition, 467, 469–70
  fatigue life data, 469

Nickel-based superalloys—*contd.*
  see also Hastelloy; Inconel . . . ;
    KSN; MA-X7; NSC-1;
    R4286; SSS . . . ; SZ
Nishimura, A., 488
Noll, W., 19
Nonholonomic property,
    interpretation of, 185
Nonlocality (of material), meaning of
    term, 20
Non-polarity, meaning of term, 81
Non-proportional loading, 296
Non-Riemannian space, analysis of
    material space based on,
    184–90
Nonuniqueness in choice of active
    slip systems, 64, 72–3
Normal distribution functions, 35–6,
    488, 493, 494, 496, 497, 498
Notch effect
  collaborative research on, 333–9
  creep crack behavior of nickel
    affected by, 371
  Inconel 617 affected by, 452–3
  mechanical factors affecting,
    304–10
  transition phenomenon
    creep–fatigue interaction
      conditions, 310–20
    creep rupture, 304, 305
    frequency dependence of, 318
Notch opening displacement (NOD)
  crack initiation observed by, 326–7
  curves for Inconel 617 superalloy,
    398
Notched specimens
  creep curves for, 308, 309
  life prediction in creep–fatigue
    interaction, 327–40
  low-cycle fatigue failure of, 320–7
  stress variation in, 304–6
NSC-1 superalloy, 467, 468, 469
Nuclear reactors
  steelmaking use of, 444, 445
  see also High-temperature gas-
    cooled reactor (HTGR);
    Very high temperature gas-
    cooled reactor (VHTGR)

Nuclear steelmaking
  characteristics of, 445
  research background of, 443–4
Nucleation theory, 29–31
Nye, J. F., 75

Objectives (of this book), 1–2
Ohara, H., 496
Ohkubo, T., 66
Ohmori, K., 27
Ohtani, R., 402, 407
Oliver's law, 47
Optical microscopy
  creep crack tip in stainless steel,
    304
  vacuum effect on creep rupture of
    copper, 382, 383, 384
Orientation function
  calculation of, 413
  relation to misorientation, 413, 414
ORNL kinematic hardening rule,
    notched specimens, 333–6,
    340
Orowan, E., 75
Ostergren, W., 299–300
Ostergren's (fatigue life prediction)
    method, 299–300, 301, 302,
    304
Oxide films, creep rate in vacuum
    affected by, 375–8
Oyane (plasticity) model, 70

Paris–Erdogan rule, 97
Parsons, W. B., 46
Partial safety factor design, 39
Particle, meaning of term, 15
Particle size
  relation to
    damage factors, 419
    number of cycles for crack
      propagation, 420–1, 422
Partitioning
  crack length, 402–3, 408
  strain partitioning method, 287
Path-independent integrals, 358, 362
  see also J-integral

Peierls, R. E., 75
Peierls–Nabarro theory, 75
Permanent softening region, 218
Petch's relation, 225
  see also Hall–Petch relation
Phenomenological yielding (of mild steel), 53
Pile-up dislocations, 94, 95, 131, 132, 218
  models for, 77
Plastic deformation
  gradient, 76
  polycrystalline materials
    strain history effect when cyclically stressed, 137–45
    typical modes of, 120, 121
Plastic work, principle of maximum, 64, 65
Plasticity
  historical account of, 61–74
  imperfect crystals, 184–237
  laws, creep of polycrystalline metals, 162–80
  polycrystalline metals
    creep affected by hydrostatic pressure, 168–79, 180
    cyclic stress conditions, 130–7
    tension conditions, 120–30
  self-consistent model for, 69, 70
  stratificational concepts of, 16
Poisson, S. D., 87
Polarity, meaning of term, 20, 81
Polycrystalline metals
  characteristics of, 82
  creep behavior of
    hydrostatic stress, effects of, 168–79, 180
    strain history, effects of, 163–8
  crystal plasticity of
    cyclic stress conditions, 130–7
    historical account of, 66–9
    tension conditions, 120–30
  fracture toughness affected by hydrostatic pressure, 359–68
  rate-dependent constitutive model for slip in, 73

Polycrystalline metals—*contd.*
  strain history effect on plastic deformation under cyclic stressing, 137–45
  work-hardening curves, historical account of, 70–3
Prager flow-rule, 165, 166
Pre-damage concept, 284, 286, 292
Pressure soaking
  changes in conceptual material space, 223–4
  dislocation and imperfection densities affected by, 220–3
  effects studied by means of continuously dislocated continuum, 220–8
  friction stress affected by, 225–7
  locking force of dislocation affected by, 225–7
Principal strain direction, 125–6
  angle measurement method, 127
  relation to misorientation, 126–7
  principal stress direction, 143–4
Principal stress axes, changing of, fatigue lives affected by, 296–303
Principal stress direction
  relation to principal strain direction, 143–4
Probability density function, meaning of term, 494
Probability distribution functions, 34
Pure metals
  load wave shape, effect on fatigue of, 400–9
  plastic compression under high hydrostatic pressures, 436–43
  plastic drawing under high hydrostatic pressures, 428–9, 431–6
  see also Aluminum; Copper; Lead; Nickel; Zinc
Push–pull tests
  crack propagation in, 252–3
  failure life in, 250–1

R4286 superalloy, 466, 467
Rate process theory, 30, 32
Read, W. T., 194
Reliability
  collaborative research on, 470–82
  engineering, 40–1, 43
  meaning of term, 494
Remaining-life prediction, 41, 104, 470
  method applied, 472–9
Research Association of Applied Geometry (RAAG, Japan), 76
Resistance factor, meaning of term, 38
Reversing torsional tests
  crack propagation in, 252–3
  failure life in, 150–1
Rice, J. R., 26, 93, 111, 245
Riedel's (screw dislocation) method, 375
Riemann–Christoffel curvature tensor, 21, 95, 187, 208
  lattice defects corresponding to, 84
Riemannian space. *See* Non-Riemannian space
Royal Aircraft Establishment (RAE, Britain), 26, 108, 109
RS-513 heat-resisting metal, 467
Rubbra, A. A., 110

S25C steel
  fluctuation of
    fatigue life of, 499
    rotating/bending fatigue limit of, 490
S35C steel
  fluctuation of
    fatigue life of, 499
    rotating/bending fatigue limit of, 490
S45C steel
  fluctuation of
    fatigue life of, 499

S45C steel—*contd.*
  fluctuation of—*contd.*
    rotating/bending fatigue limit of, 491
S50C steel, fluctuation of fatigue life of, 499
S55C steel, fluctuation of rotating/bending fatigue limit of, 491
S–N curve, 491
SA heat-resisting metal, 467
Sachs, G., 66, 69
Sachs equation, 433
Sachs (plasticity) model, 68–9, 70
Safety
  factor, 38
  interdisciplinary nature of, 40
  life, 40
  margin, 38–9
    HTGR heat exchanger, 463
Sakai, T., 499
Sawczuk (researcher), 80
SB heat-resisting steel, 467
Scale parameter, meaning of term, 18–19, 81
Scale-up, 44–7
Scanning electron microscopy (SEM)
  fracture surfaces, 8, 9, 10
  vacuum effect on creep rupture of copper, 382
Scattering functions, 34–7
Schmid, E., 63
Schmid factor, 63
  relation to
    dislocation density, 202, 204
    imperfection density 202, 204
    misorientation, 135, 136
    slip spreading of aluminum, 123
Schmid law, 63
SCM4 steels
  fluctuation of
    fatigue life of, 499
    rotating/bending fatigue limit of, 492–3
Screw dislocation
  array pile up, 386
  density, effect of surface oxide thickness on, 386–7

Screw dislocation—*contd*.
  mechanism, 75
  model at front of crack tip with oxide film, 375–6
Sedov (researcher), 47
Seeger, A., 206
Serpentine glide fracture surface, 8, 9, 364
Shahinian (researcher), 382
Shanon diagrams, 27, 28
SHAW humidity meter, 381, 395
Shear strain, relation to dislocation and imperfection densities, 206–8
Shear summation, principle of minimum, 66
Shibata, T., 496
Shimokawa (researcher), 499
Siebel equation, 441
Similitude
  law of, 45, 47–9
  types of, 47–9
Simple body, meaning of term, 19, 20
Size effect, 49–54
Slip
  bands, 62, 107
  mechanisms, history of development of, 61–6
  nonuniqueness in choice of active systems, 64, 72–3
  plane dislocation distribution, model of, 232
    hydrostatic pressure, effects of, 236–7
SM50A steel, fluctuation of strength of, 488
Smekal, A. G., 109
Softening region
  permanent, 218
  transient, 217
Somigiliana, C., 90
SSS113MA superalloy, 466, 467, 468, 469
SSS410 superalloy, 467
Stainless steels
  cyclic hardening of Mises equivalent stress amplitude, 152

Stainless steels—*contd*.
  type 304,
    biaxial low-cycle fatigue of, 293–304
    creep damage tests for, 472
    creep J-integral range versus crack propagation rate, 479
    cyclic strain-hardening behavior of, 145–62
    cyclic stressing of, 145–62
    cyclic work-hardening of, 298, 303, 333
    damage evaluation by
      X-ray diffraction, 409–16
      X-ray profile analysis, 416–24
    fatigue crack direction in, 275–80
    fatigue fracture surface of, 9, 10
    fatigue J-integral range versus crack propagation rate, 479
    low-cycle biaxial fatigue of, 255–66
    material constants used, 150
    notch effect on fatigue failure, 325–7
    notch weakening-to-strengthening transition in, 310–18, 320
    notched specimens
      hysteresis loops for, 333, 335–6
      life prediction for, 337, 338, 339
    remaining-life prediction for, 476–7
    X-ray fractography of, 420–3
  type 304HTB, 471
  type 316
    biaxial-stress effects on, 250–4
    creep damage tests for, 472
    electron probe microanalysis of, 392
    fatigue of notched specimens, 320–3
    notch weakening-to-stregthening transition in, 310 15, 320
    vacuum effects on creep rupture of, 391–4, 399–400

Stainless steels—*contd.*
   type 321HTB, 471
Statistical continuum mechanics, 15, 81
Statistical law, reliability use of, 41–2
STBA alloy steels, 471
   see also Chromium–molybdenum steels
Steam turbine rotors
   biaxial low-cycle fatigue of metal in, 267–74
   creep damage/rupture life prediction for metal in, 474
Steelmaking, nuclear reactors used, 444, 445
   see also High-temperature gas-cooled reactor (HTGR)
Steels
   See Alloy . . . ; Austenitic . . . ; Carbon . . . ; Chromium– . . . ; Heat-resisting . . . ; Mild . . . ; Stainless steels
Stochastic processes, 6, 26–8
Strain concentration factor, grain-to-grain variation of, 123, 124, 125
Strain history, effect on polycrystalline metals under stress cycling, 137–45
Strain hold-times, biaxial low-cycle fatigue lives affected by, 293–6
Strain partitioning method, 287
Strain waveforms
   creep–fatigue interaction affected by, 281–5
   types of, 296–7
Stratificational cognition, 14–18
   combined micro and macro fracture mechanics, 96
   continuum mechanics, 93–5
   elasticity, 15
   fracture, 17
   plasticity, 16
Stratificational concepts, 14–18
Stratificational hierarchy/structure, modeling as, 44

Strength and fracture of materials (SFM)
   concepts used for, 6–7
   fluctuation of, 6, 34–43, 488–94
   formation of study of, 59–60
   interdisciplinary approach necessary, 5, 59–61
   law of similitude applied to, 47–9
   phenomena involved, 5
   research approach used, 6
   scale-up in, 44–5
Stress
   concentration factor. *See* Elastic stress . . .
   corrosion cracking, aluminum alloy, 496
   dependent heterogeneous nucleation theory, 30–1
   intensity factor, 27, 92
   mean values, 51
Stress-hold (fatigue) tests
   stress wave in, 404
   transitional changes of creep propagation after reloading in, 400–6
Striation (in fracture), 8, 10, 28
Stroh, A. N., 77
Structure-sensitive properties, 6, 18, 37
Superalloys
   chemical composition of, 467
   load wave shape, effect on fatigue of, 281–92
   new developments in, 464, 466–70
   oxygen effect on creep, 449
   strain rate dependence of yield stress, 448
   temperature dependence of yield strength, 446, 448
   see also CNF; Hastelloy; Inconel . . . ; KSN; MA-X7; NSC-1; RA286; RS-513; SA; SB; SSS . . .; SZ; X-40
Surface
   analysis techniques, 379
   cracks, high-temperature biaxial fatigue

Surface—*contd*.
cracks, high-temperature biaxial fatigue—*contd*.
direction of cracks, 275–7
growth of cracks, 277–9
types of, 275, 280
dislocation model, 77, 376
energy considerations, 23–6
layer concept, 53
Survival probability, meaning of term, 494
SUS 304. *See* Stainless steel, type 304
SUS 316. *See* Stainless steel, type 316
Switzerland, creep research in, 100
SZ superalloy, 467, 468, 469

Taira, S., 99, 103
Tanaka, S., 490, 499
Taylor, G. I., 64–5, 67, 108, 109
Taylor–Bishop–Hill theory (of crystal plasticity), 66
Taylor (plasticity) model, 68, 69, 70
TDD. *See* Distributed dislocations, theory of
Tear tests
apparatus used, 359–60
information obtained from, 361
load–elongation curve for, 360
specimen configuration, 359
Tepicc-4 finite element analysis computer program, 333, 334
Terada, T., 33, 92
Terao, S., 495
Thermal fatigue
circular disk specimens, 340–53
experimental apparatus used, 345–7
hollow cylindrical specimens compared with, 351–3
specimen configuration used, 346
Thermal ratchet design procedure, 446
Time-dependent *J*-integral range, 407, 408, 478
Tokuda, A., 493, 499

Tokyo Koki Co. Ltd, test apparatus developed by, 480
Tomkins, B., 300
Tomkins's (fatigue life prediction) method, 299, 300, 302, 303
Torsion tensor, 187
lattice defects corresponding to, 84
Toughness measures, 48
Transient softening region, 217
Transition probability representation, 27, 28
Transitional probability of failure, meaning of term, 494
Transmission electron micrographs (TEM)
biaxial fatigue stainless steel specimens, 153, 154, 156–7, 158
specimen preparation for, 154, 409
type 304 stainless steel, 413
Truesdell, C., 46
Two-point dislocation correlation function, 85, 86
Type 304 stainless steel. *See* Stainless steel, type 304
Type 316 stainless steel. *See* Stainless steel, type 316

UK, creep research in, 100
USA, creep research in, 100–1
USSR, creep research in, 101

Vacuum
creep crack behavior in
theory of distributed dislocations applied to, 370–5
creep rupture behavior in
heat-resisting metals, 391–400
pure metals, 379–91
superalloys, 395–400
Variety (in fracture modes), 6, 7–11
Very high temperature gas-cooled reactor (VHTGR) applications
corrosive and mechanical behavior of alloys in, 482–5

Very high temperature gas-cooled
reactor (VHTGR)
applications—*contd*.
heat exchanger constructed, 464,
465
heat-resisting alloys tested for,
482–5
life prediction in, 458, 460–3
material strength in, 444–55
Volterra, V., 89–90
Volterra's dislocation of sixth kind,
79
Von Mises equivalent strain
notch effect represented by,
307–10
range
fatigue data plotted against, 260,
266, 298
thermal cycling, 343, 344, 345,
349, 352, 353
thermal crack propagation, 344,
345, 349, 352
Von Mises equivalent stress
amplitude
cyclic hardening of, stainless
steel, 152
relation with plastic strain
amplitude for stainless steel,
148, 149
fatigue data plotted against, 263,
266, 273
notch effect represented by, 306
Von Mises yield criterion, 67
biaxial fatigue, 146, 253, 254

Weakest-link model (for brittle
fracture), 36, 37, 51, 489
Weibull distributions, 37, 488, 489,
490, 496
Weibull, W., 37, 488
Weld-structure creep damage test
apparatus, 481
West Germany, creep research in,
100
Wöhler, A., 11, 61
Wolf, K., 109
Work-hardening. *See* Cyclic
hardening

Wu, T. T., 69, 70
Wulff's net, misorientation measured
on, 200, 201

X-40 superalloy, creep–fatigue
interaction of, 281–5
X-ray
diffraction
asterism in, 212, 213–14, 409–16
biaxial fatigue, 155
see also Laue X-ray diffraction
method
fractography, 420–3
parameters, variation within and
between grains, 127–30
profile analysis
damage evaluation by, 359,
416–24
diffraction conditions used, 417
scanning speed used, 418
residual stress measurement,
thermal cracks measured
by, 347, 350

Yamada, Y., 74
Yamamoto, S., 31–2
Yokobori, T., 11, 27, 29, 30, 60, 61,
77, 90, 109, 369, 488, 496,
497
Yokobori's theory, 31
Yukawa, H., 86

Zeller, R., 82
Zienkiewicz, O. C., 74
Zinc
compression under hydrostatic
pressure, 439, 440
creep affected by hydrostatic
stress, 170, 172, 174, 177,
180
drawing under hydrostatic
pressure, 428–9, 431, 435,
436
necking strain affected by
hydrostatic pressure, 367